Insecticides: Advances in Integrated Pest Management

Insecticides: Advances in Integrated Pest Management

Editor:

Vikas Mishra, Ph.D.

Delve Publishing

www.delvepublishing.com

Insecticides: Advances in Integrated Pest Management

Editor: Vikas Mishra, Ph.D.

708 3rd Avenue, 6th Floor
New York, NY 10017
United States of America

www.delvepublishing.com

ISBN: 978-1-68095-757-0

Library of Congress Control Number: 2016951513

About the Editor

Vikas Mishra, Ph.D.

Vikas obtained his Ph.D. from King George's Medical University, India in 2010. His interests span from bio-resources, fermentation technology to medical virology and neuroscience research. He is currently working as a research associate at the University of British Columbia.

Contents

List of Contributors

C.O. Ehi-Eromosele
Chemistry Department, School of Natural and Applied Sciences, College of Science and Technology, Covenant University, Canaan Land, Ota, Ogun State, Nigeria

O.C. Nwinyi
Department of Biological Sciences, School of Natural and Applied Sciences, College of Science and Technology, Covenant University, Canaan Land, Ota, Ogun State, Nigeria

O.O. Ajani
Chemistry Department, School of Natural and Applied Sciences, College of Science and Technology, Covenant University, Canaan Land, Ota, Ogun State, Nigeria

Charles M Benbrook
Centre for Sustaining Agriculture and Natural Resources, Washington State University

Hanem Fathy Khater
Medical and Veterinary Parasitology, Faculty of Veterinary Medicine, Benha University, Egypt

G.R. Mohammadi
Department of Crop Production and Breeding, Faculty of Agriculture and Natural Resources, Razi University, Kermanshah, Iran

Nadeem Sheikh
University of the Punjab, Quaid-i-Azam Campus, Lahore, Pakistan

Mohammad Zahangeer Alam
Department of Environmental Science, Bangabandhu Sheikh Mujibur Rahman Agricultural University, Gazipur 1706, Bangladesh

Md. Manjurul Haque
Department of Environmental Science, Bangabandhu Sheikh Mujibur Rahman Agricultural University, Gazipur 1706, Bangladesh

Md. Sirajul Islam
Farm Management Division, Bangladesh Rice Research Institute (BRRI), Gazipur 1701, Bangladesh

Emran Hossain
Administration and Common Service, Bangladesh Rice Research Institute (BRRI), Gazipur 1701, Bangladesh

Sabiha Binta Hasan
Jatiya Kabi Kazi Nazrul Islam University, Trishal, Mymensingh, Bangladesh

Shahela Binte Hasan
National University, Gazipur, Bangladesh

Md. Sakhawat Hossain
Ministry of Public Administration, Government of Bangladesh, Bangladesh

Mohammad Zahangeer Alam
Department of Environmental Science, Bangabandhu Sheikh Mujibur Rahman Agricultural University, Gazipur, Bangladesh

Alex R. Crump
Pacific Northwest National Laboratory, Richland, WA, USA

Md. Manjurul Haque
Department of Environmental Science, Bangabandhu Sheikh Mujibur Rahman Agricultural University, Gazipur, Bangladesh

Md. Sirajul Islam
Farm Management Division, Bangladesh Rice Research Institute, Gazipur, Bangladesh

Emran Hossain
Administration and Common Service, Bangladesh Rice Research Institute, Gazipur, Bangladesh

Sabiha B. Hasan
Department of Economics, Jatiya Kabi Kazi Nazrul Islam University, Mymensingh, Bangladesh

Shahela B. Hasan
Social Welfare, National University, Gazipur, Bangladesh

Md. Sakhawat Hossain
Ministry of Public Administration, Dhaka, Bangladesh

Petros Damos
Open University of Cyprus, Faculty of Pure and Applied Sciences, Department of Environmental Conservation and Management, Main OUC building: 33, Giannou Kranidioti Ave., 2220, Latsia, Nicosia, Cyprus

Lucía-Adriana Escudero Colomar
IRTA, Sustainable Plant Protection (Entomology), IRTA-Mas Badia Agricultural Experimental Station. La Tallada d'Empordà S/N. 17134, Girona, Spain

Claudio Ioriatti
Technology Transfer Centre, Fondazione Edmund Mach, Via Edmund Mach 1, 38010 San Michele all'Adige (TN), Italy

Anges Yadouleton
Entomological Research Center of Cotonou (CREC), 06 BP 2604 Cotonou, Republic of Benin

Thibaud Martin
Centre for International Cooperation in Agronomic Research for Development, CIRAD-UR HortSys, 34980 Montpellier, France

Gil Padonou
Entomological Research Center of Cotonou (CREC), 06 BP 2604 Cotonou, Republic of Benin

Fabrice Chandre
Institute of Research for Development (IRD) / Laboratory Centre against Insect Pests (LIN), WHO Collaborating Centre 911 Ave Agropolis BP 64501, 34394 Montpellier Cedex 5 France

Alex Asidi
Entomological Research Center of Cotonou (CREC), 06 BP 2604 Cotonou, Republic of Benin

Luc Djogbenou
Institute of Research for Development (IRD) / Entomological Research Center of Cotonou (CREC), 01 BP 4414 RP Cotonou, Republic of Benin

Roch Dabiré
Institute for Research in Health Science (ICES) / Centre Muraz, BP 390, Bobo-Dioulasso, Burkina Faso

Rock Aïkpon
Entomological Research Center of Cotonou (CREC), 06 BP 2604 Cotonou, Republic of Benin

Michel Boko
University of Abomey-Calavi, Department of geography, PB 526 Abomey, 01 BP 526 Cotonou, Benin

Isabelle Glitho
Lome University, Faculty of Science and Technology, Lome, Togo

Martin Akogbeto
Entomological Research Center of Cotonou (CREC), 06 BP 2604 Cotonou, Republic of Benin

S. A. Dhenge
Department of Extension Education, Post Graduate Institute, Dr. Panjabrao Deshmukh Krishi Vidyapeeth, Akola

D.M.Mankar
Department of Extension Education, Post Graduate Institute, Dr. Panjabrao Deshmukh Krishi Vidyapeeth, Akola

Barbara Nieradko-Iwanicka
Medical University of Lublin, Chair and Department of Hygiene Poland

Chris Guiver
Environment & Sustainability Institute, College of Engineering, Mathematics and Physical Sciences, University of Exeter, Penryn Campus, TR10 9FE, UK

Christina Edholm
Department of Mathematics, University of Nebraska-Lincoln, Lincoln, NE, USA

Yu Jin
Department of Mathematics, University of Nebraska-Lincoln, Lincoln, NE, USA

Markus Mueller
Environment & Sustainability Institute, College of Engineering, Mathematics and Physical Sciences, University of Exeter, Penryn Campus, TR10 9FE, UK

Jim Powell
Department of Mathematics and Statistics, Utah State University, Logan, UT, USA

Richard Rebarber
Department of Mathematics, University of Nebraska-Lincoln, Lincoln, NE, USA

Brigitte Tenhumberg
School of Biological Sciences, University of Nebraska-Lincoln, Lincoln, NE, USA

Stuart Townley
Environment & Sustainability Institute, College of Engineering, Mathematics and Physical Sciences, University of Exeter, Penryn Campus, TR10 9FE, UK

Lilian B. Calderwood
Department of Plant and Soil Science, University of Vermont, 63 Carrigan Dr., Burlington, VT 05405

Scott A. Lewins
Department of Biology, Saint Michael's College, 1 Winooski Park, Colchester, VT 05439

Heather M. Darby
Department of Plant and Soil Science, University of Vermont Extension, 278 South Main St., St. Albans, VT 05478

William R. Kuhn
Department of Biological Sciences, Rutgers University, 195 University Ave., Newark, NJ 07102

Roger R. Youngman
Department of Entomology, Virginia Tech, 216A Price Hall, Blacksburg, VA 24061

Shaohui Wu
Department of Entomology, Virginia Tech, 216A Price Hall, Blacksburg, VA 24061

Curtis A. Laub
Department of Entomology, Virginia Tech, 216A Price Hall, Blacksburg, VA 24061

Matthew H. Meisner
Department of Statistics and Center for Population Biology, University of California-Davis, Davis, California, USA

Jay A. Rosenheim
Department of Entomology and Nematology, University of California-Davis, Davis, California, USA

Ilias Tagkopoulos
Department of Computer Science, University of California-Davis, Davis, California, USA

Preface

Integrated Pest Management (IPM) is an effective and environmentally sensitive approach to pest management that relies on a combination of common-sense practices. IPM aims to suppress pest populations below the economic injury level. IPM programs use current, comprehensive information on the life cycles of pests and their interaction with the environment. Integrated pest management depends on a systematic plan to control pests with deep emphasis on the eradication or minimizing the effect of pests on crops in an effective, economical and environmentally friendly manner. With geographical variation and in accordance to the crop requirements, integrated pest management varies from site to site. Despite of site specific variations, the key components of the integrated pest management involves identification and enumeration of pest populations, impact on crop, assessment and utilization of best available pest management techniques. Development of insecticide resistance, dominance of secondary pests, effect of pest control methods on the non-target organisms, environmental impact and cost evaluation may change the course and effectiveness of the employed pest control measures and thus each of the these components requires a thorough examination in a site dependent manner. The aim of this book is to provide an overview of integrated pest management with highlights on the areas of major concerns. The book compiles case studies detailing different pest control measures and their use in different geographical, crop and entomological disciplines. Integrated pest management, alternative insect control strategies, ecological impact of insecticides as well as pesticides and drugs of forensic interest are included. The content coverage in this book should appeal the reader's concern to upkeep rational decisions about the use of pesticides. IPM emphasizes the growth of a healthy crop with the least possible disruption to agro-ecosystems and encourages natural pest control mechanisms." Entomologists and ecologists have urged the adoption of IPM pest control since the 1970s. IPM allows for safer pest control. This includes managing insects, plant pathogens and weeds. Globalization and increased mobility often allow increasing numbers of invasive species to cross national borders. IPM poses the least risks while maximizing benefits and reducing costs. Synthetic pesticides are used as required and often only at specific times in a pest's life cycle. Many newer pesticides are derived from plants or naturally occurring substances (e.g.—nicotine, pyrethrum and insect juvenile hormone analogues), but the toxophore or active component may be altered to provide increased biological activity or stability. Applications of pesticides must reach their intended targets. Matching the application technique to the crop, the pest, and the pesticide is critical. The use of low-volume spray equipment reduces overall pesticide use and labor cost.

Editor
Vikas Mishra, Ph.D.

INTRODUCTION

Integrated pest management is a balanced, tactical approach to pest control. It involves taking action to anticipate pest outbreaks and to prevent potential damage. IPM utilizes a wide range of pest control strategies or tactics. The goal of this strategy is to prevent pests from reaching economically or aesthetically damaging levels with the least risk to the environment.

IPM programs are very site-specific. IPM is based on the identification of pests, accurate measurement of pest populations, assessment of damage levels, and knowledge of available pest management strategies or tactics that enable the specialist to make intelligent decisions about control. IPM offers the possibility of improving the effectiveness of pest control programs while reducing some of the negative effects. Many successful IPM programs have reduced pesticide use and increased protection of the environment.

Pesticide use is and will continue to be significant in food and fiber production, forestry, turf and landscape maintenance, and public health. Pest management has shifted from relying heavily on pesticides to using an integrated approach based on pest assessment, decision making, and evaluation.

Why Practice IPM?

Why have pest managers shifted to IPM when chemical pesticides so often succeed at controlling pests? There are many reasons to broaden pest management beyond the use of chemicals.

IPM helps to keep a balanced ecosystem

Every ecosystem, made up of living things and their non-living environment, has a balance; the actions of one kind of organism in the ecosystem usually affect other species. Introducing chemicals into the ecosystem can change this balance, destroying certain species and allowing other species (sometimes pests themselves) to dominate. Pesticides can kill beneficial insects that consume pests, leaving few natural mechanisms of pest control.

Pesticides can be ineffective

Chemical pesticides are not always effective. Pests can become resistant to pesticides. In fact, some 600 cases of pests developing pesticide resistance have been documented to date, including many common weeds, insects, and disease-causing fungi. Furthermore, pests may survive in situations where the chemical does not reach pests, is washed off, is applied at an improper rate, or is applied at an improper life stage of the pest.

IPM can save money

IPM can avoid crop loss caused by pests and prevent unnecessary pesticide expense. Applicators can save on pesticide costs because the need for control, rather than routine application triggered by the calendar, is the basis for applying pesticides.

IPM promotes a healthy environment

We have much to learn about the persistence of chemicals in the environment and their effect on living creatures. Cases of contaminated groundwater appear each year, and disposal of containers and unused pesticides still pose challenges for applicators. Make sure that environmental impacts are considered in any pest management decisions. Using IPM strategies helps keep adverse effects to a minimum.

IPM maintains a good public image

IPM is now demanded by many sectors of our society. IPM has been implemented to grow our food, to manage turf and ornamentals, to protect home and business structures, to manage school grounds, and to protect humans, pets, and livestock health.

Components of an Integrated Pest Management Program

Planning is at the heart of an IPM program. Every crop has pests that need to be considered. If you wait until problems arise during a growing season, you'll end up relying on pesticides more and more. A good Integrated Pest Management program has three components:

1) identifying and monitoring pest problems;
2) selecting the best pest management tactics;
3) recordkeeping and evaluating the program.

By considering each of these components, applicators can set up an IPM program for insects, plant diseases, weeds, and vertebrate pests.

Pest Control Goals

There are many tactics that are effective in controlling pests, and each one (or combinations) will give pesticide applicators the best strategy for a specific goal.

Prevention

Prevention, suppression, and eradication are three approaches to maintain pest damage below economic levels. Prevention includes such things as planting weed- and disease-free seed and growing varieties of plants resistant to diseases or insects, sanitation, using cultural controls to prevent weedy plants from seeding, and choosing planting or harvesting times that minimize pest problems. Pesticides are sometimes used for pest prevention as well.

Suppression

Suppressive pest-control methods are used to reduce pest population levels. The methods chosen usually do not eliminate all pests, but reduce their populations to a tolerable level or to a point below an economic injury level; additional suppressive measures may be required if the first attempt does not achieve the management goal.

Eradication

Eradication is the total elimination of a pest from a designated area. Over larger areas eradication may be very expensive and often has limited success. Large eradication programs are usually directed at exotic or introduced pests posing an immediate area-wide public health or economic threat.

Pest Management Methods

The pest control strategy or tactics you choose depends on the nature of the pest, the environment of the pest, and economic or tolerance considerations. The combination of prevention and suppression techniques usually enhances a pest-management program. When implementing the IPM program, try to select the methods that are the most effective and the least harmful to people and the environment. Use several methods whenever possible, and be sure to use them correctly. It is also important to observe all state and federal regulations regarding the methods chosen or combined in an IPM program.

Natural Controls

Natural Controls are the measures that check or destroy pests without depending on humans for their continuance or success. Natural controls include climatic factors such as wind, temperature, sunshine, and rain. Topographic features such as rivers, lakes, and mountains can influence pest movement. Naturally occurring predators, parasites, and pathogens can regulate pest populations. When natural controls have not held pests in check, humans must intervene and apply pest-management controls. Maintaining populations of natural enemies by avoiding damaging cultural practices or the indiscriminate use of pesticides can be one of the most economical means of control. If pesticides are part of your control program, select types that are known to be less toxic to natural enemies or, if recommended, apply pesticides at lower-than-label rates to avoid harming natural enemies. Sometimes it is possible to modify certain parts of the environment, such as by planting crops or ground covers, to maintain or enhance natural enemies.

Biological Control

Most pests have natural enemies that control or suppress them effectively in some situations. Natural enemies, including pathogens and insects, are being used successfully as biological control agents to manage certain insect, mite, fungus, animal, and weed pests. Biological control is often directed against pests that are not native to a geographical area. Introduced pests often cause problems in their new locations because they lack natural enemies to help control them. Laws have been enacted that strictly control the importation of all organisms, including biological control agents, into the United States, to prevent these organisms from also becoming pests. Biological control also involves the mass release of large numbers of natural enemies into fields, orchards, greenhouses, or other locations to control specific pests. This method usually does not have long-term results, so these natural enemies must be released periodically. Several natural enemies are reared or cultured commercially. Predatory mites are used to control plant-feeding spider mites. Parasitic wasps and lacewings are used to control various insect pests. Nematodes and fungi are being studied as biological control agents for certain weeds and some insects. General predators, such as praying mantids and lady beetles, are sold with claims made for biological control.

Mechanical Control

Mechanical control involves the use of devices, machines, and other physical methods to control pests or alter their environment. Traps, screens, barriers, fences, and nets are examples of devices used to prevent pest activity or remove pests from an area.

Cultivation: Cultivation is one of the most important methods of controlling weeds. It is also used for some insects and other soil-inhabiting pests. Devices such as plows, disks, mowers, cultivators, and bed conditioners physically destroy weeds or control their growth and disrupt soil conditions suitable for the survival of some microorganisms and insects.

Exclusion: Exclusion is a mechanical control technique that consists of using barriers to prevent pests from getting into an area. Window screens, for example, exclude flies, mosquitoes, and other flying insects. Patching or sealing cracks, crevices, and other small openings in buildings can exclude insects, rodents, bats, birds, or other pests. Fences and ditches make effective barriers against many vertebrate pests. Wire or cloth mesh excludes birds from fruit trees. Sticky material painted onto tree trunks, posts, wires, and other objects prevents crawling insects from crossing.

Trapping Traps physically catch pests within an area or building. Several types of traps are commonly used. Some kill animals that come in contact with the trap; others snare animals so they can then be relocated or destroyed. Traps are either mechanical devices or sticky surfaces.

Cultural Control

The goal of cultural control is to alter the environment, the condition of the host, or the behavior of the pest to prevent or suppress an infestation. It disrupts the normal relationship between the pest and the host and makes the pest less likely to survive, grow, or reproduce. Cultural practices and sanitation are two examples of cultural control.

Cultural Practices: Many cultural practices influence the survival of pests. In agricultural crops, selection of crop plant varieties, timing of planting and harvesting, irrigation management, crop rotation, and use of trap crops help reduce populations of weeds, microorganisms, insects, mites, and other pests. Weeds also can be managed by mulching (with plastic, straw, shredded bark, or wood chips) and by using cover crops.

Sanitation: Sanitation, or source reduction, involves eliminating food, water, shelter, or other necessities important to the pest's survival. In crop production, sanitation includes such practices as removing weeds that harbor pest insects or rodents, eliminating weed plants before they produce seed, destroying diseased plant material or crop residues, and keeping field borders or surrounding areas free of pests and pest breeding sites. Animal manure management is an effective sanitation practice used for preventing or reducing fly problems in poultry and livestock operations.

Insecticide

An insecticide is a substance used by humans to gain some advantage in the struggle with various insects that are considered "pests." In the sense used here, a pest insect is considered undesirable, from the human perspective, because: (a) it is a vector that transmits disease-causing pathogens to humans (such as those causing malaria or yellow fever), or other diseases to livestock or crop plants; or (b) it causes a loss of the productivity or economic value of crop plants, domestic animals, or stored foodstuffs. The abundance and effects of almost all insect pests can be managed through the judicious use of insecticides.

However, the benefits of insecticide use are partly offset by important damages that may result. There are numerous cases of people being poisoned by accidental exposures to toxic insecticides. More commonly, ecological damage may be caused by the use of insecticides, sometimes resulting in the deaths of large numbers of wildlife.

Types of Insecticides

There are five different types of insecticides that include Stomach Insecticides, Systemic Insecticides, Contact Insecticides, Translaminar Insecticides and Fumigant Insecticides. All are either natural (organic), man-made (synthetic) formulas or preparations that are used to control or kill unwanted insects. Although insecticides have widespread use in agricultural applications, natural (organic) formulas have been used in homes for centuries. The 20th century saw the introduction of many man-made (synthetic) formulas that have been used for controlling pests in residential and workplace settings.

Stomach Insecticides

Stomach insecticides are mostly recommended and used against chewing type of insect pests because these insecticides enter into the stomach of the insect through their food.

Systemic Insecticides

The most common application for systemic insecticides is through soil drenching. The insecticide is introduced into the soil where it is absorbed by plant roots. It then moves up through the plant to external areas (leaves, twigs, fruits, branches), where it lays on the plant surface area and is poisonous to any insects that come chewing on the plant.

Contact Insecticides

Contact insecticides act like "bullets" aimed at a target and can only kill insects that are hit by its application. Common household insect spray acts much like a contact insecticide. Because this type must directly hit the insect, it is ineffective as a preventative insecticide.

Translaminar Insecticides

Translaminar insecticides have special characteristic as compared to other insecticides because why these are applied on the leaves of the plants these easily cross the lamina of the leaf and moves to other surface of the leaf and therefore it is mostly recommended against sucking type insect pests of different crops.

Fumigant Insecticides

Fumigant insecticides are most of the time used against indoor insects because these enter the body of the insect or target organism through their spiracles or tracheae in gaseous form.

Keep These Tips in Mind When Using Insecticides

- Practicing Integrated Pest Management can significantly reduce the amount of insecticides needed to control many insect problems.
- Using more than one insecticide product in the same location can increase or decrease each one's effectiveness. It may also result in a greater risk to health and/or the environment.
- Broad-spectrum insecticides are effective against all insects, even the good ones. Other insecticides target certain insects. Using a targeted insecticide minimizes the risk to beneficial or non-target insects.
- Some insecticides work immediately to kill insects while others may need some time to take effect.
- Insect growth regulators like pyriproxyfen and methoprene do not kill insects; they make it impossible for exposed insects to molt (grow) or lay eggs properly.
- Insecticidal baits can be used instead of spraying large areas, especially for social insects like ants. This can decrease the risk of exposure, but remember to place baits where children and pets won't have access.

CHAPTER 1 INTEGRATED PEST MANAGEMENT

C.O. Ehi-Eromosele, O.C. Nwinyi and O.O. Ajani

INTRODUCTION

Integrated Pest Management (IPM) has had a varied history, with different definitions. It has been implemented under an array of different connotations (Lewis *et al*., 1997). The term was earlier used as "integrated control" by Bartlett (Bartlett, 1956) and was further elaborated on by Stern and co-workers (Stern *et al*., 1959). In reference to the concept of integrating the use of biological and other controls in complementary ways, the term was later broadened to embrace coordinated use of all biological, cultural, and artificial practices (van den Bosch and Stern, 1962). The term "IPM," under various authors have advocated for the principle of incorporating, the full array of pest management practices with production objectives in a total systemic approach. Nonetheless, there is no universally agreed definition of IPM.

Alastair (2003) conceived IPM as a method of rationalizing pesticide use to prevent or delay the resurgence of pest populations, that had become resistant to pesticides, and to protect beneficial insects. Today, concerns about pesticide residues in the food chain and in the environment have led to alternative definitions that exclude the use of conventional pesticides. Nevertheless, there are broad agreements on the core principles of IPM. These include IPM being:

- An integrated scheme due to the methods of control that are seen as component technologies rather than alternatives.
- emphasizes pest management, within a balanced system whereas control strategy suggests direct intervention with little concern for sustainability (Alastair, 2003).

Citation: C.O. Ehi-Eromosele, O.C. Nwinyi and O.O. Ajani (2013). Integrated Pest Management, Weed and Pest Control - Conventional and New Challenges, Dr. Sonia Soloneski (Ed.), InTech, DOI: 10.5772/54476.

In principle, IPM may be defined as a flexible and holistic system. This views the agro-ecosystem as an interrelated whole, that utilizes a variety of biological, cultural, genetic, physical, and chemical techniques that hold pests below economically damaging levels with a minimum disruption to the cropping ecosystem and the surrounding environment (Malena, 1994). In the definition of FAO (2012), integrated pest management (IPM) is an ecosystem approach to crop production and protection that combines different management strategies and practices to grow healthy crops and minimize the use of pesticides. IPM therefore, utilizes the best mix of control tactics for a given pest problem when compared with the crop yield, profit and safety of other alternatives (Kenmore *et al.,* 1985). Other definitions of IPM according to the United States Environment Protection Agency (2012), involves an effective and environmentally sensitive approach to pest management that relies on a combination of common-sense practices. IPM could be a broad based ecological approach to structural and agricultural pest control that integrates pesticides/herbicides into a management system incorporating a range of practices for economic control of a pest. In addition, IPM, attempts to prevent infestation, to observation of patterns of infestation when they occur, and to intervene (without poisons) when one deems necessary. Finally, IPM is the intelligent selection and use of pest control actions that ensures favourable economic, ecological and sociological consequences (Sandler, 2010).

Succinct History of IPM

IPM is not a new philosophy. The concept has been around since the 1920's when cotton pest management program was developed. Under this scheme, insect control was "supervised" by qualified entomologists, and insecticide applications were based on conclusions reached from periodic monitoring of pest and natural-enemy populations. This was viewed as an alternative to calendar-based insecticide programs. Supervised control is based on a sound knowledge of the ecology and analysis of the projected trends in pest and natural-enemy populations. In supervised control, (integrated control) the best mix of chemical and biological controls is sought and identified for a given insect pest. The chemical insecticides are used in a manner that is least disruptive to biological control. The chemical controls are applied only after regular monitoring indicates that a pest population had reached an economic threshold level. Thus, such treatment is required to prevent the population from reaching an economic injury level where economic losses would exceed the cost of the artificial control measures.

Typically, the main aim of IPM programmes is on agricultural insect pests (IPM Guidelines, 2009). Although, originally developed for agricultural pest management, IPM programmes are now developed to encompass diseases, weeds, and other pests that may interfere with the management objectives of sites such as residential and commercial structures, lawn and turf areas, and home and community gardens. IPM programs use current, comprehensive information on the life cycles of pests and their interaction with the environment. This information, in combination with available pest control methods, is used to manage pest damage by the most economical means, and with the least possible hazard to people, property, and the environment.

The IPM approach can be applied to both agricultural and non-agricultural settings, such as the home, garden, and workplace. IPM takes advantage of all appropriate pest management options including, the judicious use of pesticides.

IPM – A PHILOSOPHY

A Pest Management Strategy

Integrated Pest Management (IPM) is a philosophy that involves the management of a pest instead of controlling or eradicating a pest. It requires a greater knowledge of the pest, crop and the environment. Therefore, its strategy focuses on harnessing inherent strengths within ecosystems and directing the pest populations into acceptable bounds rather than toward eliminating them. This strategy avoids undesirable short term and long term ripple effects and will ensure a sustainable future (Lewis *et al.*, 1997).

IPM programs should be operated with "pest management objectives" rather than "pesticide management objectives". Integrated pest management is a comprehensive long term pest management program based on knowledge of an ecosystem that weighs economic, environmental, and social consequences of interventions (Flint and van den Bosch, 1981). The foundation for pest management in agricultural systems should be an understanding and shoring up of the full composite of inherent plant defenses, plant mixtures, soil, natural enemies, and other components of the system. These natural "built in" regulators are linked in a web of feedback loops that are renewable and sustainable. The use of pesticides and other "treat-the-symptoms" approaches are unsustainable and should be the last option rather than the first line of defense. A pest management strategy should always start with the question "Why is the pest a pest?". It should also seek to address underlying weaknesses in ecosystems and/or agronomic practice(s) that have allowed organisms to reach pest status (Lewis *et al.*, 1997).

An Integrated Process

Integration or compatibility among pest management tactics is central to Integrated Pest Management. Simply mixing different management tactics does not constitute IPM. Mixing the tactics arbitrarily may actually aggravate pest problems or produce other unintended effects. IPM recognizes there is no "cure-all" in pest control (dependence on any one pest management method will have undesirable effects). Reliance on a single tactic will favor pests that are resistant to that practice. In IPM, integrated control seeks to identify the best mix of chemical and biological controls for a given insect pest. The term "integrated" is thus synonymous with "compatibility."

Understanding Pest Biology and Ecology

The determination of the correct cause of pest problem (understanding pest biology) and ecology is essential in manipulating the environment to the crop's advantage and to the detriment of the pest.

Acceptable Pest Levels

IPM recognizes that eradication of a pest is seldom necessary or even desirable, and generally not possible. The primary objective in pest management is not to eliminate a pest organism but to bring it into acceptable bounds (Lawal *et al.*, 1997). The emphasis is on control, not eradication. IPM holds that wiping out an entire pest population is often impossible, and the attempt can be expensive and environmentally unsafe. IPM programmes initial task is to establish acceptable pest levels, called action thresholds, and apply controls where the thresholds are crossed. These thresholds are pest and site specific, meaning that it may be acceptable at one site to have for instance a weed such as white clover, but at another site it may not be acceptable. By allowing a pest population to survive at a reasonable threshold, selection pressure is reduced. This stops the pest gaining resistance to chemicals produced by the plant or applied to the crops. If many of the pests are killed, then any that have resistance to the chemical will form the genetic basis of the future, more resistant, population. By not killing all the pests there are some un-resistant pests left that will dilute any resistant genes that appear (Wikipedia, 2011).

IPM A Continuum, Not an End

Agriculture is a dynamic system that continually changes to changing crop production practices. IPM must continually change to meet pest management challenges. IPM is a continuum that will change with time. Every farmer practices some type of IPM, as long as they make progress to better its management. As new pest control techniques are discovered, the producer and crop advisor must adapt their pest control program to reflect these changes. What is considered a good IPM program today may be considered a chemical intensive program in a few years. Additionally, some good advice to the producer and crop advisor is to try the new changes on a limited scale, while becoming comfortable with the suggested practices before wide-scale changes are made.

IPM PROCESS

IPM is applicable to all types of agriculture and sites such as residential and commercial structures, lawn and turf areas, and home and community gardens. The process includes:

Proper Identification Pest Damage and Responsible Pests

Identification must be the first objective. When the identity of a pest is not known, then, a strategy built to control the pest cannot be transferred from one site to another, primarily, because the pest species or strain (biotype) might behave differently. Thus, a solid foundation must be built on systematic, taxonomy, etiology, and spatial distribution (Irwin, 1999). Cases of mistaken identity may result in ineffective actions. If plant damage is due to over-watering, it could be mistaken for fungal infection, since many fungal and viral infections arise under moist conditions. This could lead to spray costs, but the plant would be no better off.

Pest and Host Life Cycles Biology

Understanding crop growth and development is an underlying principle of IPM. We cannot just focus on the pest. The interactions between crop and pest (as well as the environment) are very important. To deplore an efficient IPM programme, literature and other data sources about the pest, the pest's life cycle, host range, distribution, movement, and basic biology will have to be researched. At the time you see a pest, it may be too late to do much about it except maybe spray with a pesticide (Metcalf and Luckmann, 1994). Often, there is another stage of the life cycle that is susceptible to preventative actions. For example, weeds reproducing from last year's seed can be prevented with mulches and pre-emergent herbicide. Also, learning what a pest needs to survive allows you to remove these.

Monitor or Sample the Environment for Pest Populations

After the pest has been correctly identified, monitoring must begin before it becomes a problem. Sampling and monitoring methodologies must be designed and tested to provide the ability for assessing instantaneous and dynamic aspects of the pest's density, activity, or incidence (Irwin, 1999). Understanding how the environment affects pest and crop development is very important. Understanding interactions with the environment allows crop advisors to react to changing conditions. Environmental influences like drought stress influences pest management recommendations. When a crop is under stress it can be less capable of dealing with the stress caused by insects that extract plant sap (e.g. aphids, leafhoppers) and this stress may slightly lower the economic threshold. Weed populations which would not normally cause an economic loss may do so under drought conditions when they compete with the crop for limited water.

The weather is notorious for affecting pest development and survival. Certain weather patterns may affect weed seed germination and explain why certain weeds are more abundant during wet fall or springs.

Establish Action Threshold (Economic, Health, And Aesthetic)

The question here is: how many are too many or how much can be tolerated? In some cases, there are standardized numbers of pests that can be tolerated. Soybeans are quite tolerant of defoliation, so if there are a few caterpillars in the field and their population may not be increasing dramatically; thus, no urgent action may be necessary. Conversely, there is a point at which an action must be taken to control cost. For instance the farmer can control cost at the point when the cost of damage by the pest is more than the cost of control. This is an economic threshold. Tolerance of pests varies according to the health hazard (low tolerance) or merely a cosmetic damage (high tolerance in a non-commercial situation). Different sites may also have varying requirements based on specific areas. For instance, white clover may be perfectly acceptable on the sides of a tee box on a golf course, but unacceptable in the fairway where it could cause confusion in the field of play (Purdue University, 2006).

Choose an Appropriate Combination of Management Tactics

The word 'integrated' in IPM initially referred to the simultaneous use or integration of any number of tactics in combination, with focus on maintaining a single pest species below its economic injury level. Although, in theory a single strategy results from the simultaneous integration of several tactics, in practice, the integration actually occurs in a step-wise, time-delayed fashion. Several of the tactics are compatible, but some are not. Certainly the tactics of biological control, habitat manipulation, and legal control go alongside. The tactic of host resistance can stand alone or be combined with the other tactics just mentioned. Chemical control is generally compatible with host resistance. Thus, a management strategy integrates one or several compatible tactics into a single package (Irwin, 1999).

Evaluate and Record Results

Evaluation is often one of the most important steps in Integrated Pest Management (Bennett *et al.*, 2005). It is the process of reviewing an IPM program and the results it has generated. Asking the following questions is useful: Did the steps one took effectively control the population? Was this method safe enough? Where there any expected side effects? What is the next step? Understanding the effectiveness of the IPM program allows the site manager to make modifications to the IPM plan prior to pests reaching the action threshold and requiring action again.

PEST MANAGEMENT TACTICS

There are different pest management tactics to suppress pests. They include host resistance, chemical, biological, cultural, mechanical, sanitary and mechanical controls. The primary pest management tactic involves maximization of built-in pest reduction features of an ecosystem. Molecular or genetic mechanisms are potentially manifested in a number of these more specific tactics. Each category, discussed below, employs a different set of mechanisms for suppressing populations.

Chemical Control

The therapeutic approach of killing pest organisms with toxic chemicals has been the prevailing pest control strategy for over 50 years. Safety problems and ecological disruptions continue to ensue (Wright, 1996), and there are renewed appeals for effective, safe, and economically acceptable alternatives (Benbrook, 1996). Synthetic chemical pesticides are the most widely used method of pest control. The four major problems encountered with conventional pesticides are toxic residues, pest resistance, secondary pests, and pest resurgence (Lewis, 1997). The use of natural pesticides and organophosphates that are more environmentally friendly are encouraged and synthetic pesticides should only be used as a last resort or only used as required and often only at specific times in a pests life cycle.

Biological Control

This involves the use of other living things that are enemies of a pest in order to control it. Sometimes, the term "biological control" has been used in a broad context to encompass a full spectrum of biological organisms and biologically based products including pheromones, resistant plant varieties, and autocidal techniques such as sterile insects. IPM is mainly aimed at developing systems based on biological and non-chemical methods as much as possible.

Host Plant Resistance

This involves breeding varieties with desirable economic traits, but less attractive for pests, for egg laying and subsequent development of insect, disease or nematode. It also involves withstanding the infestation/infection or the reduction of pests to level that they are not large numbers during the plant growth period (Sharma, 2007).

Cultural Measures

This involves practices that suppress pest problems by minimizing the conditions that favour their existence (water, shelter, food). Some of these factors are intrinsic to crop production while making the environment less favourable for survival, growth and reproduction of pest species. If followed in an appropriate manner, the cultural practices can provide significant relief from pests. The selection of appropriate site for the cultivation of field crops and fruit trees can reduce future infestation from insect pests. The culture should be selected in such a manner that it should be suitable for growing in the area and tolerant to important pests diseases of the area.

Mechanical Control

This is the use of machinery and other tools to control pests. It involves agricultural practices like tillage, slash and burn, and hand weeding. The pruning of infested parts of fruits and forest trees and defoliation in certain crops help reduce the pest population. Chaffing of sorghum/maize stalks and burning of stubbles kills maize borer.

Sanitary Control

Preventive practices are important part of an IPM programme. These include cleaning field equipment (i.e., tillage equipment, haying equipment, etc.), planting certified seeds and quarantine of infested crops or farmlands. These are methods used to prevent the introduction of a pest into the field.

Natural Control

Natural control involves the enhancement of naturally occurring pest management methods to combat pests like using beneficial insects and diseases. Here, insecticides will only be used when they are economically feasible and it is apparent that natural enemies will not control the pests.

IPM: A MULTI-DISCIPLINARY APPROACH

IPM is a management intensive philosophy which stresses a multidisciplinary approach. Pests interact with each other, the crop, and the environment. Similarly, pest and crop management disciplines must work together to develop control recommendations that reflect these interactions.

For example, management of the Soybean aphid includes entomologists who study the insect and their damage to soybean, agronomists that identify crop stage which are most vulnerable to soybean aphid damage, plant pathologists who study the viruses transmitted by aphid feeding, and soil scientists who study the aphids interaction with nutrient deficiencies.

BENEFITS OF AN IPM PROGRAMME

The benefits of Integrated Pest Management are immense directly to farming and indirectly to society.

- Integrated Pest Management (IPM) protects environment through elimination of unnecessary pesticide applications. In IPM, pesticides are used at the smallest effective dose when other methods of pest control have failed. Also, they are used in bringing a pest organism to acceptable bounds with as little ecological disruption as possible.
- IPM improves profitability. Since IPM programme applies the most economical management pest tactics, profitability is ensured for the grower or farmer.
- It reduces risk of crop loss by a pest. Applying pest management and monitoring tactics will also ensure the reduction of crop loss or damage by pests.
- Long term sociological benefits of IPM would also emerge in areas of employment, public health, and well being of persons associated with agriculture.

DISADVANTAGES OF AN IPM PROGRAMME

In spite of the numerous benefits of IPM stated so far, there are also some drawbacks to it:

An IPM Program Requires a Higher Degree of Management

Making the decision not to use pesticides on a routine or regular basis requires advanced planning and therefore, a higher degree of management. This planning includes attention to field histories to anticipate what the pest problems might be, selecting crop varieties which are resistant or tolerant to pest damage, choosing tillage systems that will suppress anticipated pest damage while giving the crop the greatest yield potential.

IPM Can Be More Labor Intensive

Consistent, timely and accurate field scouting takes time. However, it is this information that is necessary and is the corner stone of IPM programs. Without this information you cannot make intelligent management decision.

Success Can Be Weather Dependant

Weather can complicate IPM planning. For example you might want to lower herbicide rates and use row cultivation to manage weed pressure. However an extended wet period may reduce (or eliminate) the effectiveness of row cultivation. Therefore, good IPM planners will have a alternate plan for when these problems arise.

REFERENCES

1. Alastair, O. (2003) Integrated Pest Management for Resource-Poor African Farmers: Is the Emperor Naked? World Development Vol. 31, No. 5, pp. 831–845
2. Barlett, B. R. (1956) Natural predators. Can selective insecticides help to preserve biotic control Agric. Chem. 11, 42–44.
3. Benbrook, C. M. (1996) Pest Management at the Crossroads (Consumer's Union, Yonkers, NY).
4. Bennett, G.W., Owens, J.M., Corrigan, R.M. (2005) Truman's Scientific Guide to Pest Management Operations, 6th Edition, Purdue University, Questex, USA. pp. 10-12,
5. Ehler, L. E.; Bottrell D. G. (2000) The illusion of integrated pest management. Issues in Science and Technology 16, 61-64.
6. Flint, M. L. and van den Bosch, R. (1981) Introduction to Integrated Pest Management (Plenum, New York).
7. Food and Agricultural Organisation of the United Nations http://www.fao.org/agriculture/crops/core-themes/theme/pests/ipm/en/
8. IPM Guidelines . UMassAmherst: Integrated Pest Management, Agriculture and Landscape Program 2009 http://www.umass.edu/umext/ipm/publications/guidelines/index.html.
9. Irwin, M.E. (1999) Agricultural and Forest Meteorology 97: 235–248
10. Kenmore, P.E., Heong, K.L., Putter, C.A (1985) Political, social and perceptual aspects of integrated pest management programmes. In:Lee, B.S., Loke, W.H., Heong, K.L. (Eds.), Integrated Pest Management in Asia. Malaysian Plant Protection Society, Kuala Lumpur, pp. 47- 66
11. Lewis, W. J., van Lenteren, J. C., Sharad C. Phatak, and Tumlinson, J. H. (1997) Proc. Natl. Acad. Sci. USA 94: 12243–12248
12. Malena, C. (1994) Gender issues in IPM in African agriculture. Socio-economic series no. 5. Chatham: Natural Resources Institute.
13. Purdue University Turf Pest Management Correspondence Course, Introduction, 2006
14. Robert L. M., William H. L. (1994). Introduction to Insect Pest Management. New York: John Wiley and Sons, Inc. pp. 266.
15. Sandler, Hilary A. (2010). "Integrated Pest Management". Cranberry Station Best Management Practices 1 (1): 12–15.
16. Stern, V. M., Smith, R. F., van den Bosch, R. and Hagen, K. S. (1959) The integration of chemical and biological control of the spotted alfalfa aphid (the integrated control concept). Hilgardia 29:81-101.
17. Sharma, B.K. (1997) Environmental Chemistry, Krishna Ltd., Delhi pp. soil-114
18. United States Protection Agency "Pesticides and Food: What Integrated Pest Management Means". 2011. http://www.epa.gov/pesticides/food/ipm.htm.
19. United States Environmental Protection Agency "Integrated Pest Management (IMP) Principles". 2012. http://www.epa.gov/pesticides/factsheets/ipm.htm.
20. van den Bosch, R and Stern, V. M. (1962) The Integration of Chemical and Biological Control of Arthropod Pests Annu. Rev. Entomol. 7: 367–387.
21. Vereijken, P., Edwards, C., El Titi, A., Fougeroux, A. and Way., M. (1986) Proceedings of Workshop on Integrated Farming. Bull. I. O. B. C./W. P. R. S./IX 2, 1–34.
22. Wikipedia, The Free Encyclopaedia http://en.wikipedia.org/wiki/Integrated_pest_management
23. Wijnands, F. G. and Kroonen-Backbier, B. M. A. (1993) In Modern Crop Protection: Developments and Perspectives, ed. Zadoks, J. C. (Wageningen Pers, Wageningen, The Netherlands), pp. 227–234.
24. Wright, L. 1996. Silent sperm. (A reporter at large.) The New Yorker (Jan. 15):42,48,50-53,55. www.cias.wisc.edu/curriculum/.../Notes_for_IPM_powerpoint.doc

CHAPTER 2

IMPACTS OF GENETICALLY ENGINEERED CROPS ON PESTICIDE

Charles M Benbrook

Keywords: Herbicide-resistant crops, Herbicide-tolerant soybeans, Glyphosate, 2,4-D, Bt crops, Genetically engineered corn, Roundup Ready crops, Biotechnology and pesticide use, Glyphosate resistant weeds

BACKGROUND

Public debate over genetically engineered (GE) crops is intensifying in the United States (U.S.), driven by new science on the possible adverse health impacts associated with herbicide-resistant (HR) crop pesticide use, and the rapid spread of glyphosate-resistant weeds. Still, many experts and organizations assert that GE crops have reduced, and continue to reduce herbicide, insecticide, and overall pesticide use. Fortunately, high quality and publically accessible U.S. Department of Agriculture (USDA) pesticide use data are available and can be used to track changes in pesticide use on crops containing GE traits. Moreover, the impacts of these traits on U.S. pesticide use trends are substantial and obvious, especially in recent years as a result of the growing number and geographical spread of glyphosate-resistant (GR) weeds.

Stable reductions in insecticide use in *Bt*-transgenic corn are also now in jeopardy as a result of the emergence of corn rootworm (CRW) populations resistant to the Cry 3Bb1 toxins expressed in several corn hybrids [1, 2]. To combat this ominous development, some seed and pesticide companies are recommending a return to use of corn soil insecticides as a resistance management tool. There is a degree of irony in such recommendations, given that the purpose of Cry 3Bb1 corn was to eliminate the need for corn soil insecticides.

The emergence of herbicide-resistant genetically engineered crops in 1996 made it possible for farmers to use a broad-spectrum herbicide, glyphosate, in ways that were previously impossible. From 1996 through 2011, 0.55 billion hectares of HR corn (*Zea mays*), soybeans (*Glycine max*), and cotton (*Gossypium hirsutum*) were grown in the U.S. [Additional file 1: Table S7]. In 2011, an estimated 94% of the soybean area planted, 72% of corn, and 96% of cotton were planted to HR varieties, respectively, while about 65% of corn and 75% of cotton hectares in the U.S. were planted to *Bt* varieties [Additional file 1: Table S6].

Glyphosate-resistant, Roundup Ready (RR) crops now comprise the overwhelming majority of HR crops. RR crops were rapidly adopted because they provided farmers a simple, flexible, and forgiving weed management system, especially compared to systems reliant on the low-dose, persistent herbicide chemistries on the market in the late 1990s, such as imazethapyr (43% soybean hectares treated in 1996) and chlorimuron-ethyl (14% treated). From 1996 through 2008, HR crops resistant to herbicides other than glyphosate either disappeared from the market (e.g. bromoxynil HR cotton), or have been planted on relatively few hectares (e.g. glufosinate HR, LibertyLink cotton and corn).

Net reductions in pesticide use, encompassing changes in both herbicide and insecticide kilograms/pounds applied, are among the purported claims of GE crops [3, 4, 5]. Analysts assessing the impacts of *Bt* crops on insecticide use report reductions, or displacement, in the range of 25% to 50% per hectare [6]. A more recent study reports a 24% reduction [5]. On GE and non-GE corn since 1996, the volume of insecticides applied has declined, because of the pesticide industry-wide trend toward more biologically active insecticides applied at incrementally lower application rates.

Citation: Charles M Benbrook: Impacts of genetically engineered crops on pesticide. Environmental Sciences Europe 2012 24:24, doi:10.1186/2190-4715-24-24.

The corn rootworm (CRW) has been the target of the majority of corn insecticide applications the last several decades. The average corn insecticide application rate in 1996 was about 0.76 kilograms of active ingredient per hectare (kgs/ha) (0.7 pounds/acre) and is less than 0.2 kgs/ha today (0.18 pounds a.i./acre) [Additional file 1: Table S12]. The two contemporary corn soil insecticide market leaders – tebupirimiphos and tefluthrin – are applied at average rates around 0.13 kgs/ha (0.12 pounds/acre). In 1996, the market leaders were chlorpyrifos and terbufos, insecticides applied at rates above 1.12 kgs/ha (1.0 pounds/acre) [Additional file 1: Table S12]. Obviously, planting*Bt* corn in 2011 reduced insecticide use less significantly compared to land planted to *Bt* corn in the late 1990s.

Few comprehensive estimates have been made of the impacts of HR crops on herbicide use. The USDA has not issued a new estimate in well over a decade; the USDA's Economic Research Service (ERS) reported an 3.7 million kg (8.2 million pound) decrease in pesticide use in 1998 as a result of GE corn, soybeans, and cotton [7], an estimate that is comparable to the present study's estimate of a 4.4 million kg (9.6 million pound) reduction [Additional file 1: Table S15]. A series of unpublished simulation studies have been carried out by the National Center for Food and Agriculture Policy (NCFAP). In a report covering crop year 2005, NCFAP projected that HR corn, soybean, and cotton reduced total herbicide use by 25.6 million kgs, compared to hectares planted to non-HR varieties [6]. Sankula's herbicide use estimates are based on observations of mostly university experts regarding "typical" herbicide use rates on farms planting HR versus non-HR varieties. The rates incorporated in Sankula's estimates often differ from those published for the same year by USDA's National Agricultural Statistics Service (NASS) [8]. NASS reported that an average 1.5 applications of glyphosate were made on HR soybeans in 2005, while Sankula assumes only 1.18 applications. Sankula's estimate of total herbicide use on RR soybeans in 2005, 1.15 kgs/ha (1.03 pounds/acre), is less than the NASS figure for glyphosate alone, 1.23 kgs/ha (1.1 pounds/acre). If true, Sankula's data suggests that essentially no other herbicides were applied to RR soybeans in 2005, when in fact the average soybean hectare in 2002 was treated with 1.66 herbicides according to NASS data.

This paper quantifies the impacts of GE crops on the kilograms of pesticides applied per hectare and across all GE hectares, drawing upon publicly accessible USDA data. The pesticide use impacts of the six major, commercial GE pest-management traits are modeled and then aggregated over the 16years since commercial introduction. While most of the pesticide use data incorporated in the model were originally reported by U.S. government agencies in pounds of active ingredient, and/or pounds of a.i./acre, results are reported herein in SI units (kilograms of active ingredient and kg/ha). Some key results are also reported in pounds/acre. Convert kilograms to pounds by multiplying by 2.205, and pounds to kgs by multiplying by 0.454. To convert from kg/ha to pounds/acre, multiply by 0.893; to convert from pounds/acre to kg/ha, multiply by 1.12.

RESULTS AND DISCUSSION

Farmers planted 0.55 billion hectares (1.37 billion acres) of HR corn, soybeans, and cotton from 1996 through 2011, with HR soybeans accounting for 60% of these hectares [Additional file 1: Table S7]. In terms of overall herbicide use per hectare based on NASS data, substantial increases have occurred from 1996 through 2011. In soybeans, USDA reported herbicide applications totaling 1.3 kgs/ha (1.17 pounds/acre) in 1996, and 1.6 kgs/ha (1.42 pounds/acre) in 2006, the last year soybeans were surveyed by USDA. In cotton, herbicide use has risen from 2.1 kgs/ha (1.88 pounds/acre) in 1996 to 3.0 kgs/ha (2.69 pounds/acre) in 2010, the year of the most recent USDA survey. In the case of corn, herbicide use has fallen marginally from 3.0 kgs/ha (2.66 pounds/acre) in 1996 to 2.5 kgs/ha (2.26 pounds/acre) in 2010, largely as a result of lessened reliance on older, high-rate herbicides.

Compared to herbicide use rates per hectare on non-HR hectares, HR crops increased herbicide use in the U.S. by an estimated 239 million kgs (527 million pounds) in the 1996–2011 period, with HR soybeans accounting for 70% of the total increase across the three HR crops. Rising reliance on glyphosate accounted for most of this increase.

In light of its generally favorable environmental and toxicological properties, especially compared to some of the herbicides displaced by glyphosate, the dramatic increase in glyphosate use has likely not markedly increased human health risks. Because glyphosate cannot be sprayed on most actively growing, non-GE plants, residues of glyphosate in food have been rare, at least until the expansion ~ 2006 in the number of late-season glyphosate applications on wheat and barley as a harvest aid and/or to control escaped weeds. Presumably as a result of such uses, 5.6% of 107 bread samples tested in 2010 by the U.K. Food Standards Agency contained glyphosate residues [9]. Three samples had 0.5 parts per million of glyphosate [9], a relatively high level compared to the other pesticides found in these bread samples.

Budget pressures have forced the U.S. Department of Agriculture to reduce the number of crops included in its annual NASS pesticide use survey. Soybean pesticide use has not been surveyed since 2006, about when the spread of glyphosate-resistant weeds began to significantly increase herbicide use in selected areas. Herein, total herbicide use on HR hectares is projected to rise 13.5% from 2006–2011 (about 2.7% annually), compared to a 6.6% (1.3% annually) increase on conventional soybean hectares. By way of contrast, the NASS-reported glyphosate rate of application per crop year on the average hectare of soybeans increased 8.9% per annum from 2000–2006 (see Table 1). So, despite the significant and widespread challenges inherent in managing glyphosate-resistant weeds in the 2006–2011 period, a substantial decrease is projected in the rate of increase in glyphosate applications per hectare of HR soybeans. The justification for this projected fall in the rate of increase is recognition by farmers that further increases in glyphosate use will likely not prove cost-effective, coupled with positive responses by farmers to the near-universal recommendation that corn-soybean farmers incorporate into their spray programs herbicides that work through modes of action other than glyphosate's [10, 11, 12, 13, 14, 15].

Table 1 Projected rates of change in herbicide use since the most recent USDA survey, relative to recent annual percent changes in rates

	2010-2011	2005-2010	Per Year 2005-2010
Corn			
Total Herbicides	2%	10.2%	2.0%
Glyphosate	2.5%	12.9%	2.6%
Soybeans	**2007-2011**	**2000-2006**	**Per Year 2000-2006**
Total Herbicides	3.2%	35.2%	5.9%
Glyphosate	3.3%	53.4%	8.9%
Cotton	**2010-2011**	**2007-2010**	**Per Year 2007-2010**
Total Herbicides	2.2%	3.1%	1.0%
Glyphosate	-1%	-10.3%	-3.4%

Since 1996, about 317 million trait hectares (782 million trait acres) have been planted to the three major *Bt* traits –*Bt* corn for European corn borer (ECB) and CRW, and *Bt* cotton. *Bt* corn and cotton have delivered consistent reductions in insecticide applications totaling 56 million kgs (123 million pounds) over 16 years of commercial use.*Bt* corn reduced insecticide use by 41 million kgs (90 million pounds), while *Bt* cotton displaced 15 million kgs (34 million pounds) of insecticide use.

Taking into account applications of all pesticides targeted by the traits embedded in the three major GE crops, pesticide use in the U.S. was reduced in each of the first six years of commercial use (1996–2001). But in 2002, herbicide use on HR soybeans increased 8.6 million kgs (19 million pounds), driven by a 0.2 kgs/ha (0.18 pounds/acre), increase in the glyphosate rate per crop year, a 21% increase. Overall in 2002, GE traits increased pesticide use by 6.9 million kgs (15.2 million pounds), or by about 5%. Incrementally greater annual increases in the kilograms/pounds of herbicides applied to HR hectares have continued nearly every year since, leading to progressively larger annual increases in overall pesticide use on GE hectares/acres compared to non-GE hectares/acres. The increase just in 2011 was 35.3 million kgs (77.9 million pounds), a quantity exceeding by a wide margin the cumulative, total 14 million kg (31 million pound) reduction from 1996 through 2002.

Total pesticide use has been driven upward by 183 million kgs (404 million pounds) in the U.S. since 1996 by GE crops, compared to what pesticide use would likely have been in the absence of HR and *Bt* cultivars. This increase represents, on average, an additional ~0.21 kgs/ha (~0.19 pounds/acre) of pesticide active ingredient for every GE-trait hectare planted. The estimated overall increase of 183 million kgs (404 million pounds) applied over the past 16 years represents about a 7% increase in total pesticide use.

There are two major factors driving the upward trend in herbicide use on HR hectares compared to hectares planted to non-HR crops: incremental reductions in the application rate of herbicides other than glyphosate applied on non-HR crop hectares, and second, the emergence and rapid spread of glyphosate-resistant weeds. The first factor is driven by progress made by the pesticide industry in discovering more potent herbicidal active ingredients effective at progressively lower rates of application.

Twenty-seven percent of U.S. soybean hectares in 1996 were treated with pendimethalin at an average rate of 1.1 kgs/ha and another 22% were sprayed with trifluralin at a rate of 0.99 kgs/ha, while the market leader (imazethapyr) was applied to 43% of hectares planted at a rate of 0.07 kgs/ha [16]. By 2002 the combined percentage of soybean hectares treated with these two high-dose herbicides had dropped from 49% to 16% [17], and just 5% were treated in 2006 [18]. Between 1996 and 2006, the number of registered soybean herbicides applied at rates below 0.11 kgs/ha increased from nine to 17. As a result, the amount of herbicides applied to conventional crops has steadily fallen since 1996. In contrast, glyphosate is a relatively high-dose herbicide that is usually applied at a rate between 0.67 to 0.9 kgs per hectare.

Resistant Weeds

The emergence and spread of glyphosate-resistant weeds is the second, and by far most important factor driving up herbicide use on land planted to herbicide-resistant varieties. Glyphosate resistant (GR) weeds were practically unknown before the introduction of RR crops in 1996. The first glyphosate-resistant weed (*Lolium rigidum*) emerged in Australia in 1996 from canola, cereal crop, and fence line applications [19]. In the mid-1990s, as the first glyphosate-resistant crops were moving toward commercialization and gaining market share, Monsanto scientists wrote or were co-authors on several papers arguing that the evolution of GR weeds was unlikely, citing the herbicide's long history of use (~20 years) and relative absence of resistant weeds [20, 21].

Other scientists, however, challenged this assertion [22]. Dr. Ian Heap, long-time manager of the international database on resistant weeds, warned in a 1997 conference presentation that to limit glyphosate selection pressure in Roundup Ready cropping systems, the herbicide would need to be used in conjunction with proven resistance-management practices and with non-chemical weed control methods [23]. A 1996 report by Consumers Union stated that HR crops are "custom-made" for accelerating resistance and called for the Environmental Protection Agency (EPA) to revoke approval of HR crops when and where credible evidence of resistance emerges [24].

Today, the Weed Science Society of America (WSSA) website lists 22 GR weed species in the U.S. [19]. Over two-thirds of the approximate 70 state-GR weed combinations listed by WSSA have been documented since 2005, reflecting the rapidly spreading nature of the GR-weed problem.

According to the WSSA, over 5.7 million hectares (14 million acres) are now infested by GR weeds, an estimate that substantially underestimates the actual spread of resistant weeds [16, 22], [and personal communication, Dr. Ian Heap]. Dow AgroSciences carried out a recent survey on the percent of crop acres/hectares in the U.S. impacted by glyphosate-resistant weeds [25]. Findings from the survey were provided to USDA in support of Dow AgroSciences's petition for deregulation of 2,4-D herbicide-resistant corn, and suggest that around 40 million hectares (100 million acres) are already impacted by glyphosate-resistant weeds, an estimate that Heap considers inflated [personal communication]. The true extent of spread in the U.S. likely lies around the midpoint between the WSSA and Dow AgroSciences estimates (i.e., 20–25 million hectares), and by all accounts, will continue to rise rapidly for several years.

Why have GR weeds become such a serious problem? Heavy reliance on a single herbicide – glyphosate (Roundup) -- has placed weed populations under progressively intense, and indeed unprecedented, selection pressure [10]. HR crops make it possible to extend the glyphosate application window to most of the growing season, instead of just the pre-plant and post-harvest periods. HR technology allows multiple applications of glyphosate in the same crop year. The common Midwestern rotation of HR corn-HR soybeans, or HR soybeans-HR cotton in the South, exposes weed populations to annual and repetitive glyphosate-selection pressure.

These factors trigger a perfect storm for the emergence of GR weeds. Research has traced the resistance mechanism in Palmer amaranth (*Amaranthus palmeri*) to 5-enolpyruvylshikimate-3-phosphate synthase (EPSPS) gene amplification. Resistant weed populations from Georgia contained 5-fold to 160-fold more copies of the EPSPS gene, compared to susceptible plants [26]. Moreover, EPSPS gene amplification is heritable, leading Gaines et al. to warn that the emergence of GR weeds "endangers the continued success of transgenic glyphosate-resistant crops and the sustainability of glyphosate as the world's most important herbicide."

Resistant Palmer amaranth (*Amaranthus palmeri*) has spread dramatically across southern states since the first resistant populations were confirmed in 2005, and already poses a major economic threat to U.S. cotton production. Some infestations are so severe that cotton farmers have been forced to leave some crops unharvested.

Responding to Resistance

GR weed phenotypes are forcing farmers to respond by increasing herbicide application rates, making multiple applications of herbicides, applying additional herbicide active ingredients, deep tillage to bury weed seeds, and manual weeding. In recent years the first three of the above responses have been the most common. Each response increases the kilograms of herbicides applied on HR crop hectares. All five interventions increase costs. Moreover, if 2,4-D and dicamba herbicide-resistant corn and soybeans are fully deregulated by the U.S. government, there will be growing reliance on older, higher-risk herbicides for management of glyphosate-resistant weeds.

Based on an upward trajectory in the planting of 2,4-D HR corn reaching 55% of corn hectares planted by 2019, coupled with an average of 2.3 applications (the label allows three) and an average rate of 0.94 kgs/ha (0.84 pounds/acre) (the label allows 1.12 kgs/ha (1.0 pounds/acre)), 2,4-D use on corn in the U.S. would increase over 30-fold from 2010 levels [Additional file 1: Table S19]. Such a dramatic increase could pose heightened risk of birth defects [27, 28] and other reproductive problems [29], more severe impacts on aquatic ecosystems [30], and more frequent instances of off-target movement and damage to nearby crops and plants. Moreover, the efficacy of 2,4-D corn may well prove short lived, since a population of 2,4-D resistant waterhemp (*Amaranthus tuberculatus*) has now been confirmed in Nebraska [31], and there are already at least eight other weeds resistant to 2,4-D [19].

GR weeds typically emerge first on a few isolated fields, but their pollen, genes, and seeds can travel widely and spread quickly, especially if glyphosate continues to be relied on heavily [11]. No substantial change in the intensity of glyphosate use in the U.S. is expected in the foreseeable future; nearly all corn, soybean, and cotton cultivars now carry a RR gene. The seed industry has no plans to grow and sell more non-HR seed, and indeed is moving in the opposite direction by developing more stacked, multiple HR varieties. The share of total national corn, soybean, and cotton hectares impacted by GR weed populations is likely to grow and will, as a result, increase both the number of different herbicides applied, as well as the total kgs of herbicides applied.

As argued by many weed scientists and extension specialists, integrated weed management systems, coupled with markedly lessened reliance on RR technology are now essential to extend the useful life of RR technology [10, 12,14, 32]. Without major change, a crisis in weed management systems is likely, triggering possibly ominous economic, public health, and environment consequences.

Higher Costs Triggered by Resistant weeds and HR Technology

Weed management costs per hectare increase by 50% to 100% or more in fields infested with glyphosate-resistant weeds, as evident in a series of case studies submitted to the USDA by Dow AgroSciences in support of its petition to the USDA seeking deregulation of 2,4-D herbicide-resistant corn [25]. In soybean production in Arkansas, for example, Dow AgroSciences compared the average cost/acre of the top-five, most popular herbicide programs in Roundup Ready soybeans in fields without resistant weeds, compared to the average of two recommended programs in fields infested with glyphosate-resistant Palmer amaranth. Herbicide costs rise 2.7-fold (from $16.29 to $44.34 per acre) [23], [Table thirty, page 93]. In Illinois soybean production, the increase in herbicide costs is estimated at 64% ($19.21 to $31.49 per acre) [23], [Table thirty-two, page 95], while in Iowa corn production, the increase is 67% ($19.23 to $32.10 per acre) [23], [Table thirty-six, page 99].

The markedly higher cost/hectare of herbicide-resistant seeds must be added to the higher herbicide costs noted above to more fully reflect the added costs associated with HR technology. The cost of a bushel of conventional, not-GE soybean seed increased during the GE-crop era from $14.80 in 1996 to $33.70 in 2010, while a bushel of GE soybean seed cost, on average, $49.60 in 2010 (all seed price data derived from USDA data) [33]. Accordingly, the cost of GE soybean seed in 2010 was 47% higher per bushel than non-GE seed. In the case of corn, conventional seed prices rose from $26.65 per acre planted in 1996 to $58.13 in 2010. The average cost of GE corn seed per acre in 2010 was $108.50, with some GE cultivars selling for over $120 per planted acre. Hence, GE corn seed costs per acre were about double the cost conventional seed.

Public Health Concerns

Heightened risk of public health impacts can be expected in the wake of more intensive herbicide use, especially applications later in the season on herbicide-resistant crop varieties. While current risk assessment science suggests that glyphosate is among the safer herbicides per hectare treated in terms of human health risks, both the frequency of human exposures and levels of exposure via food, drinking water, and the air have no doubt risen in the U.S. in recent years. Two-thirds to 100% of air and rainfall samples tested in Mississippi and Iowa in 2007–2008 contained glyphosate [34].

The likely approval and use of herbicide-resistant crops in the U.S. engineered to survive applications of multiple herbicides adds tricky new dimensions to herbicide-risk assessments. Applications later in the growing season will be more likely to lead to residues in silage or forage crops. As a result, herbicide residues in milk, meat, or other animal products might become more common. The jump in herbicide volumes applied during June and July will increase the risk of drift and herbicide movement via volatilization, possibly exposing people via the air, water, or crops grown in the proximity of treated fields. Risks from the drift and volatilization of 2,4-D and dicamba are of special concern, given that these two herbicides have triggered thousands of non-target crop damage episodes over the last 20 years in the U.S. Indeed, for several years, 2,4-D has been the leading cause of crop damage episodes investigated by State departments of agriculture [35].

Environmental Impacts Linked to HR Technology

A long list of environmental effects can be triggered, or made worse, by the more intensive herbicide use required to keep pace with weeds in farming systems heavily reliant on herbicide-resistant crops. Glyphosate has been shown to impair soil microbial communities in ways that can increase plant vulnerability to pathogens [36, 37, 38], while also reducing availability of certain soil minerals and micronutrients [39]. Landscapes dominated by herbicide-resistant crops support fewer insect and bird species; e.g., a study in the American Midwest reported a 58% decline in milkweed and an 81% drop in monarch butterflies from 1999 to 2010 [40]. Heavy use of glyphosate can reduce earthworm viability [41] and water use efficiency [42]. Several studies have documented reductions in nitrogen fixation in herbicide-resistant soybean fields sprayed with glyphosate [43, 44]. Transgene flow from herbicide-resistant crops can occur via multiple mechanisms and can persist in weedy relatives [45].

Individually, these environmental impacts appear, for the most part, of the same nature and in the same ballpark as the risks associated with other herbicide-based farming systems, but collectively they raise novel concerns over long-term, possibly serious impacts on biodiversity, soil and plant health, water quality, aquatic ecosystem integrity, and human and animal health.

Bt corn and Cotton Impacts and Prospects

While *Bt*-transgenic corn and cotton have displaced an estimated 56 million kgs (123 million pounds) of insecticides since 1996, every plant in a *Bt* corn or cotton field is manufacturing within its cells one or more forms of the natural bioinsecticide *Bacillus thuringiensis*. The rate of synthesis of *Bt* Cry protein endotoxins is roughly proportional to the rate of plant growth. As plants mature and enter senescence, *Bt* endotoxin expression falls.

Few published estimates are available of *Bt* endotoxin expression levels in contemporary corn cultivars. Nguyen et al. projected that a hectare of *Bt*-corn for CRW control expressing the Cry3Bb1 gene in MON88017 corn produces 905 grams of Cry3Bb1 per hectare (0.8 pounds per acre) [46]. The amount of *Bt* Cry proteins produced by a hectare of *Bt* corn for ECB and CRW control are calculated in [Additional file 1: Tables S20–S22], with key results shown in Table 2 for specific corn events, traits, and endotoxins. [Additional file 1 Tables S23–25] cover *Bt* cotton events. Expression level data reported by companies in regulatory documents were used to calculate per hectare production of specific endotoxins. [Additional file 1: Tables S21 and Table S24 contain the expression level data for*Bt* corn and cotton events, and [Additional file 1: Table S22 and Table S25] report the volumes of *Bt* Cry proteins produced per hectare and acre based on contemporary seeding rates.

Table 2 *Bt* cry protein synthesis in major GE corn cultivars

	Cry Protein	Cry/Shoot	Cry/Root	Cry/Plant	Plants per hectare	Cry Toxin	Plants per Acre	Cry Toxin
						kg /ha		**lb/acre**
MON 810	Cry1Ab	1193	496	1689	79,040	0.133	32,000	0.119
MON 88017	Cry3Bb1	14915	4030	18945	79,040	1.497	32,000	1.333
MON 89034	Cry1A.105	2826	620	3446	79,040	0.272	32,000	0.242
MON 89034	Cry2Ab2	4553	496	5049	79,040	0.399	32,000	0.355
TC 1507	Cry1F	1207	165	1372	79,040	0.108	32,000	0.097
DAS 59122	Cry34Ab1	26376	2647	29023	79,040	2.294	32,000	2.042
DAS 59122	Cr35Ab1	5825	567	6392	79,040	0.505	32,000	0.45
SmartStax Corn								
MON 88017	Cry3Bb1	7536	2015	9551	79,040	0.755	32,000	0.672

MON 89034	Cry1A.105	2983	651	3634	79,040	0.287	32,000	0.256
MON 89034	Cry2Ab2	4553	558	5111	79,040	0.404	32,000	0.36
TC 1507	Cry1F	1413	185	1598	79,040	0.126	32,000	0.112
DAS 59122	Cry34Ab1	24649	2623	27272	79,040	2.156	32,000	1.918
DAS 59122	Cr35Ab1	5275	586	5861	79,040	0.463	32,000	0.412
SmartStax Total						**4.191**		**3.73**

Major contemporary *Bt* corn events targeting the ECB synthesize nearly as much or more insecticidal Cry protein per hectare than the weighted-average rate of conventional insecticides applied on a hectare planted to *Bt* corn for ECB control (about 0.15 kgs insecticide per ha; 0.13 pounds/acre in 2010 [Additional file 1: Table S11]). MON810, the Cry protein in Monsanto's original Yieldgard corn, expresses 0.2 kgs/ha of endotoxin, whereas Syngenta's *Bt* 11 synthesizes 0.28 kgs/ha [Additional file 1: Table S22]. Newer events for ECB control like Monsanto's Genuity VT Double PRO (MON 89034) produce Cry 1A.105 and Cry 2Ab2 endotoxins totaling 0.62 kgs/ha. The Dow AgroSciences-Pioneer Hi-Bred Herculex I (TC1507) event expresses the least endotoxin – 0.1 kg *Bt* endotoxin per hectare – just below the rate of insecticides applied.

In the case of *Bt* corn targeting the CRW, every hectare planted in recent years expresses substantially greater volumes of *Bt* endotoxins than the ~0.2 kgs of insecticides applied on the average hectare for CRW control (0.19 pounds/acre [Additional file 1: Table S12]). MON 88017 expresses 0.62 kgs/ha of Cry 3Bb1, while DAS 59122–7 expresses two Cry proteins totaling 2.8 kgs/ha, 14-fold more than the insecticides displaced [Additional file 1: Table S22]. SmartStax GE corn synthesizes six Cry proteins, three targeting the ECB, and three the CRW. Total Cry protein production is estimated at 4.2 kgs/ha (3.7 pounds/acre), 19-times the average conventional insecticide rate of application in 2010.

Should Bt Endotoxins Count as Insecticides Applied?

Entomologists are divided on the question of whether the *Bt* produced by transgenic plants should be counted as "insecticides applied." The case for doing so is strong, despite the obvious differences in how Cry proteins enter corn agroecosystems. When a field of corn is sprayed with a foliar *Bt* insecticide, the amount of toxin sprayed per hectare should be counted when computing total insecticide use. The primary difference between the *Bt* Cry proteins in a *Bt*-transgenic plant, and a field of non-GE plants sprayed with foliar *Bt*, is that in the later case, the toxin is present predominantly on plant tissue surfaces, whereas in the former *Bt*-crop case, the toxin is inside plant cells. This distinction does not greatly matter from the perspective of the overall load of pesticides in the environment, although the presence of pesticides inside plants, as opposed to on their surface, alters relative risk profiles across non-target organisms.

It should also be noted that, in general, the systemic delivery of *Bt* Cry proteins poses more significant risks to animals and humans ingesting *Bt* crops than applications of *Bt* insecticides via liquid sprays. Systemic delivery also enhances a range of environmental and ecological risks [47] compared to foliar *Bt* use patterns that result in rapid breakdown of *Bt* Cry proteins as they are exposed to sunlight and rainfall.

Most corn insecticides are applied in ways that expose active ingredients to destructive abiotic and biotic forces that tend to break down the chemicals to generally less toxic forms. Granular soil insecticides applied via boxes on corn planters tend to break down within weeks as a result of soil microbial activity. Because properly applied granular insecticides are buried in the soil, exposure to non-target organisms is limited, although poorly operated or calibrated planting equipment can result in grains of insecticide remaining on the soil surface, posing a serious potential risk to some bird species. A significant portion of the foliar insecticides applied per hectare for ECB control never hit its plant target, and a portion of the insecticide that does land and lodge on plant tissues is washed off within hours, days, or weeks during rainfall events. This is why insecticide residues are rarely detected in corn grain and silage at harvest time, and why conventional insecticide applications on corn pose little or no human dietary risk.

By virtue of their altered environmental fate and risk profile, all systemic pesticides should be counted when measuring pesticide use, and hence so too should the *Bt* proteins manufactured in *Bt*- transgenic crops. If *Bt*-transgenic plants produced proteins that disrupted insect morphology, feeding behavior, or reproduction, the absence of a toxic mode of action would strengthen the argument that *Bt* Cry proteins are not functionally equivalent to insecticides, and hence should not be counted as insecticides applied. *Bt*-crop technology that limits*Bt*-endotoxin expression to only those tissues that are under active attack, and then only during times when insects are actively feeding, would also support the view that *Bt* crops are compatible with IPM.

CONCLUSIONS

Today's pest-management related GE traits have proven popular and commercially profitable for the biotech-seed industry, but over-reliance has set the stage for resistance-driven problems in both herbicide-resistant and *Bt*-transgenic crops. Largely because of the spread of glyphosate-resistant weeds, HR crop technology has led to a 239 million kg (527 million pound) increase in herbicide use across the three major GE-HR crops, compared to what herbicide use would likely have been in the absence of HR crops. Well-documented increases in glyphosate applications per hectare of HR crop account for the majority of this 239 million kg increase.

While *Bt* corn and cotton have reduced insecticide applications by 56 million kgs (123 million pounds), resistance is emerging in key target insects and substantial volumes of *Bt* Cry endotoxins are produced per hectare planted [corn, Additional file 1: Tables S20–S22, cotton, Additional file 1: Tables S23–S25], generally dwarfing the volumes of insecticides displaced. Documenting the full range of impacts on the environment and public health associated with the *Bt* Cry proteins biosynthesized inside *Bt*-transgenic plants remains a challenging and largely ignored task, especially given the recent move toward multiple *Bt* protein, stacked-trait events.

Overall, since the introduction of GE crops, the six major GE technologies have increased pesticide use by an estimated 183 million kgs (404 million pounds), or about 7%. The spread of GR weeds is bound to trigger further increases, e.g., the volume of 2,4-D sprayed on corn could increase 2.2 kgs/ha by 2019 (1.9 pounds/acre) if the USDA approves unrestricted planting of 2,4-D HR corn [Additional file 1: Table S19]. The increase in herbicides applied on HR hectares has dwarfed the reduction in insecticide use over the 16 years, and will almost surely continue to do so for several more years.

Estimating the impacts of GE crops on pesticide use is growing more complex because of gaps in NASS pesticide use data collection for the three major crops, increases in the average number of traits per GE-crop hectare planted, the registration of HR crops engineered to resist herbicides other than glyphosate, massive disruption in weed communities, and the presence of one to three, or even more, glyphosate-resistant weeds in many crop fields. It is difficult to project what the distribution, population levels, and phenotypes of weeds would have been over the last 16 years in the absence of HR technology. Inevitably, weed management systems and technology would have evolved along other trajectories in the absence of HR crops these last 16 years, resulting in heightened reliance on both pre-plant and post-emergence applications of multiple, low-dose herbicides.

A majority of American soybean, maize, and cotton farmers are either on, or perilously close to a costly herbicide and insecticide treadmill. Farmers lack options and may soon be advised, out of necessity, to purchase HR crop cultivars resistant to multiple active ingredients and to treat *Bt* corn with once-displaced corn insecticides. The seed-pesticide industry is enjoying record sales and profits, and the spread of resistant weeds and insects opens up new profit opportunities in the context of the seed industry's current business model. Regulators cannot restrict the use of a previously approved HR technology because it increases pesticide use and triggers resistance, nor have U.S. government agencies turned down an application for a new HR or *Bt*-transgenic trait because of the likelihood it would accelerate the spread of resistant weeds or insects. Whether the USDA has the statutory authority to deny a petition for HR crop deregulation (i.e., approval) on the grounds of worsening problems with resistant weeds is a contested issue in ongoing litigation.

Profound weed management system changes will be necessary in the three major GE crops to first stabilize, and then hopefully reduce herbicide use, the costs of weed management, and herbicide-related impacts on human health and the environment. Weed management experts are largely in agreement that the percent of cropland area planted to glyphosate-based HR seeds must decline dramatically (e.g., by at least one-third to one-half) for farmers to have a realistic chance at success in preventing resistance [10, 12, 14]. Unfortunately, there appears little interest across the seed-biotech industry in increasing production of non-Roundup Ready or not-*Bt* transgenic seed. Since the decisions made by the seed industry in any given year determine the traits offered by the industry to farmers in next crop season, the seed industry must act first in order for farmers to turn the corner toward more sustainable weed and insect pest management systems. The many important ramifications of this practical reality – that the seed industry must act first -- have yet to be fully appreciated by farmers, weed management experts, and policy makers in the U.S.

Regulators in the U.S. have thus far done little to prevent the emergence and spread of resistant weeds, while several resistance-management interventions have been imposed as part of the approval of *Bt* crops. In addressing weed resistance, the hands-off regulatory posture in the U.S. reflects, in part, the basic authorities granted to the EPA and USDA in federal law. Both agencies regard weed resistance as an efficacy-economics challenge that can best be addressed by the private sector consistent with market forces. The need for novel policy interventions will grow in step with the emergence and spread of resistance weeds and evidence of adverse economic, environmental, and public health consequences triggered by markedly increasing reliance on older, higher-risk herbicides.

METHODS

The model calculates the impact of HR and *Bt*-transgenic crop varieties on pesticide use annually from 1996 through 2011, and aggregates results over this 16-year period. The model is composed of 16 tables [Additional file1: Tables S1–S16]. Nine additional tables, [Additional file 1: Tables S17–S25] address changes in pesticide use, the spread of resistant weeds, and the quantity of *Bt* endotoxins produced per hectare by today's major corn and cotton*Bt* traits.

The model was developed using the units of measure typical in USDA-NASS surveys (pounds of active ingredients, acres planted); the Additional files are available in pounds and acres units only. In this paper, metric units are used to report results, although selected key results will be reported in both units of measure.

[Additional file 1: Table S1] records average per acre herbicide and insecticide use data, drawing on pesticide use data compiled annually by the USDA's NASS. These surveys record the percent of crop acres treated with specific active ingredients, average one-time rates of application, the average number of applications, the rate per crop year (average rate multiplied by the average number of applications), and total pounds applied.

In the case of herbicides, [Additional file 1: Table S1] reports total herbicide, all glyphosate, and "Total Herbicides Minus Glyphosate." "All Glyphosate" aggregates the multiple chemical forms of glyphosate surveyed by NASS, and calculates average rates of application and number of applications, weighted by frequency of use. The same procedure is used to calculate average pounds/acre applied of other herbicides of interest for which NASS reports use data for multiple chemical forms (e.g. 2,4-D, dicamba). [Additional file 1: Table S2] includes national acres planted to each crop, average pesticide use rates, and total pounds applied per acre and overall herbicide, insecticide, and herbicide + insecticide volumes applied.

[Additional file 1: Tables S3–S6] record the percent of national acres planted to a crop variety expressing each of the six, major commercial GE traits. The USDA's ERS provides data on the percent of total national corn [Additional file 1: Table S3], soybean [Additional file 1: Table S4], and cotton hectares [Additional file 1: Table S5] that were planted to each GE crop trait for 1996–2011. Percent acres planted to all six GE traits by year are presented in [Additional file 1: Table S6]; there is a high level of confidence in these data.

[Additional file 1: Table S7] reports acres planted to each of the six traits, multiplying the percent acres planted to each trait in ST 6 by total acres planted to each crop in [Additional file 1: Table S2]. [Additional file 1: Tables S8–S10] calculate, for the three HR crops, the estimated difference in average herbicide use on HR hectares versus land planted to conventional, non-GE varieties. [Additional file 1: Tables S11–S13] report the basis for calculating the pounds of insecticides displaced by the planting of *Bt* corn and cotton traits. [Additional file 1: Table S14] integrates all of the average per acre pesticide use rates by crop, trait and year, and reports the estimated difference between per acre rates on GE versus non-GE acres. [Additional file 1: Table S15] converts the differences in rates per acre to differences in pounds applied nationally by crop, trait, and year, and over the 16-year period. [Additional file 1: Table S16] provides details on glyphosate use from NASS surveys over the 1996–2010 period, and is the source of data on glyphosate use in other Additional files.

Assumptions, Projections, and Calculations

A series of assumptions, projections, and calculations are embedded in the model in order to estimate total herbicide and insecticide use on GE versus not-GE hectares. Table 3 outlines model assumptions and Table4describes the projections embedded in the model's calculations.

Table 3 Data sources and assumptions required to quantify the impact of GE crops on pesticide use in the U.S., 1996-2011

Parameter	Source	Supplemental table impacted	Basis and explanation
National Pesticide Use per Acre/Hectare	NASS-USDA	1, 2	Best publicly available estimates of annual per acre herbicide and insecticide use
Annual Gaps in NASS Survey Data by Crop	Interpolated	1, 2	Changes in total herbicide, glyphosate, and insecticide use occur linearly/annum when there are gaps in NASS pesticide use surveys
Annual Application Rates of "Other Herbicides on HR Hectares"	(See Table 4)	8, 9, 10	Trends by crop on HR acres track changes in total herbicide use, as reported by NASS; changes from year to year are gradual
Bt Cry Proteins Produced by *Bt* Corn and Cotton Plants	Projected (see text, Additional files)	20-25	Trait-specific expression levels by tissue taken from documents submitted by technology developers; used to quantify volume of each*Bt* endotoxin produced by plants per acre/hectare based on typical planting density
Insecticide Use on *Bt*Corn	(Details in Table 4)	11, 12	Insecticide displacement as a result of planting *Bt* corn corrected for hectares not likely to have been treated in the absence of *Bt* corn cultivars
Insecticide Use on *Bt*Cotton	NASS-USDA	13	Budworm/bollworm control insecticide displacement on hectares planted to *Bt* cotton is 100%

Table 4 Projections required quantifying the impact of GE crops on insecticide use in the US, 1996–2011

Parameter	**Supplemental table(s) impacted**	**Basis for setting value**	**Basis and explanation**
Corn			
Share of Insecticide Applications Targeting the European Corn Borer (ECB) Versus Corn Rootworm (CRW)	11, 12	Guidance from extension IPM specialists and land grant university spray guides	Some insecticides applied exclusively for control of ECB, others for control of CRW; and some target both. The percent hectares treated with a given insecticide are apportioned relative to target pests: ECB, CRW, or other insects.
"Other Insecticides" Applied in 2010 for ECB Control	11	NASS data on "Other Insecticides" applied in 2010	NASS reported 237,000 pounds of "Other Insecticide" use in 2010; 30% of these "Other Insecticides" applied to corn in 2010 projected to target the ECB.
"Other Insecticides" Applied in 2010 for CRW Control	12	NASS data on "Other Insecticides" applied in 2010	NASS reported 237,000 pounds of "Other Insecticide" use in 2010; 60% of "Other Insecticides" applied to corn in 2010 projected to target the CRW.
Cotton			
Share of Insecticide Applications Targeting the Budworm/Bollworm Complex	13	Guidance from extension IPM specialists and land grant university spray guides	Some insecticides applied exclusively or partly for control of the budworm/bollworm complex, others for other insects; percent hectares treated with a given insecticide is apportioned relative to target insects.

NASS surveyed corn, soybean, and cotton pesticide use in most years from 1996–2010. None of the crops were surveyed in 2008; cotton was last surveyed in 2007 and 2010; corn was surveyed in 2005 and 2010; and soybeans have not been surveyed since 2006. In estimating the impacts of GE crops on pesticide use from 1996–2011, average application rates per crop year were interpolated in years with no data, when NASS had surveyed a previous and subsequent year, based on the assumption of linear change in the intervening years.

It is assumed that changes in the volume of herbicides other than glyphosate applied on the average HR hectare tracks changes in total herbicide use, and also changes gradually from year-to-year. With few exceptions, these patterns of change in herbicide use are evident in all crops surveyed by USDA. Significant annual changes in total herbicide use, as well as non-glyphosate applications, are almost always linked to an increase or decrease in acres treated with one or more relatively high-dose herbicides applied at or around 1 pound/acre, compared to use of herbicides applied at rates less than 0.5 pound/acre (several are sprayed at rates below 0.05 pounds/acre).

The volumes of *Bt* Cry endotoxins produced per acre/hectare of *Bt* corn and cotton are not included in the estimates of changes in insecticide use on acres/hectares planted to *Bt* cultivars, although the volumes are surprisingly significant compared to the volume of insecticides applied on treated acres/hectares (see "Discussion"). In the case of insecticide use on *Bt* corn, the volume of insecticide use displaced per acre/hectare is adjusted in light of the likely percent of *Bt* corn acres/hectares that would have been treated with an insecticide in the absence of *Bt* cultivars. Multiple analysts have reported substantial planting of *Bt* corn as insurance against possible insect feeding damage, on acres/hectares that farmers would not prophylactically apply insecticides [4, 13]. In a January 2010 survey, 73.3% of 518 farmers surveyed at regional extension meetings in Illinois reported that they planted *Bt*corn "Knowing That Anticipated Damage Levels Were Low" [48]. USDA has surveyed corn insecticide use 14 times since 1991. The total area treated with an insecticide has fallen in the range 31% +/− 5% in all years, with the average around 33%.

It is assumed that farmers planting *Bt* cotton do not spray conventional insecticides against the budworm/bollworm complex of insecticides, leading to 100% displacement of such applications. This assumption likely overestimates displacement marginally, especially in recent years where isolated populations of less susceptible or resistant populations have emerged.

Table 3 describes the basis for projecting a number of missing values over the 1996-2011-time period. In the years since the last NASS survey, pesticide rates were projected based on recent trends and changes in weed pressure.

In the case of corn, total herbicide and glyphosate use trends from 2005–2010 are projected to continue unchanged through 2011, despite the accelerating emergence and spread of resistant weeds in the Midwest. The rapid rate of increase in total herbicide and glyphosate use/acre in soybean production systems from 2000–2006 (5.9% and 8.9%/annum) is projected to decline to an average increase of 3.2% and 3.3% per annum in 2007–2011. Reductions in annual rates of increase reflect the decision by many HR soybean farmers to follow the advice of weed management specialists [10, 11] to diversity the modes of action included in herbicide-based control programs. The rate of increase in total herbicide use on HR cotton from 2010 to 2011 is projected at about twice the annual rate, 2007–2010, whereas the rate of decline in per hectare glyphosate use is projected to fall from −3.4% to −1% per annum as farmers increase rates and/or frequency of applications of glyphosate in regions where resistant weeds are now posing serious management challenges.

Estimating herbicide use on conventional and HR hectares

NASS surveys do not report pesticide use on GE and conventional crop hectares separately.

The volume of herbicides applied to HR hectares can be approximated by adding NASS-reported glyphosate use per crop year to an estimate of the volume of herbicides other than glyphosate (hereafter, "other herbicides") applied on HR hectares. The volume of "other herbicides" applied on HR hectares is estimated based on the average number of non-glyphosate herbicides applied per hectare, coupled with the average rate per application of non-glyphosate herbicides. In addition, the rate of "other herbicides" on HR hectares is adjusted to reflect changes from year to year in overall herbicide use and glyphosate application rates. For example in recent years, "other herbicides" have been applied to around one-half of HR soybean hectares at an average rate of ~0.34 kgs/ha (~0.3 pounds/acre), resulting in an average ~0.17 kgs/ha (~0.15 pounds/acre) of "other herbicide" applications on all HR hectares (0.5 × 0.34).

The shares of total crop hectares in a given year planted to conventional and HR crop varieties is compiled by the USDA's ERS [Additional file 1: Tables S3–S5] and can be used in a weighted-average formula to calculate the kgs of herbicides applied on non-HR hectares –

$$\text{THACropx}=[(\%\text{HPHTx})\text{x}(\text{HAHTx})]+[(\%\text{HPCONx})\text{x}(\text{HACONx})] \qquad (1)$$

Where,

THA Crop_x="Total Herbicides Applied" (kgs active ingredient/hectare in a crop year);

% HPHT_x=Percent national "Hectares Planted to HR" cultivars;

HAHT_x="Herbicides Applied on HR" hectares (kg a.i./crop year);

% HPCON_x=Percent national "Hectares Planted to Conventional" non-HR hectares; and

HACON_x="Herbicides Applied on Conventional" hectares (kgs a.i./crop year).

The first four of the above-five variables are reported or can be derived from USDA data; the fifth can be calculated by solving the above equation for HACON_x. For each HR crop and year combination, the impact of HR cultivars on average herbicide use is calculated by subtracting HAHT_x from HACON_x. This difference is then multiplied by the HR hectares planted, to calculate the impact of HR crops on herbicide use in a given year. Increases or decreases in the volume of herbicides applied as a result of the planting of HR crops are then aggregated across all years (1996–2011) and the three HR crops.

In the case of *Bt* transgenic corn, the average rate of application of insecticides targeting the ECB and the CRW must be calculated. This process is complicated by the fact that several insecticides are applied for control of the ECB and CRW, as well as other insects. Pesticide labels, treatment recommendations in university spray guides, and experts in corn Integrated Pest Management (IPM) were consulted in carrying out this step [Additional file 1: Tables S11, S12].

Average rates of insecticide application across all corn hectares treated per crop year are then calculated, weighted by portions of total hectare treatments. This weighted-average rate of insecticide application on hectares treated for ECB control declines from 0.24 kgs/ha (0.21 pounds/acre) of active ingredient in 1996 to 0.15 kgs/ha (0.13 pounds/acre) in 2010. In the case of CRW insecticides, the rate falls from 0.76 kgs/ha in 1996 to 0.2 kgs/ha in 2010.

The next step in calculating the pounds of insecticides displaced by the planting of *Bt* corn is to estimate the portion of hectares planted to *Bt* corn for ECB and/or CRW control that would have been treated with an insecticide if the corresponding *Bt* crop had not been planted. Doing so requires a set of assumptions and projections.

Historically, USDA data shows that before the advent of *Bt* corn, 10% +/− 3% of national corn hectares were treated for ECB control, while 27% +/− 4% were treated for CRW control. Yet by 1998 (third year of commercial sales), 19% of corn hectares were planted to a *Bt* cultivar targeting the ECB – about double the historic share of hectares treated with an insecticide for this pest. Today, close to two-thirds of corn hectares are planted to *Bt* for ECB cultivars, some six-times the historic rate. In the case of *Bt* corn for CRW, by the fifth year of commercial sales, 2007, the share of corn hectares planted to CRW hybrids was 25.6%, roughly equaling the historic share of hectares treated with CRW insecticides (27% +/− 4%). In 2011, 60% of corn hectares were planted to a CRW hybrid, double the historic share of corn hectares treated with a CRW insecticide.

The impact of *Bt* corn on the volume of insecticide displaced per hectare should be adjusted downward to account for hectares that would, in all likelihood, not have been treated. In the case of *Bt* corn targeting the ECB, the likely share of hectares planted to *Bt* corn that would have been sprayed for ECB control begins at 90% in 1997, the first year of commercial planting, and drops incrementally to 45% in 2007.

This percent is left unchanged from 2008–2010, despite the increase in corn hectares planted to *Bt* corn for ECB from 49% to 65%, because of reported increases in insect pest pressure in major corn producing regions [49]. The result is the projection that in 2011, insecticide applications were displaced on 10.9 million hectares of corn (27 million acres) planted to *Bt* hybrids for ECB control (45% of the 65% of corn hectares planted to *Bt* for ECB hybrids). These 10.9-million hectares are 29% of total corn hectares planted, and is about three-times the historic level of insecticide applications for ECB control.

In the case of *Bt* corn for CRW control, the percent of hectares planted that displaces insecticide use begins at 95% in 2003, the first year of commercial sales, and declines to 55% in 2011. In 2011, 57% of corn hectares were planted to a *Bt* CRW hybrid, and hence *Bt* corn for CRW displaced insecticide use on 31% of national hectares planted. This estimate assumes that any hectare planted to a *Bt* corn for CRW control was not also treated with a CRW insecticide. In addition, 9.4% of corn hectares were sprayed for CRW control with an insecticide. Accordingly, about 40% of corn hectares were either sprayed for the CRW or planted to a *Bt* variety for CRW control, well above the 27% +/− 4% level treated with insecticide over the last 20 years.

The historically high, projected level of CRW treatment is justified, in part, by the emergence in the late 1990s of a variant of the CRW that learned to overwinter in soybean fields, thus undermining the efficacy of corn-soybean rotations in reducing CRW populations [50]. Recent, historically high corn prices have also increased the frequency of continuous corn, a management factor that surely has increased CRW pressure.

Bt cotton targets the budworm/bollworm complex, but does not affect other insect pests, including the boll weevil, plant bugs, white flies, and stinkbugs. Applications of broad-spectrum insecticides are typically made on essentially 100% of planted cotton hectares to control the budworm/bollworm complex and other insects. *Bt* cotton will reduce the use of insecticides on the budworm/bollworm complex, but will only indirectly impact applications of insecticides targeting other insects.

[Additional file 1: ST 13] reports the basis for estimating the pounds of insecticides displaced by each acre planted to *Bt* cotton. University insect management guides and experts were consulted to estimate the portion of hectares treated with each cotton insecticide that targeted the budworm/bollworm complex, versus other insects. The number of acres treated with each insecticide is calculated from NASS data, as well as the share of total acres treated. Average insecticide use rates are then calculated, weighted by each active ingredient's share of insecticide acre treatments targeting the budworm/bollworm complex. The weighted average cotton insecticide application rate falls modestly from 0.46 kgs/ha (0.41 pounds/acre) in 1997 to 0.27 kgs/ha (0.24 pounds/acre) in 2010–2011.

Table 4 summarizes the basis for projections required to estimate the volume of insecticide use displaced by the planting of a hectare to *Bt* corn or cotton cultivars.

AUTHOR CONTRIBUTION

Charles Benbrook (CB) developed the model, carried out the analysis, and wrote the paper. The author read and approved the final manuscript.

AUTHOR INFORMATION

CB is a Research Professor, Center for Sustaining Agriculture and Natural Resources, Washington State University, Pullman, Washington, USA. CB has studied the impacts of agricultural biotechnology since the mid-1980s in a variety of positions including Executive Director, Board on Agriculture, National Academy of Sciences/National Research Council, and as Chief Scientist, The Organic Center.

ABBREVIATIONS

AI: Active ingredient

Bt **:** Bacillus thuringiensis

CRW: Corn rootworm

ECB: European corn borer

EPSPS: Enolpyruvylshikimate-3-phosphate synthase

EPA: Environmental Protection Agency

ERS: Economic Research Service

GE: Genetically engineered, genetic engineering

GR: Glyphosate resistant

ha: Hectare

HR: Herbicide Tolerant

IPM: Integrated Pest Management

kgs: Kilograms

NASS: National Agricultural Statistics Service

NCFAP: National Center for Food and Agriculture Policy

RR: Roundup Ready

SI: International System of Units

ST: Supplemental Table

THA: Total hectares

US: United States

USDA: United States Department of Agriculture

WSSA: Weed Science Society of America.

DECLARATIONS

Acknowledgements

Thanks to the analysts in the U.S. Department of Agriculture's (USDA) National Agricultural Statistics Service and the Economic Research Service (ERS) for compiling the data essential to carry out this work. Dr. Merritt Padgitt of the ERS (retired) carried out a special tabulation of USDA survey data on soybean pesticide use that was used to calibrate the model. Valuable assistance was provided in developing and refining the model and underlying dataset by Karen Benbrook and Karie Knoke. Dr. Margaret Mellon, Dr. Jane Rissler, and Dr. Doug Gurian-Sherman of the Union of Concerned Scientists (UCS) contributed to the conceptual development of the model. Mr. Bill Freese, Center for Food Safety, provided helpful suggestions and data on resistant weeds. Dr. Robert Kremer, Dr. Michael Gray, Dr. Matt Liebman, and Dr. Michael Owen are among the land grant pest management scientists that provided guidance as the model was developed. Funding to support the development of the model was provided by the Institute for Agriculture and Trade Policy, Consumers Union, UCS, and The Organic Center.

COMPETING INTERESTS

The author declares he has no competing interests.

REFERENCES

1. Gassmann AJ, Petzold-Maxwell JL, Keweshan RS, Dunbar MW: Field-evolved resistance to Bt maize by western corn rootworm. PLoS One 2011, 6: e22629. doi:10.371/journal.pone.0022629.
2. Berry I: Illinois researcher confirms rootworm resistance to Monsanto corn trait. Wall Street Journal.: Dow Jones Newswire; 2012. published 17 August 17.
3. Federoff NV, Battisti DS, Beachy RN, Cooper PJM, Fischhoff DA, Hodges CN, Knauf VC, Lobell D, Mazur BJ, Molden D, Reynolds MP, Ronald PC, Rosegrant MW, Sanchez PA, Vonshak A, Zhu J-K: Radically rethinking agriculture for the 21st century. Science 2010, 327:833–834.
4. Carpenter J: GM crops and patterns of pesticide use. Science 2001, 292:637–638.
5. Brookes G, Barefoot P: Global impact of biotech crops environmental effects, 1996–2010. GM Crops and Food: Biotech in Ag and the Food Chain 2012, 3:1–9.
6. Sankula S: Quantification of the impacts on US agriculture of biotechnologyderived crops planted in 2005, National Center for Food and Agricultural Policy, 1–110. 2006 [http://www.ncfap.org/documents/2005biotechExecSummary]
7. Economic Research Service: Genetically engineered crops: has adoption reduced pesticide use?. 2000 [http://webarchives.cdlib.org/sw1610ws5g/http:/ ers.usda.gov/publications/agoutlook/aug2000/ao273f.pdf]
8. USDA National Agricultural Statistics Service: Agricultural Chemical Usage Field Crop Summary, 2003–2005; [http://usda.mannlib.cornell.edu/MannUsda/ viewDocumentInfo.do?documentID=1560]
9. United Kingdom Health and Safety Executive, Expert Committee on Pesticide Residues in Food (PRiF), Pesticide Residues Committee: Pesticide residues monitoring report; third quarter report 2010, quarter ended September
10. Published 10 March 2010. [http://www.pesticides.gov.uk/Resources/ CRD/igrated-Resources/Documetns/Other/2010_Q3_Report.pdf]
11. Mortensen DA, Egan JT, Maxwell BD, Ryan MR, Smith RG: Navigating a critical juncture for sustainable weed management. BioScience 2012, 62:75–84.
12. Owen MDK: Weed species shifts in glyphosate-resistant crops. Pest Manag Sci 2008, 64:377–387.
13. Duke SO: Comparing conventional and biotechnology-based pest management. J Agric Food Chem 2011, 59:5793–5798.
14. Steffey K, Gray M: IPM and the integrated control concept: progress after 50 years in the commercial corn and soybean landscape? Bulletin. University of Illinois Extension, 2009, No. 1, Article 5.
15. Harker KN, O'Donovan JT, Blackshaw RE, Beckie HJ, Mallory-Smith C, Maxwell BD: Our view. Weed Sci 2012, 60:143–144.

16. Hartzler B, et al: Preserving the value of glyphosate. Iowa State University:; 2004. http://www.weeds.iastate.edu/mgmt/2004/preserving.shtml

17. National Agricultural Statistics Service: Agricultural Chemical Usage Field Crop Summary, 1997. [http://usda.mannlib.cornell.edu/MannUsda/ viewDocumentInfo.do?documentID=1560]

18. National Agricultural Statistics Service: Agricultural Chemical Usage Field Crop Summary, 2003. [http://usda.mannlib.cornell.edu/MannUsda/ viewDocumentInfo.do?documentID=1560]

19. National Agricultural Statistics Service: Agricultural Chemical Usage Field Crop Summary, 2007. [http://usda.mannlib.cornell.edu/MannUsda/ viewDocumentInfo.do?documentID=1560]

20. Weed Science Society of America: International survey of herbicide resistant weeds. 2012 [http://www.weedscience.org/In.asp]

21. Padgett SR, Kolacz KH, Delanney X, Re DB: Development, identification, and characterization of a glyphosate-tolerant soybean line. Crop Sci 1995, 35:1451–1461.

22. Bradshaw LD, Padgette SR, Kimball SL, Wells BH: Perspectives on glyphosate resistance. Weed Tech 1997, 11:189–198.

23. Gressel J: Fewer constraints than proclaimed to the evolution of glyphosate-resistant weeds. Resis Pest Manag 1996, 8:2–5.

24. Heap IM: The occurrence of herbicide-resistant weeds worldwide. Pestic Sci 1999, 51:235–243.

25. Benbrook C, Groth E, Hansen M, Halloran J, Benbrook K: Pest Management at the Crossroads. Yonkers, New York: Consumers Union; 1996.

26. Blewett TC: Comments on behalf of Dow AgroSciences LLC on Supplemental information for petition for determination of nonregulated status for herbicide resistant DAS-40278-9 Corn. Economic and agronomic impacts of the introduction of DAS-40278-9 corn on glyphosate resistant weeds in the U.S. cropping system; 2011:1–202.

27. Gaines TA, Zhang W, Wang D, Bukun B, Chrisholm ST, Shaner DL, Nissen SJ, Patzoldt WL, Tranel PJ, Culpepper AS, Grey TL, Webster TM, Vencill WK, Sammons RD, Jiang J, Preston C, Leach JE, Westra P: Gene amplification confers glyphosate resistance in amaranthus palmeri. Proc Nat Acad Sci 2010, 107:1029–1034.

28. Gary VF, Harkins ME, Erickson LL, Long-Simpson LK, Holland SE, Burroughs BL: Birth defects, season of conception, and sex of children born to pesticide applicators living in the red river valley of Minnesota, USA. Environ Health Perspect 2002, 110:441–449.

29. Arbuckle TE, Lin ZQ, Mery LS: An exploratory analysis of the effect of pesticide exposure on the risk of spontaneous abortion in an Ontario farm population. Environ Health Perspect 2001, 109:851–857.

30. Schreinemachers DM: Birth malformations and other adverse perinatal outcomes in four U.S. Wheat-producing states. Environ Health Perspect 2003, 111(9):1259–1264. July.

31. Rohr JR, McCoy KA: A qualitative meta-analysis reveals consistent effects of Atrazine on freshwater fish and amphibians. Environ Health Perspect 2009, 118(1):20–32. 2010 January.

32. Bernards ML, Crespo RJ, Kruger GR, Gaussoin R, Tranel PJ: A waterhemp (Amaranthus tuberculatus) population resistant to 2,4-D. Weed Sci 2012, 60:379–384.

33. Owen MDK: Weed resistance development and management in herbicide-tolerant crops: experiences from the USA. Furverbraucherschultz und lebensmittelsicherheit 2011, 6(Suppl 1):85–89.

34. Benbrook CM: The magnitude and impacts of the biotech and organic seed price premium, The Organic Center. 2009. http://www.organic-center.org/ reportfiles/SeedPricesReport.pdf.

35. Chang F-C, Simcik MF, Capel PD: Occurrence and fate of the herbicide glyphosate and its degradate aminomethylphosphonic acid in the atmosphere. Environ Tox and Chem 2011, 30:548–555. doi:10:1002/35c.431.

36. Association of American Pest Control Officials (AAPCO): 2005 AAPCO pesticide drift enforcement survey. 2005. http://aapco.org/documents/surveys/DriftEnforce05Rpt.html.

37. Kremer RJ, Means NE: Glyphosate and glyphosate-resistant crop interactions with rhizosphere microorganisms. Europ J Agronomy 2009, 31:153–161. doi:10.1016/j.eja.2009.06.004.

38. Fernandez MR, Zentner RP, Basnyat P, Gehl D, Selles F, Huber D: Glyphosate associations with cereal diseases caused by Fusarium spp. in the Canadian Prairies. Europ. Agronomy. 2009, 31:133–143. doi:10.1016/j.eja.2009.07.003.

39. Cakmak I, Yazici A, Tutus Y, Ozturk L: Glyphosate reduced seed and leaf concentrations of calcium, manganese, magnesium, and iron in nonglyphosate resistant soybean. Europ J Agronomy 2009, 31:114–199. doi:10:1016/j.eja.2009.07.001.

40. Zobiole LHS, de Oliveira RS, Huber DM, Constantin J, de Castro C, de Oliveira FA, de Oliveira A: Glyphosate reduces shoot concentrations of mineral nutrients in glyphosate-resistant soybeans. Plant Soil 2009, 328:57–69. doi:10.1007/s11104-009-0081-3.

41. Pleasants JM, Oberhauser KS: Milkweed loss in agricultural fields because of herbicide use: effects on the monarch butterfly population. Insect Conservation and Diversity 2012, doi:10.1111/j.1752-4598-2012.00196.x.

42. Casabé N, Piola L, Fuchs J, Oneto ML, Pamparato L, Basack S, Gimenez R, Massaro R, Papa JC, Kesten E: Ecotoxicological assessment of the effects of glyphosate and chlorpyrifos in an Argentine soya field. J Soils Sediments 2007, 7:232–239. doi:10.1065/jss2007.04.224.

43. Zobiole LHS, de Oliveira R, Kremer RJ, Constantin J, Bonato CM, Muniz AS: Water use efficiency and photosynthesis of glyphosate-resistant soybean as

affected by glyphosate. Pesticide Biochem and Physiol 2010, 97:182–193.

44. Zobiole LHS, de Oliveira RS, Kremer RJ, Constantin J, Yamada T, Castro C, de Oliveira FA, de Oliveira A: Effect of glyphosate on symbiotic N2 fixation and nickel concentration in glyphosate-resistant soybeans. Appl Soil Ecol 2010, 44:176–180.

45. King AC, Purcell LC, Vories ED: Plant growth and nitrogenise activity of glyphosate-tolerant soybean in response to glyphosate applications. Agron J 2001, 93:179–186.

46. Warwick SI, Légeré A, Simard MJ, James T: Do escaped transgenes persist in nature? The case of an herbicide resistance transgene in a weedy Brassica rapa population. Mol Biol 2007, 17:1387–1395. doi:10.1111/j.1365-294X.2007.03567.x.

47. Nguyen HR, Jehle JA: Expression of cry3Bb1 in transgenic corn MON88017. J Agric Food Chem 2009, 57:9990–9996.

48. Stotzky G: Persistence and biological activity in soil of inserted proteins from Bt and of bacterial DNA bound on clay and humic acids. J Environ Qual 2000, 29:691.

49. Gray ME: Relevance of traditional Integrated Pest Management (IPM) strategies for commercial corn producers in a transgenic agroecosystem: a bygone era? J Agric Food Chem 2011, 59:5852–5858.

50. Gray ME: Additional reports of severe rootworm damage to Bt corn received: questions and answers, The Bulletin University of Illinois Extension; 2011. No. 22, Article 2.

51. Pierce CMF, Gray ME: Population dynamics of a Western corn rootworm (Coleoptera: Chrysomelidea) variant in east central Illinois commercial maize and soybean fields. J Econ Entomol 2007, 100(4):1104–1115. Aug.

CHAPTER 3

ECOSMART BIORATIONAL INSECTICIDES

Hanem Fathy Khater

INTRODUCTION

Pest insects can damage agricultural crops, consume and/or damage harvested food, or transmit diseases to humans and animals. The past 30 years has witnessed a dramatic reemergence of epidemic vector-borne diseases throughout much of the world (Atkinson, 2010). Prior to the development and commercial success of synthetic insecticides in the mid-1930s to 1950s, botanical insecticides were the foremost weapons against insect pests. The synthetic insecticides (organochlorines, organophosphates, carbamates and later the pyrethroids and neonicotinoids) are characterized by efficacy, speed of action, ease of use, and low cost. Accordingly, they drove many natural control methods, such as using of botanicals, predators, and parasitoids to near obscurity. Twenty years after synthetic insecticides were overzealously entrenched in 'modern' agricultural production; they induce widespread environmental contamination, toxicity to non-target organisms, development of resistance against insecticides, and negative effects on animal and human health (Pretty, 2009). Consequently, there is an urgent need to explore and utilize naturally occurring products for combating pests. The terms "biorational pesticide" and "biopesticides" are gaining popularity in the current climate of environmental awareness and public concern. Both terms are derived from two words, "biological" and "rational", referring to pesticides that have limited or no adverse effects on the environment, non-target organisms including humans. Biorational insecticides include: biochemicals insecticides (botanicals, insect growth regulators, insect pheromones, photoinsecticides, and inorganics); biological insecticides, using of natural enemies such as parasitoids, predators, nematodes, and pathogens (virus, bacteria, fungi, or protozoa); and transgenic insecticides (genetically modified plants or organisms). Natural enemies play an important role in limiting potential pest populations and they are more likely to survive in case of application of ecofriendly biopesticides. Approaches to the biological control of insects include: conservation of existing natural enemies; introducing new natural enemies and establishing a permanent population (called "classical biological control"); and mass rearing and periodic release, either as a seasonal introduction of a small population of natural enemies, or a massive, "inundative" release. In developing countries, biopesticides offer unique and challenging opportunities for exploration and development of their own biorational insecticides. Nanotechnology has become one of the most promising new technologies in the recent decade for protection against insect pests. Such technology will revolutionize agriculture including pest management in the foreseeable future. Integrated pest management (IPM) is the use of all available means to maintain pest populations below levels that would cause economic loss while minimally impacting the environment. Several tactics could be utilized in IPM programs as chemical, cultural, physical, and biological control (Vreysen et al. 2007). The introduction of more effective biorational products through IPM programs will reduced rates of chemical pesticides and prevent, or at least delay the development of resistance in target pests to both chemical pesticides and biopesticide toxins. Flourishing of organically produced food in the developed world facilitates greater farmer acceptance of biopesticides

Citation: Hanem Fathy Khater (2012). Ecosmart Biorational Insecticides, Insecticides - Advances in Integrated Pest Management, Dr. Farzana Perveen (Ed.), InTech, DOI: 10.5772/27852.

as the sales of organically produced food are increasing at a significantly faster rate than sales of any other food commodity. Consequently, biorational insecticides will dominate the market of pesticides in the near future. Here, I am concerned about control of insects and arachnids (ticks and mites) of agricultural, medical, and veterinary importance and referred to them as insects or insect pests. The words "biorational" and "biopesticide" as well as "pesticides" and "insecticides" are used interchangeably throughout this chapter. Finally, I review current biorational insecticides and their mode of actions, uses, commercial products, and safety concerns.

BIOCHEMICAL CONTROL

Botanical Insecticides

The practice of using plant derivatives or botanical insecticides in agriculture dates back at least two millennia in ancient Egypt, India, China, and Greece. In Europe and North America, the documented use of botanicals extends back more than 150 years, dramatically predating discoveries of the major classes of synthetic chemical insecticides beginning in the 1940s.

Traditional Botanical Insecticides

Pyrethrum

Pyrethrum is one of the oldest and safest insecticides. The ground, dried flowers of Tanacetum cinerariaefolium (Asteraceae) were used in the early 19th century to control body lice during the Napoleonic Wars. Pyrethrum contains three esters of chrysanthemic acid and three esters of pyrethric acid. Among the six esters, those incorporating the alcohol pyrethrolone, namely pyrethrins I and II, are the most abundant and account for most of the insecticidal activity. Technical grade pyrethrum, the resin used in formulating commercial insecticides, typically contains from 20% to 25% pyrethrins (Casida & Quistad, 1995). Recently, Australia produces almost one-half of the world supply and produces a technical grade material comprising 50% pyrethrins by weight. Pyrethrins affect the insect on contact, creating disturbances in the nervous system which eventually result in convulsions and death. Pyrethrin acts on insects with phenomenal speed causing immediate paralysis, notably in flying insects, some of which are immobilized within 1 s. It blocks voltage-gated sodium channels in nerve axons. The mechanism of action of pyrethrins is qualitatively similar to that of DDT and many synthetic organochlorine insecticides. Pyrethrums are mixed with a synergist such as piperonyl butoxide (PBO) to increase insect mortality and to extend their shelf life. In purity, pyrethrins are moderately toxic to mammals, but technical grade pyrethrum is considerably less toxic (Casida & Quistad, 1995). Major uses of pyrethrum are for structural pest control, in public health, and for treatment of animal premises. Pyrethrins have limited use outdoors as they are especially labile in the presence of the UV component of sunlight (Ware & Whitacre, 2004). Pyrethrum products represent 80% of the total market of global botanical insecticides (Isman, 2005) and are favored by organic growers because of their low mammalian toxicity and environmental nonpersistence making it among the safest insecticides in use. For more information about pyrethrum, see Taylor (2001), Collins (2006), and Gilbert & Gill (2010).

Other Traditional Botanicals

A handful of other plant materials have seen limited commercial use as insecticides and their uses are in decline, such as sabadilla, a powder based on the ground seeds of the South American plant Schoenocaulon officinale ; Wood of the Caribbean tree Ryania speciosa; Quassia amara, a small tree from Brazil; woodchips and ground bark of this species have been used traditionally as an insecticide, as have plant parts from the related tree, Ailanthus altissima; rotenone, an isoflavonoid obtained from the roots or rhizomes of tropical legumes in the genera Derris, Lonchocarpus, and Tephrosia. Rotenone is used as insecticide and mainly a fish poison to paralyze fish, causing them to surface and be easily captured, but there is a growing concern about its safety and its relation to Parkinson's disease (Betarbet et al., 2000). For more niceties about traditional botanical insecticides, see Ware & whitecare (2004), Isman (2005, 2006, 2010), Isman & Akhtar (2007), Gilbert & Gill (2010), Kumar et al (2010), Dubey (2011), and Mehlorn (2011).

Newer Botanical Insecticide

"Neem" Neem (Azadirachta indica A. Juss: Meliaceae) is a large, evergreen, hardy tree, native to the Indian sub-continent and well known their as the' Botanical Marvel', It is an old and new insecticide. The Indians used neem, from prehistoric times, primarily against household and storage pests, and to some extent against pests related to field crops. In addition, they traditionally burn neem leaves in the evening to repel mosquitoes. It is effective against more than 500 species of insects and arthropods. Neem has attracted global attention recently due to its potential as a source of natural drugs and as environment-friendly pesticides, see Schmutterer (1995), Kumar (2002), Isman et al. (2011), and Mehlhorn (2011) for more fine points.

Chemical composition

Neem seeds are a rich storehouse of over 100 tetranortriterpenoids and diverse nonisoprenoids. The neem tree contains more than 100 different limonoids in its different tissues (Isman et al., 1996). Many of them are insect feed deterrents. The highly oxygenated azadirachtin ($C_{35}H_{44}O_{16}$), a nortriterpenoid belonging to the lemonoids, is the most biologically active constituent of neem. Azadirachtin has shown bactericidal, fungicidal, and insecticidal properties, including insect growth regulating qualities (Ware & Whitacre, 2004). It is systemic in nature, absorbed into the plant and carried throughout the tissues, being ingested by insects when they feed on the plant. Thus, it is effective against certain foliage-feeders that cannot be reached with spray applications. In general, chewing insects are affected more than sucking insects and insects that undergo complete metamorphosis are also generally affected more than those that do not undergo metamorphosis (Dubey, 2011).

Mode of Action

The effects of azadirachtin on insects include feeding and oviposition deterrence, growth inhibition, and fecundity and fitness reductions (Schmutterer, 1990). Azadirachtin is a common example of a natural plant defense chemical affecting feeding, through chemoreception (primary antifeedancy), that consists in the blockage of the input from receptors that normally respond to phagostimulants, or from stimulation of specific deterrent cells or both (Dethier, 1982) and through a reduction in food intake due to toxic effects if consumed (secondary antifeedancy), where food intake is reduced after application of azadirachtin in ways which bypass the mouth part chemoreceptors (Mordue & Blackwell, 1993). The antifeedant effect is highly variable among pest species, and even those species initially deterred are often capable of rapid desensitization to azadirachtin (Bomford & Isman, 1996). Azadirachtin is a tetranortriterpenoid, structurally similar to insect hormones "ecdysones", its biological activity as ecdysone-blocker thus disturbing insect growth. This substance interferes with synthesis of the insect molting hormone, a-ecdysone, as well as other physiologically active neuropeptides in insects, producing a wide range of physiological and behavioral effects, such as anorexia. It also leads to sterility in female insects due to its adverse effects on ovarian development, fecundity, and fertility. For more information about the mode of action of neem, see Isman and Akhtar (2007) and Insect growth regulators below.

Safety

Azadirachtin is nontoxic to mammals. Different neem products were neither mutagenic nor carcinogenic, and they did not produce any skin irritations or organic alterations in mice and rats, even at high concentrations. The pure compound azadirachtin, the unprocessed materials, the aqueous extracts and the seed oil are the most safe to use as an insecticide to protect stored seeds for human consumption (Boeke et al., 2004). Ecologically, azadirachtin is non toxic to fish (Wan et al., 1996), natural enemies and pollinators (Naumann & Isman, 1996), birds, other wild life, and aquatic organisms as azadirachtin, breaks down in water within 50–100 h. It is harmless to non-target insects (bees, spiders, and butterflies). The effect of azadirachtin on natural enemies is highly variable (Hohmann et al., 2010, Kumar et al., 2010). Environmentally, azadirachtin induce no accumulations in the soil, no phytotoxicity and accumulation seen in plants, and no adverse effect on water or groundwater (Mehlhorn, 2011). Neem is sensitive to light and the half-life of azadirachtin is only one day (Kleeberg, 2006), leaving no residues on the crop and therefore are preferred over chemical pesticides. Azadirachtin is classified by the Environmental Protection Agency (EPA) in class IV.

Risk Factors

The most critical adverse effects are reproduction disturbances, although these are often reversible. (Boeke et al., 2004). Neem pollen induces allergenic effect to some individuals (Karmakar & Chatterjee, 1994). Moreover, the oil can turn rancid (De Groot, 1991) and is easily contaminated with aflatoxins, so contaminated neem seeds with aflatoxin should not be picked from the ground but seeds that are greenish yellow in color should be picked from the trees or swept regularly under the tree (Gunasena & Marambe, 1998). Ecto-endo parasitoids vulnerable to neem but soil application could reduce negative side effects compared to plant spraying and hence improve selectivity (Kumar et al., 2010). Treating the host with neem before parasitism was less deleterious to wasp emergence, especially for Trichogrammatoidea annulata (Hohmann et al., 2010). For more details about safety of neem, see Boeke et al. (2004), Mehlhorn (2011), Kumar et al. (2010), and Homanni et al. (2010).

Production

In order to produce and use efficacious neem pesticides, Saloko et al. (2008) reviewed some points that should be notes: neem leaf extracts are less effective than seed extracts due to lower azadirachtin content; neem preparations should be kept away from sunlight to avoid photodegradation of active ingredients by UV light, and formulations are better applied at dusk when sun is weak; sun screens such as Para Amino Benzoic Acid (PABA) could be added to reduce the photo-oxidation of azadirachtin by UV light. Neem seeds contain 0.2% to 0.6% azadirachtin by weight, so solvent partitions or other chemical processes are required to concentrate to be 10% to 50% as in the technical grade material used for commercial production. World wide, there are over 100 commercial neem formulations such as Margosan-O, Bio-neem, Azatin, , Neemies, Safer's ENI, Wellgro, RD-Repelin, Neemguard, Neemark, and Neemazal. Formulations include emulsifiable concentrates (ECs), suspension concentrates (SCs), ultra low volume (ULV) formulations and granular formulations. The chemistry of azadirachtin was reached in 2007 and its synthesis was completed, see Morgan (2009). Azadirachtin and botanical preparations based on neem seed extracts are environmentally friendly pesticide and virtually non-toxic to mammals and wildlife, making them among the safest of all insecticides that used for integrated pest management and organic farming, For more details about neem, see Collins (2006), Isman & Akhtar (2007), Saloko et al. (2008), Gilbert & Gill (2010), Dubey (2011), and Regnault-Roger (2011).

Essential Oils

Aromatic oils obtained through steam distillation of many plant families, ex. Myrtaceae, Lamiaceae, Asteraceae, Apiaceae, and Rutaceae are highly targeted for anti-insect activities against several insect orders. Approximately 3000 essential oils are known, and 10% of them have commercial importance in the cosmetic, food, and pharmaceutical industries. They are generally recognized as safe, GRAS, by the US Food and Drug Administration. Complete essential oils are more effective than individual constituents or even a combination of constituents.

Essential oil chemistry

The volatile components of essential oils can be classified into four main groups: terpenes, benzene derivatives, hydrocarbons, and other miscellaneous compounds. The major constituent of some oils are 8-cineole from rosemary (Rosmarinus officinale) and eucalyptus from (Eucalyptus globus); eugenol from clove oil (Syzygium aromaticum); thymol from garden thyme (Thymus vulgaris); and menthol from various species of mint (Mentha species). More information about essential oil chemistry is given by Isman (2006) and Tripathi et al. (2009).

Mode of action

Aromatic plants produce many compounds that act as ovicidal, larvicides, adulticides, insect arrestants and repellents or act to alter insect feeding behavior, growth and development, ecdysis (molting) and behavior during mating and oviposition. Essential oils are lipophilic in nature and interfere with basic metabolic, biochemical, physiological, and behavioral functions of insects. Commonly, essential oils can be inhaled, ingested or skin absorbed by insects. The rapid action against some pests is indicative of a neurotoxic mode of action, and there is evidence for interference with the neuromodulator octopamine (Enan, 2005) or GABA-gated chloride channels (Priestley et al., 2003). Several essential oil compounds have been demonstrated to act on octopaminergic system of insects. Octopamine is a neurotransmitter, neurohormone, and circulating neurohormone – neuromodulator (Hollingworth, et al., 1984) and its disruption results in total break down of nervous system in insects. The lack of octopamine receptors in vertebrates likely accounts for the profound mammalian selectivity of essential oils as insecticides. Eugenol mimicked octopamine in increasing intracellular calcium levels in cloned cells from the brain of Periplaneta americana and Drosophila melanogaster (Enan, 2005). Consequently, octopaminergic system of insects represents a biorational target for insect control. Plant volatile oils have long been known to affect the behavioural responses of pests, with the monoterpenoid components appearing most useful as insecticides or antifeedants (Palevitch & Craker, 1994). LMW terpenoids may be too lipophilic to be soluble in the haemolymph after crossing the cuticle, and proposed a route of entry through the tracheae (Veal, 1996). Most insecticides bind to receptor proteins in the insect and, in doing so; they interrupt normal neurotransmission, which lead to paralysis and subsequently death. Recent evidence suggests that low-molecular-weight (LMW) terpenoids may also bind to target sites on receptors that modulate nervous activity. Ionotropic, ┘ -aminobutyric acid, GABA receptors, the targets of organochlorine insecticides lindane and dieldrin, are modulated by LMW terpenoids with vastly different structures (Priestley et al., 2006). Valuable appraisals about the mode of action are those of Price & Berry (2006), Isman (2006, 2010); Tripathi et al. (2009); and Dubey (2011). Some essential oils have larvicidal effect and the capacity to delayed development and suppress adults emergences and induce abnormalities during development of insects of medical and veterinary importance, Fig (1-5) (Khater, 2003; Shalaby & Khater, 2005; Khater & Shalaby, 2008; Khater & Khater (2009); Khater et al., 2009, 2011).

Plant- based repellents: Some plant-based repellents are comparable to, or even better than synthetics; however, essential oil repellents tend to being short-lived in their effectiveness due to their volatility. Nerio (2010) review some splendid ideas for improvement of repellency of essential oils. Repellency assays with essential oils were done for Diptera species, especially mosquitoes and to a lesser extent to coleopteran insects related to losses in stored food. Plants with strong smell, such as French marigold and coriander act as repellents and can protect the corps nearby. Several essential oil- producing plants have been widely studies, such as Cymbopogon spp., Eucalyptus spp., Ocimum spp., the osage orange (hedgeapple) (Maclura pomifera), and catnip (Nepeta cataria). Several plant oils or their constituents have been commercialized as insect repellents in the past decade, such as soybean, lemon grass, cinnamon, and citronella. Neem oil, from A. indica, when formulated as 2% in coconut oil, provided complete protection (i.e. no confirmed bites) for 12 hours from Anopheles mosquitoes (Sharma et al., 1993). Essential oils have pronounced In vitro and In vivo pediculicidal activity as the number of lice infesting water buffaloes in Egypt was significantly reduced 3, 6, 4, and 6 days after treatment with the essential oils of camphor (Cinnamomum camphora), peppermint (Mentha piperita), chamomile (Matricaria chamomilla), and onion (Allium cepa), respectively. Surprisingly, the same oils repelled flies (Musca domestica, Stomoxys calcitrans, Haematobia irritans, and Hippobosca equine) infecting buffaloes for almost 6 days posttreatment. No adverse effects were noted on either animals or pour-on operators after exposure to the applied oils (Khater et al., 2009).

Metabolites reliable for repellent activity: Nerio et al. (2010) reviewed the repellent activity of essential oils which contributed to some metabolites, such as monoterpenes (┘-pinene, cineole, eugenol, limonene, terpinolene, citronellol, citronellal, camphor, and thymol) against Mosquitoes (Yang et al., 2004) sesquiterpenes, ┘-caryophyllene, repellent against A. aegypti; phytol, a linear diterpene alcohol, against Anopheles gambiae; and phenylethyl alcohol, ┘-citronellol, cinnamyl alcohol, geraniol, and ┘-pinene, isolated from the essential oil of Dianthus caryophyllum, aginst ticks (Ixodes ricinus). In addition, cineole, geraniol and piperidine found in bay leaves (Laurus nobilis, Lauraceae) possess repellent properties towards cockroaches. Repellents may have an increasingly important role in eliminating insects from certain environments and essential oils could play a major role in new repellent technology. Valuable review on the repellent activity of essential oils are those of Tripathi et al. (2009), Isman (2010), Kumar et al. (2010), Nerio et al. (2010), Dubey (2011), and Maia & Moore (2011).

Fumigant

Today, The used fumigants, for instance, phosphine, methyl bromide, and DDVP (2,2- dichlorovinyl dimethyl phosphate) do have adverse effets. Phosphine is the major cause of suicidal deaths in India. Methyl bromide has ozone-depleting potential (UNEP, 2000) and DDVP has a possible human carcinogen potential (Lu, 1995). Thus, there is an urgent need for development of safe alternative that have the potential to replace the toxic fumigants against pests attacking grains, dry stored food, and other agricultural products. The active principles are monoterpenes, sesquiterpenes and their biogenically related phenols. In addition to direct toxicity to insects, many of these substances are eterrents or repellents. Essential oils of Artemisia species, Anethum sowa, Curcuma long, and Lippia alba. Clove, rosemary, thyme, eucalyptus and various mint species have demonstrated contact and fumigant toxicity to a wide spectrum of insects, including human head lice (Toloza et al., 2008). Isolates like d-limonene, carvones and 1,8-cineole have been well documented as fumigants. The exact mode of action of these oils as fumigant is unknown, but the oils mainly act in the vapour phase via respiratory system. Physical properties of essential oils such as high boiling point, high molecular weight and low vapor pressure are barriers for application in large scale fumigation. For more details about fumigants, see Tripathi et al., (2009), Isman (2010), and Dubey (2011).

Commercialization

Although essential oils are effective when freshly applied, their protective effects usually dissipate relatively quickly. In their review, Nerio et al. (2010) discussed methods to access repellency effects, the synergistic phenomena of such oils and some novel ideas to increase the repellent efficiency. Some fixative materials such as liquid paraffin, vanillin, salicyluric acid, mustard, and coconut oils have been used. Formulations based on creams, polymer mixtures, or microcapsules for controlled release, resulted in an increase of repellency duration. Still, essential oils can be incorporated with polymers into sheets and attractant adhesive films with essential oils were prepared to control insects in agriculture and horticulture. Novel ideas are needed to be explored for better commercialization of essential oil- based pesticides. Several essential oil constituents are already in use as an alternative to conventional insecticides, such as Green Ban® (containing oils of citronella, cajuput, lavender, safrole free sassafrass, peppermint, and bergaptene free bergamot oil); Buzz Away® (containing oils of citronella, cedarwood, eucalyptus, and lemongrass); ValeroTM, a miticide/fungicide for use in grapes, berry crops, citrus, and nuts; and CinnamiteTM, an aphidicide /miticide/fungicide for glasshouse and horticultural crops. The last two products are based on cinnamon oil, with cinnamaldehyde (30% in EC formulations) as the active ingredient. In addition, d-limonene is an active ingredient of commercially available flea shampoos, plus pulegone and citronellal are used as mosquito repellents

Safety of Essential Oils

Currently, the US Environmental Protection Agency (US EPA) has registered citronella, lemon, and eucalyptus oils as insect repellent ingredients for application on the skin. Using essential oils or some of their products could cause dermatitis, they should be rubbed on a small portion of skin to determine if there will be an allergic reaction before treating your whole body. The most attractive aspect of using essential oils and/or their constituents for pest control is their favorable mammalian toxicity because many essential oils and their constituents are commonly used as culinary herbs and spices. Many of the commercial products including essential oils are included on the GRAS list fully approved by FDA and EPA in USA for food and brevarage consumption (EPA, 1993). Some of the purified terpenoid constituents of essential oils are moderately toxic to mammals, but, with few exceptions, the oils themselves or products based on oils are mostly nontoxic to mammals, birds, and fish. Although natural enemies are susceptible via direct contact, predators and parasitoids reinvading a treated crop one or more days after treatment are unlikely to be poisoned by residue contact as often occurs with conventional insecticides. Owing to their volatility, the oils and their constituents are environmentally nonpersistent, with outdoor half lives of\24 h on surfaces, in soil and in water (Isman el al., 2011). There is no harvest restrictions or worker re-entry restrictions for treated crops; they are compatibile with biological control agents and indigenous natural enemies of pests, and they bring about reduce risks to honeybees and other foraging pollinators. For additional information about safety of essential oils, see Isman (2006, 2010), Tripathi et al. (2009), Nerio et al. (2010), and Regnault-Roger (2011). Because many conventional pesticide products fall into disfavour with the public, botanical-based pesticides should become an increasingly popular choice for pest control.

Insect Growth Regulators

Insect growth regulators (IGRs) are chemical compounds that alter growth and development in insects. They don't directly kill insects, but interfere with the normal mechanisms of development, resulting in insects dying before they reach adulthood. IGRs are classified into two general categories based on mode of action: chitin synthesis inhibitors and substances interfering with the action of insect hormones.

Chitin Synthesis Inhibitors

Chitin synthesis inhibitors (CSIs) affect the ability of insects to produce new exoskeletons when molting. They act on the larval stages by inhibiting or blocking the synthesis of chitin which represent 30-60% of the insect exoskeleton structure. They also increase egg mortality. CSIs include conventional benzoylureas, triazine/pyrimidine derivatives, and buprofezin.

Benzoylphenylurea

Typical effects benzoylureas or benzoylphenylurea (BPUs) on developing larvae are the rupture of malformed cuticle or death by starvation. BPUs act as ovicides, reducing the egg laying rate or hindering the hatching process by inhibiting embryonic development or failure of hatchability. Commertial porducts of BPUs include diflubenzuron (Dimilin®, Adept®, Micromite®); triflumuron (Alsystin®); teflubenzuron (Nomolt®, Dart®), hexaflumuron (Trueno®, Consult®); chlorfluazuron (Atabron®); flufenoxuron (Cascade®); and flucycloxuron (Andalin®). Among the newer benzoylureas only hexaflumuron (1993) and novaluron (2001) have been registered by EPA. Studies with diflubenzuron, the most investigated BPU, revealed that it alters cuticle composition, especially inhibition of chitin, resulting in abnormal endocuticluar deposition that affects cuticular elasticity and firmness, and cause abortive molting. Diflubenzuron (Dimilin® El –Delta Company, Egypt) is highly effective in controlling mosquitoes, Culex pipiens than house flies, Musca domestica. LC 50 values were 1.26 and 1000 ppm, respectively. All treated late 3rd and early 4th larvae of C. pipiens (concetnraions: 0.04 - 40 ppm) were eventually died as Dimilin® prolonged the larval durations (11.9 days vs. 4 days in the control group) and increased larval abnormalities (46.7%). Such abnormalities were larvae with transparent cuticle, splitting of cuticle, and pharate pupae (Fig. 4). It induces pupal abnormalities as well (Fig. 5). Treatment of of M. domestica with the same product (at 1ppm) induced larval and pupal malformations reached 23.3 and 56.5%, respectively, and reduce adult emergence (66.7%). Abnormalities of M. domestica include small, shrunken, macerated larvae and larvae with week cuticle as well as distorted puparia and failure of adult eclosion (Khater, 2003) (Fig. 1-3).

Triazine/pyrimidine derivatives

Cyromazine: Cyromazine (Larvadex ®, Trigard®), a triazine, is a potent CSI and it is selective toward dipterous species and fed to poultry or sprayed to control flies on animals, in manure of broiler and egg producing operations. It controls blowfly infesting sheep and persist for up to 13 weeks (O'Brien & Fahey, 1991) after a single pour-on application, or longer if applied by dip or shower. Moreover, it is used as a leafminers spray in vegetable crops and ornamentals. Cyromazine may inhibit growth or expansion of the body wall (or both) sufficiently to prevent normal internal growth, producing the observed symptoms and leading to abnormal development. The presence of three resistant house fly populations to cyromazine in Brazilian poultry farms strongly suggests that the operational aspects of larvicide use are important for the development of resistance. Cyromazine is applied as a feed-through, both in Brazil and in the USA, where resistance has already been documented. However, in Denmark, where it was approved only as a topical manure spray, no case of resistance has yet been detected (Pinto & do Prado, 2001).

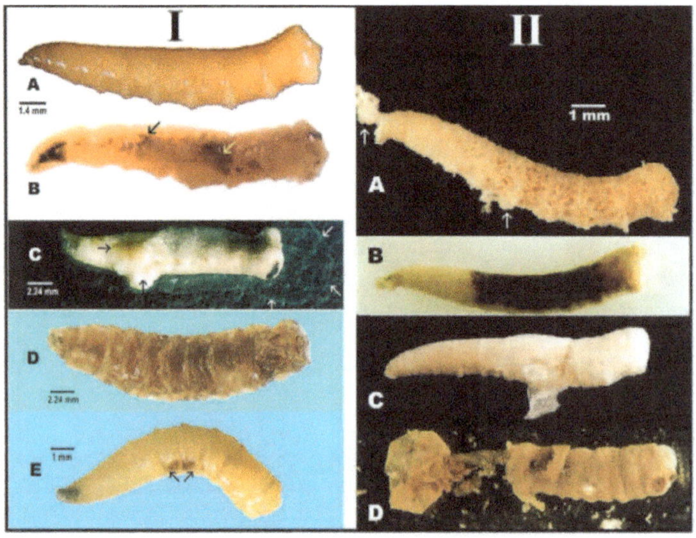

Figure. 1. Morphological malformations of larvae of house flies. I. A. Normal larva. B-E. Malformed larvae, treated with essential oils and insect growth regulators showing signs of pigmentation. C. Macerated larva with week transparent cuticle. II. Larvae infected with fungi. A. Red Pin -point pigments all over the larval body with apparent fungal growth (arrow heads). B. Larva with diffuse blackish pigmentation. C. Larva with an ulcer in the middle. D. Ulcerated and macerated larva with white nodules and fungal growth.

Dicyclanil: Dicyclanil (ZR ®, ComWin ®), a pyrimidine derivative, is highly active against dipteran larvae and available as a pour-on formulation for blowfly control in sheep in Australia and New Zealand providing up to 20 weeks' protection (Bowen et al., 1999). On the whole,

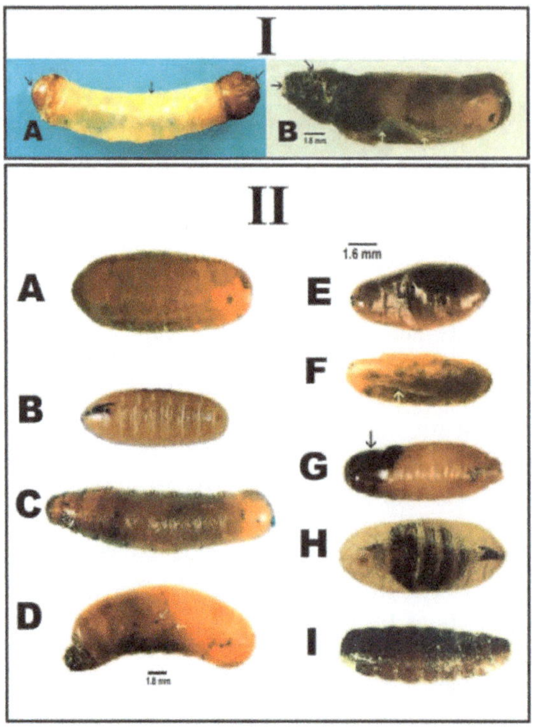

Figure. 2. Pupal abnormalities of house flies: I. Larval pupal- intermediates: A. The anterior and posterior ends as pupae (left and right arrows), while the rest of the body as larva (middle arrow). B. Fungal growth anteriorly (left arrow) and cracked pupal

case at the middle (lower arrows). II. Pupae: A. Normal pupa. B. Small pupa with visible cephalopharyngeal skeleton. C. Larviform pupa, pigmented with small dark spots at the intersegmental regions. D. Cshaped pupa with anterior constriction and small patches of black pigments (arrow heads). E. Distorted puparia. F. Pupa with a groove (arrow head). G. Blackish posterior end (arrow head). H. Transparent puparia. I. Hyphal growth appears on a small puparia. Adapted form Khater (2003).

Figure. 3. Morphological abnormalities of house flies. I. (A-F) Failure of adult eclosion. II. Adults (A. Normal adult, B. Crumpled adult with poorly developed wing and legs, C. Small adult, D. Elongated adult with deformed wing, abdomen, and legs. Adapted from Khater (2003).

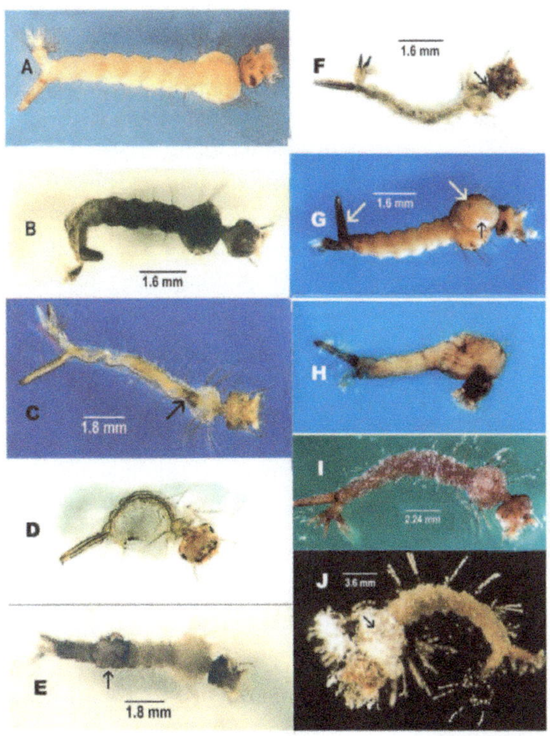

Figure. 4. Morphological abnormalities of mosquito larvae: A. Normal larva, B. Pigmented larva, C-F. Larvae with deformed cuticles, G. Larva with an opaque swelling on the thorax and black coloration at the posterior end, H. Pharate pupa (prepupa), I, J. larvae show symptoms of mycosis. Adapted from Khater (2003).

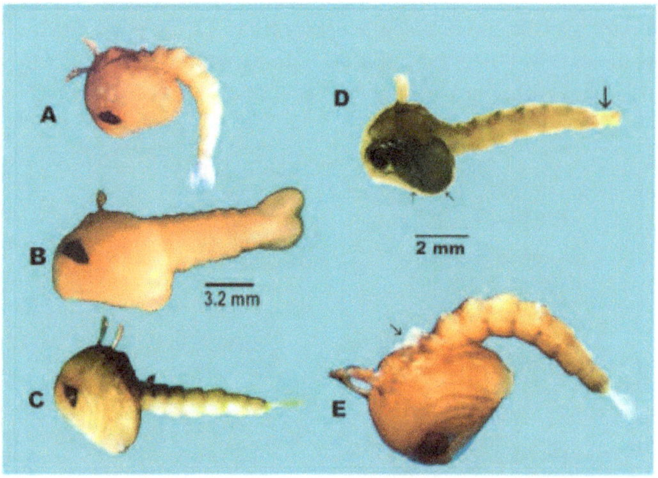

Figure. 5. Pupal abnormalities of mosquitoes: A. Normal pupa. B. Albino pupa. C. Elephantoid pupa, with enlarged cephalic region and extended abdomen. D. Black cephalothorax (lower arrows) and extended abdomen with transparent posterior end (upper arrow) and no anal gills. E. Pupa with apparent fungal growth (arrow head). Adapted form Khater (2003).

most CSI compouns are very potent against a variety of different pests, especially lepidopterous insects and whiteflies. They are harmless or exert little adverse effect on bees, predators, or parasitoids (Tomlin, 2000) which renders them acceptable for inclusion in IPM programs.

Substances Interfering With the Action of Insect Hormones

Growth and development of insects are regulated by hormones: prothoracicotrophic hormones (PTTH) (brain hormone), ecdysteroids, and juvenile hormones (JH). PTTH controls the secretion of the molting hormone (ecdysone) from the prothoracic gland. Ecdysone is responsible for cellular programming and, together with JH, initiating for the molting process. High secrection of JH from the corpora allata allows the epidermis to be programmed for a larval molt, or else it is programmed for metamorphosis. JH is virtually absent in the pupae, but is present in adults to serve some functions in reproduction. JH suppresses pupation and induces vitellogenesis during the reproductive stage of the insect (Eto, 1990).

Ecdysteroid Agonist

Bisacylhydrazine (BSH), a newer class of IGRs, is ecdysone agonists or disruptors with molting hormone activity. BSHs include chromafenozide (Virtu ®), Tebufenozide (Mimic ®, Confirm ®), Halofenozide (Mach-2 ®), and Methoxyfenozide (Intrepid ®). They have a greater metabolic stability than the insect steroid molting hormone 20-hydroxyecdysone (20E) In vivo (Retnakaran et al., 1995). They are toxic after ingestion or exposure to higher doses of topical application. Ingestion of BSH creates hyperecdysonism in susceptible insect, including molting events. The effect starts with feeding inhibition within 3-14 h (Retnakaran et al., 1997), which is very important for preventing further crop damage. In the mean while, larvae become moribund, slip their head capsule, and the hind gut may be extruded in extreme cases. Synthesis of new cuticle begins and apolysis of the new cuticle form the old one takes place. The new cuticle is not sclerotized or tanned. Consequently, the food intake by larva is prevented as the mouthparts become soft and mushy. Larval death is due to incomplete molting, starvation, and desiccation due to hemorrhage. Both Tebufenozide and methoxyfenozide, selectively toxic to Lepidopteran pests, have been classified by US EPA as reduced risk pesticides. It worth to mention that azadirachtin in neem tree acts as ecdysone blocker and disrupt insect growth, see botanical insecticides above. In general, BSHs have insect selectivity, reduced risk ecotoxicology, and mammalian profile as they have new mode of action, therefore, they are ideally integrated in IPM programs.

Juvenile Hormone Analogues

The juvenile hormone analogues (JHA) mimic the activity of naturally occurring JHs and prevent metamorphosis to the adult stage. JH is one of the most pleiotropic hormones known and functions in various aspects such as embryogenesis, molting and metamorphosis, reproduction, diapause, communication, migration/ dispersal, caste differentiation, pigmentation, silk production, and phase transformation. The major function of JH is the maintenance of larval status or the so- called juvenilizing effect, See Tunaz (2004). JHAs are highly effective at the beginning stage of metamorphosis and embryogenesis, for instance, freshly ecdysed last larval instars, freshly ecdysed pupal instars, and deposited eggs. Thus embryogenesis is disrupted when young eggs are treated with JHAs. Application to early last instar larvae would result in the development of supernumerary instars, whereas treatment at the later stage would result in abnormal pupation and development of larval-pupal mosaics or intermediates (Khater, 2003). Treatment of of M. domestica and C. pipiens with pyriproxyfen (Sumilarv ® Somitomo Co.) interferes with normal metamorphosis and results in various developmental abnormalities including larval pupal intermediates that do not survive (Fig. 1-5); Lower concentrations significantly prolonged larval and papal durations, 10.5 and 11 days (at conc. 0.1 ppm) and 9.9 and 15.1 days (at conc. 4 ppm), respectively, than those of the control (3, 5.7 days and 3.9, 1.1 days, respectively) (Khater, 2003). Methoprene (Altosid ®, Apex ®, Pharorid ®, Precor ®), the first compound introduced into the market, is a terpenoid compound with very low mammalian toxicity and is regularly used for flea control. It is sensitive to light. Thus, It has been used extensively and successfully in indoor environments and on pets in the form of collars, shampoos, sprays, and dips and also as a feed through larvicide for hornfly, Haematobia, control on cattle (Graf, 1993). Other IGRs available for use against household and agricultural pests are pyriproxyfen and fenoxycarb. Fenoxycarb (Logic®, Award®, Comply®, Torus®) is a carbamate stomach insecticide that has also JH-type effects when contacted or ingested by various arthropod pests, e.g., ants, roaches, ticks, chiggers and many others. Some JHAs are of plant origin, such as Juvenoids, isolated form plants, the "paper factor" form the balsam fir (Abies balsamea) and Juvocimenes from the sweet basil plant (Ocimum basilicum) (Bowers & Nishida, 1980). Morphogenetic control is effective for controlling insects that are pests as adults, like most insects of medical and veterinary importance, ex. mosquitoes, nuisance flies, and fire ants. They are not effective against lepidopteran agricultural pests because the larval stage is responsible for plant destruction. Many JH analogs (or mimics) (JHAs) are attractive candidates for pest control because of the ease of their synthesization and their pest selectivity than those of other peptide and steroid hormones (Eto, 1990).

Antijuvenile Hormones

Antijuvenile hormones, anti- JHs, are very effective pest control agents as they prevent JH production, facilitate JH degradation, or destroy corpus allatum. Intoxicated newly emerged larvae with anti- JH would create miniature pupae, thus abbreviating the distructive part of insect life cylce. Some of these plant-derived substances actually serve to inhibit the development of insects feeding and protecting the host plant. These are referred to broadly as antijuvenile hormones, more accurately, antiallatotropins, or precocenes. Although the mode of action of the precocenes is still unclear, it is known that they depress the level of juvenile hormone below that normally found in immature insects. As anti-JHs is competing with JH in binding to the JH receptors or to the JH carrier proteins, injuring the corpora allata cells, or interfering with JH biosynthesis (Leighton et al., 1981), other JHAs may also function as anti-JHs, such as ETB [ethyl 4-(2- pivaloyloxybutyloxy)-benzoate], which showed JH agonist and antagonist activities in Manduca sexta larvae as it compete with JH at the receptor site and become feedback inhibitors of JH biosynthesis (Staal, 1986).

Safety

Several IGRs are registered by the EPA, such as methoprene (Altosid ®) which is used as a grain protectant, as mosquito growth regulator; Precor ® for indoor control of dog and cat fleas; Hydroprene (Gentrol ® , Mator ®) for use against cockroaches, and stored grain pests; and kinoprene (Enstar II ®), which is effective against aphids, whiteflies, mealybugs, and scales (both soft and armored) on ornamental plants and vegetable seed crops grown in greenhouses and shadehouses. Although some insects acquired resistance against some IGR- based products, IGRs are typically "safer" to use around humans, pets, and natural enemies than conventional insecticides and acaricides. IGRs are effective when applied in very minute quantities and generally have few or no effects on humans and wildlife. They are, however, nonspecific, since they affect not only the target species, but other arthropods as well. For more information, see Taylor (2001), Tunaz (2004), Ware & Whitacare (2004), Collins (2006), and Gilbert & Gill (2010).

Pheromones

Pheromones are a class of semiochemicals that insects and other animals release to communicate with other individuals of the same species. Such behavioral chemicals, range from small hydrophobic molecules to water-soluble peptides, leave the body of the first organism, pass through the air (or water) and reach the second organism, where they are detected by the receiver. Signalling chemicals play an essential role in arthropod life cycles. They provide the means whereby mates, host and oviposition sites are located and recognized (Mordue Luntz, 2003). Pheromones may signal various information. Longlasting pheromones allow marking of food sources or territorial boundaries. Other signals are very short-lived and provide an immediate message, such as a brief period of reproductive readiness or short-term warning of danger.

Uses of Pheromones

There are five principal uses for sex pheromones: population monitoring, mass trapping of insects, movement studies, detection of exotic pests and, mating disruption. Such disturbance is very important in reducing the population density of pests as synthetic pheromone is dispersed into the field and the false odor plumes attract males away from females that are waiting to mate, thus reduce the population density of the pests. In contrast to the previous benefits, the high degree of selectivity may be a barrier to large-scale implementation where secondary pests become a problem as insecticides are eliminated (Walker & Welter, 2001). Other obstacles include the following: the lack of an identified pheromone for some pest species, high development and production costs, requirements for specialized application techniques or equipment, and the need to supplement the pheromone program in high pest-pressure situations (Welter et al., 2005). Large-scale implementation projects have yielded significant reductions in pesticide use while maintaining acceptably low crop-damage levels. Pheromones manipulate the behavior of insect pests. With these non-toxic and biodegradable chemicals, insects can be lured into traps or foiled into wasting energy that they normally need for locating food and mates. Pheromones are species-specific chemicals that affect insect behavior and they are not toxic to insects or other non- target organisms, creating opportunities for the biological control of other pest species. They are attractive in extremely low doses and used to bait traps or confuse a mating population of insects. Other advantages include negligible health risks, no accumulation in wildlife or groundwater, limited impacts on other management practices; manage insecticide resistance, and a more rapid registration process. Pheromones can play an important role in integrated pest management for medical or veterinary, structural, agricultural, landscape, and forest pest problems. Pheromone programs are most effective with low to moderate population densities. Because of some difficulties with high populations of pests, these programs should not be viewed as stand-alone strategies but rather as one tactic within a suite of integrated pest management options (Welter et al.,

2005). For more information about pheromones, see; Cooping & Menn (2000), Witzgall (2001), Leal et al. (2003), Bray et al. (2009), and van Emden & Service (2011).

Photosenitizers

Development of new, ecologically safe technologies to control insect pest populations is of great importance. Photosensitizers are activated by illumination with sunlight or artificial light sources, have been shown to be accumulated in significant amounts by a variety of insects when they are administered in association with suitable baits. The subsequent exposure of such insects to UV/visible light leads induces lethal photochemical reactions and death. The most famous photosensitizers are xanthenes (e.g. phloxin B) and porphyrins (e.g. haematoporphyrin) which appear to be endowed with the highest photoinsecticidal activity. In particular, porphyrins absorb essentially all the UV/visible light wavelengths in the emission spectrum of the sun; hence they are active at very low doses. Photoactive compounds usually used for photosensitization might be effective as pesticide agents, with low impact on the environment, being non-toxic and not mutagenic, see Ben Amor & Jori (2000), Mangan & Moreno (2001), Ragaei & Khater (2004), Lukšienė etal.(2007), and Awad et al. (2008) for more fine points.

Inorganics and Organic Acid

Several inorganic substances are well know with their insecticidal effect, such as potassium silicate, diatomaceous earth (DE, diatomite or kieselgur), mineral oils, sulfur, boric acid, sodium borate, silica gels, kaolin clay, and soap spray. For more details about inorganic and their uses, see Ware & Whitacare (2004), and Collins (2006). It worth to mention that peracetic acid, an organic acid, $C_2H_4O_3$, has strong acaricidal effect which was discovered for the first time (Khater & Ramadan, 2007). PPA had a great potential as acaricide against the cattle tick, Boophilus annulatus, and the fowl tick, Argas persicus, In vitro. Two minutes after treatment with 0.5%, PAA induced 100% mortality of both tick species and LC50 values for cattle and fowl ticks, after treatment for 30 min, were 0.06 and 0.05%, respectively. Following treatment with 0.25%, the LT50 values were 0.02 and 3.12 min, respectively. Furthermore, the detrimental effect of PAA against cattle tick extended beyond the adult stage, it significantly prolonged the preoviposition period, shortened the oviposition period, and decreased the mean number of the layed eggs, and such parameters were 14.75 and 6.57 days, and 457.50, respectively, after treatment with 0.25% of PAA. Therefore, PAA is highly effective when used at lower doses and short exposure time. The high speed of killing ticks is very important for avoidance of the hazard ensued by pathogen transmission in the course of delayed mortality caused by the currently used acaricides (Khater & Ramadan, 2007). PAA is highly effective against lice, In vitro and soft tick (In vivo), (Khater, H.F., Unpublished data).

BIOLOGICAL CONTROL

Biological control is the reduction or protection of pest populations by natural enemies which include four categories: microbes or pathogens (such as viruses, bacteria, protozoa, and fungi); entomopathogenic nematodes; predators (such as lady beetles and lacewings); and parasitoids (wasps and some flies). Protozoa, predators, and parasitoids are out of the scope of this chapter. Natural enemies are responsible for natural suppression of pest population. The first successful large scale microbial control application using coidiospores of the fungus Metarhizium anisopliae was carried out in the Russian Ukraine against the beet weevil, Bothynoderes punctiventris (Metchnikoff, 1879).

Virus

Entomopathogenic viruses are obligate disease-causing organisms that can only reproduce within a host insect. Among the fifteen or more families of viruses of invertebrates, it is mainly those having virus particles (virions) occluded within a proteinaceous matrix, an occlusion body (OB), have been used successfully in controlling pest populations. Such families are Entomopoxviridae (Entomopoxviruses, EPVs), Reoviridae (cypoviruses, CPVs), and Baculoviridae (Baculoviruses, BVs) (Lacy & Kaya, 2007). Only BVs have been used as pesticides (Szewczyk et al., 2009, 2011). In the past, the classification of the family Baculoviridae was based on virus morphology. It was divided into two genera: the Nucleopolyhedrovirus (NPVs) and the Granulovirus (GVs). A new division on the basis of comparison of genomic sequences indicate that virus phylogeny followed more closely the classification of the hosts than the virion morphological traits. Accordingly, family Baculoviridae contains four genera: Alphabaculovirus (lepidopteranspecific NPVs), Betabaculovirus (lepidopteran-specific GVs), Gammabaculovirus (hymenopteran-specific NPVs), and Deltabaculovirus (dipteran-specific NPVs) (Jehle et al., 2006). Although wider use of BVs as commercial insecticides was restricted because of their slow killing action and difficulties in large scale production, a very successful project was carried out in Brazil (Moscardi, 1999); over 2.0 million hectare of soybean had been already controlled annually by velvetbean caterpillar BV. Consequently, many countries have increased the area of fields and forests protected by BV pesticides.

Life Cycle and Viral Stability

Virual infection begins in the insect's digestive system after consumption of plant material with viral particles. The virus spread though out the body but the digestive system is among the last part to be destroyed, so the insects usually continue to feed until they die leading to economic loss. See Lacey & Kaya (2007) and McNiel (2010) for more information about biology and ecology of viruses. The virus manipulates the behavior of infected larvae for its own dissemination. For example, larvae of the cabbage moth, Mamestra brassicae infected with NPV moved up to five times more than their healthy counterparts during the middle stages of infection (Vasconcelos et al., 1996b). Adult insects can also disperse viruses via vertical and horizontal transmission. Understanding the biology and ecology of viruses is crucial for optimizing pest control strategies of exotic and genetically modified organisms, see Lacy & Kaya (2007) and Szewczyk et al. (2009, 2011) for more details. UV protectants are very important for stability of the viral product, stilbene fluorescent brighteners (e.g. Phorwite AR ®, Blankophor ®, and others) induce the best results. Plant metabolites as peroxidases generate free radicals

(Hoover et al., 1998) inactivate BVs. As a result, addition of free radical scavengers such as mannitol or enzyme superoxide dismutase to BV preparations can reduce such inactivation (Zhou et al., 2004). Tillage buries virus particles in the soil, thus good agricultural practices can reduce viral persistence between seasons.

Products

Some BVs are produced as commercial products, mainly for caterpillars, such as Gemstar LC (NPV of Heliothis/ Helicoverpa spp. e.g., corn earworm, tobacco budworm, cotton bollworm); Spod-X LC (NPV of Spodoptera spp. e.g., beet armyworm); CYD-X and Virosoft CP4 (GV of Cydia pomonella, the codling moth); and CLV LC (NPV of Anagrapha falcipera, the celery looper). Successful infections can perpetuate the disease outbreak making repeat applications within a season unnecessary.

Safety

Members of BVs are regarded as safe to vertebrates. Their specificity is usually very narrow, often limited to single insect species. Regarding beneficial arthropods, immature larvae of parasitoids in infected hosts may die not due to virus infection, but relatively to premature loss of the host or to variation in quality of the host. Virus can provide safe, effective and sustainable control of a variety of insect pests as a part of a varied IPM programs, see ElHusseini (2006) and Szewczyk et al. (2009). Entomopathogenic viral agents are ideal for pest management purposes due to their specificity, dispersal capacity, and self propagation. Viruses are the most environmentally acceptable components of direct and integrated management regimes. Accordingly, there is global interest in use of viral agents. Valuable reviews on entomopathogenic viruses are those of Szewczyk et al. (2009, 2011), Lacey & Kaya (2007), McNiel (2010), Ahemed et al. (2011), and Singh et al. (2011).

Bacteria

The insecticidal bacterium, Bacillus thuringiensis (Bt) is a widely occurring gram-positive, spore-forming soil bacterium that produces parasporal, proteinaceous, crystal inclusionbodies during sporulation. Bt has been the most successful commercial microbial insecticide, and also has been the subject of the overwhelming majority of genetic engineering studies to improve efficacy (Federici, 2010; Lacey & Kaya, 2010; Mehlhorn, 2011). Bt is actually a complex of bacterial subspecies that occur in soil, leaf litter, leaf surfaces, insect feces, and as a part of the flora in the midguts of many insect species. There are several insecticides based on various sub-species of Bacillus thuringiensis Berliner (Bt), such as B thuringiensis israelensis (Bti), with activity against mosquito larvae, black fly (simuliid), fungus gnats, and related dipterans species; B thuringiensis kurstaki (Btk) and B thuringiensis aizawai (Bta)with activity against lepidopteran larval species; B thuringiensis tenebrionis (Btt), with activity against coleopteran adults and larvae; and B thuringiensis japonensis (Btj) strain buibui, with activity against soil-inhabiting beetles.

Mode of Action

The insecticidal properties of Bt are largely a function of the presence of extra-chromosomal plasmids in the cell. These carry genes such as cry genes that encode a diverse array of these protein crystalline inclusion bodies which are toxic to insects. Upon ingestion by an insect, the crystal proteins (Cry) are solubilised and the insect gut proteases convert the original protoxin into smaller toxins. These hydrolysed toxins bind to the insect's midgut cells at high-affinity and specific receptor binding sites where they interfere with the potassium ion dependent, active amino acid symport mechanism decreasing absorption of minerals and nutrition from midgut and finally death of the columnar cells. This disruption causes the formation of large cation-selective pores that increase the water permeability of the cell membrane. A large uptake of water causes cell swelling and eventual rupture, disintegrating the midgut lining. Different toxins bind to different receptors with different intensities and this explains the selectivity of different Bt strains in different insect species (Baum et al., 1999; Cooping & Menn, 2000; Lacey & Kaya, 2007; Federici, 2010). Delayed larval mortalities (2–48 h) is caused by the crystal inclusions which have to be ingested and then processed within the insect's gut. Biological control products may contain the endotoxins plus live bacterial cells. The toxin stops feeding, this action hinder further damage caused by the feeding larva, and do not directly kill insects, but young larvae may starve to death and may die from bacterial infection over a longer period. Some commercial products contain Bti crystal proteins and spores, such as Bactimos and VectoBac (Valent BioSciences). VectoBac, Bti, (12 AS, Wady El- Niel for agricultural development Co. Egypt) is highly effective against C. pipence than M. domestica, LC50 values were 1x10-5 and 3.86 x103 spores/ml, respectively, and LC 90 values were 0.04 and 37.28 x 103 spores/ ml, respectively. Survived mosquito larvae died as pupae (Khater, 2003). Bti is a practical substitute to organophosphate insecticides, but it is unsuitable for application to environmentally sensitive water bodies. Bacillus sphaericus control mosquito larvae, particularly Culex and anopheline spp., especially those breeding in polluted water. It controls also black fly, Simulium sp., the vector of river blindness disease. Bs is effectively controlled C. pipiens in Egypt (Ragaei et al., 2004). It is widely used in Europe and Africa. Registered B. sphaericus product is Vectolex CG (Valent BioSciences). Recent studies in Kenya have shown that at least in some areas, biting rates by Anopheles gambiae can be reduced by more than 90% by using a combination of existing commercial formulations of Bti and Bs (Filinger & Lindsay, 2000).

Bacterial Toxins

The use of insect-specific toxins from Bt and Bs is forming an increasingly component of biological control strategies. The protein crystals (protoxins) contain several toxins which are classified according to their insecticidal activity and molecular relationship into four major gorups (Hofte and Whiteley, 1989): Cry I toxins, active against larvae of lepidopterans; Cry II toxins, active against larvae of lepidopterans and dipterans; Cry III toxins, active against larvae of coleopterans (Chrysomelidae); and Cry IV toxins, active against larvae of nematoceran flies. Cry toxins bind to glycoprotein or glycolipid receptors. The toxins can also be further categorized into subclasses, ex. CryIA etc. The genes of the toxins are named as the toxins but without capital letters and in italic (e.g., *cry*IV). Cytotoxins (Cyt) found in strains toxic to larvae of nematoceran flies and coleopterans. The CytA protein binds to lipids and does not exhibit the specific binding mechanism which the Cry proteins do (Hofte and Whiteley, 1989). Cyt proteins are thought to have a similar mode of action to that of Cry proteins, with the exception that they directly bind to the microvillar lipid bilayer. Similar to Bti, the toxicity of Bs is due to protein endotoxins produced during sporulation and assembled into a parasporal body. The main toxin of Bs is a binary toxin (BinA and BinB) which are proteolytically activated in the mosquito midgut to release peptides (43 and 39 kDa, respectively) to form the binary toxin. The toxins bind to microvilli of the midgut epithelium, trigger hypertrophy and lysis of cells, leading to destruction of midgut and death of mosquito larva.

Resistance

Despite decades of use, there is no reported resistance among mosquito populations probably due to biochemical properties of Cyt protein. Bti produces a parasporal body that contains four major endotoxins, Cry4Aa, Cry4Ba, Cry11Aa, and Cyt1Aa. Both Cry and Cyt are must be ingested to yield active toxins in the midget of insect. Cyt protein synergizes the toxicity of the Bti Cry proteins resulting in delayed the phenotypic expression of the evolution of resistance. Such synergistic interactions are also known to occur between the Bti Cry proteins. When Cyt protein combined with the Bs Bin toxin, it can also overcome resistance to Bti. For detailed information, see Wirth et al. (2005) and Federici (2010). As a rule, bacterial control agents are less likely to provoke resistance because their mode of action is more complex (Wirth et al., 2005; Federici, 2010). However, resistance had developed against Btk in some pests, e.x., the stored grain pest, Plodia interpunctella, and the diamondback moth, Plutella xylostella. In most cases, resistance has been associated with a recessive or partially recessive trait(s) and appears to be linked to a single gene. The combination of protoxins in Bti is very important in reducing the likelihood of resistance. In contrary, when the gene encoding a single toxin protein was cloned into a microorganism and then fed to larval mosquitoes, resistance was induced within a few generations; see transgenic organisms below and Mehlhorn (2011) for more details. For avoidance of resistance, products with different arrays of endotoxins should be alternated (Schuster & Stansly, 2006).

Safety

The U.S. Environmental Protection Agency (USEPA) categorizes the risk posed by Bt strains to nontarget organisms as minimal to nonexistent. After using of Bt products for more than 50 years, it can be concluded that Bt belong to the most environmentally safe products as they kill target organisms and usually do not harm nontarget organisms, such as beneficial insects, plants, or humans. The safety record for occupational exposure to Bt based biopesticides with regard to human health is considered good (Barfod, 2010). Some Bt subspecies produce other types of toxins besides Cry proteins as ß-exotoxin and ˩ -toxin. Both toxins are toxic to vertebrates and they are banned in Bt products, see Lacey & Kaya (2007) and Makonde et al. (2010) for more particulars about bacterial toxins. Bt and BS are the most used biocontrol agents against nuisance, pest or vector species. They are strongly incorporated in IPM programs because they are highly efficient, easily be mass-produced, easy to handle, stable when stored, cost-effective, pest specific, and safe to people and the environment. For more information about bacteria as a biological control agent, see Lacey & Kaya (2007), Gilbert & Gill (2010), Ahmed et al. (20 11), Mehlhorn (2011), Singh et al. (2011), and van Emden & Service (2011).

Entomophathogenic Fungi

Entomophathogenic fungi are important natural control agents that limit insect populations. Survival of fungi requires a delicat balance of interactions between fungi, host, and the environment. Most fungi cause insect diseases spread by means of asexual spores called conidia which vary greatly in their ability to survive adverse environmental conditions, desiccation, and ultraviolet radiation.

Life Cycle

In general, insects get infected when they come into contact with spores on the bodies of dead insects, on the surface of plants, in the soil, or in the air as windborne particles. High humidity is usually required for germination of conidia on the insect cuticle and production of germ tube that allow the fungus to penetrate the cuticle often at joints or creases where the insect's protective covering is thinner. In contrast to virus and bacteria, fungi do not have to be ingested to cause infections. The fungus multiplies within the host insect and kill it. Many fungi produce toxins to increase the speed of kill or prevent competition from other microbes. Death is due to toxin produced by the fungus or fungal multiplication. Under favorable conditions, the fungus grows out of the cadavers, usually at thinner areas and form conidiophores or analogous structures and sporulates. Some species go into a resting stage which survive periods of adverse conditions before forming or releasing spores. Soil incorporating fungi usually avoid the adverse effects of ultraviolet radiation and desiccation, but other microorganisms that act as competitors or antagonists often alter pathogen effectiveness. Fungal pathogens differ in the range of hosts. Many important fungi attack eggs, immature, and adults of a variety of insect species. Others are more specific to immature stages or to a narrow range of insect species. Several fungi are used to control insects such as Leptolegnia spp., Coelomomyces spp., Hirsutella thompsonii, Nomuraea rileyi, and Vericillium lecanii. The most famous entomophathogenic fungi will be disused below.

Oomycetes (water molds)

Lagenidium giganteum was first described by Couch (1935) from a combined collection of copepods and mosquito larvae, Culex and Anopheles, in North Carolina, USA. It has a wide geographical distribution: North America, Europe, Africa, Asia, Antarctica (Federici, 1981). The only species of the genus Lagenidium is known to be pathogenic to mosquito larvae is L. giganteum (formerly: L. culicidum). As a facultative parasite, L.giganteum can grow as pathogen on mosquito larvae, or as a saprophyte in aquatic environments

(Sur et al., 2001). This aquatic fungus is highly infectious to larvae of several mosquito genera. It cycles effectively in the aquatic environment even when mosquito density is low, but its effectiveness is limited by high temperatures. The fungus is not effective for mosquitoes in brackish or organically rich aquatic habitats (Merriam and Axtell, 1982). L. giganteum has a wide geographical distribution: North America, Europe, Africa, Asia, Antarctica (Federici, 1981). Fungal reproduction is both asexual (zoospores) and sexual (oospores) (Federici, 1981). In order to infect mosquito larvae, biflagellate zoospores must be formed. Motile zoospores are the asexual stage of the fungus. Oospores, the sexual stage of L.giganteum, can also be used as inoculum. They are dormant propagules, resistant to desiccation and mechanical abrasion and stable for at least seven years, which allows multivoltine persistence of the fungus in some habitats (Kerwin et al., 1994). Lagenidium spp. was isolated form Egypt for the first time form Culex pipiens larvae infesting a polluted creak in Miet El- Attar, Benha, Egypt, by Khater (2003). Such fungus was propagated on SDA medium and Peptone yeast glucose (PYG) for sporulation. Five concentrations (5 x105- 1.6 x10^7) were used to infect C. pipiens late 3rd larval stage In vitro and the biological parameters had been followed up till emergence of adults. LC50 and LC90 values were 2.79 x106 and 3.94 x108 spores/ ml, respectively. At the lowest concentration, larval duration reached 8.3 days (3.6 days in the control group) and morphological changes (symptoms of mycosis) was 66.7%, (Fig. 4,5). All survived larvae died as pupae; consequently, the isolated fungus inhibited adult emergence. As this fungus has the ability to be self- propagated, it could be an effective control agent for the vector of Bancroftian filariasis and Rift valley fever virus in Egypt (Khater, 2003). The fungus has caused high mortalities in mosquito populations, see Schotle et al. (2004) for a review. Results from a small scale field trial in North Carolina indicated that L. giganteum recycled for an entire season despite periodic scarcity of hosts and short-term drought with infections ranging from 0-100% (Jaronski & Axtell, 1983). A large-scale field trial in Californian rice fields, using mycelium from either 20 or 30 liters of fermentation beer per hectare resulted in 40%-90% infection of Culex tarsalis and Anopheles freeborni sentinel larvae (Kerwin & Washino, 1987).

The Relative Potency

The relative potency indicated that Vectobac (Bti), and Lagenidium spp. were 12x1011 and 4.3 times, respectively, more effective than the fungal product Biosect®, Beauveria bassiana, on the basis of LC50 values (Khater, 2003). L. giganteum is compatible with the bacterial agents Bti and Bs Meyer and Neide when used against Culex quinquefasciatus (Orduz & Axtell, 1991), with the fungus having the distinct advantage over Bti in that it is able to recycle in stagnant water, infecting multiple and overlapping generations of mosquitoes (Legner, 1995). In field trials in which Laginex 25 was compared with Vectobac-12AS (Bti), Laginex reduced C. quinquefasciatus larvae by 100% for 22 days whereas Vectobac-12AS required retreatment by the 10th day (Hallmon et al., 2000).

Production

Unfortunately, mass production yields of oospores remain orders of magnitude below that of the less stable mycelial (asexual, presporangial) stage, and continued problems with spore activation have prevented large-scale field tests (Kerwin and Petersen, 1997). For solving this problem, improving oospore yields would be much more useful than zoospores in large-scale operational mosquito control programs. In contrast to the previous obstacles, Laginex® is a L.giganteum-based product by AgraQuest (California, USA) until 1999. It is effective against larvae of most pest mosquito species; remains infective in the environment through dry periods but it is unable to survive high summertime temperatures; the kind of spore used was not mentioned. Shathele (2009) made an excellent potential for investigations on the use of newly developed isolate (Saudi-1 and Saudi-2) media for an isolated strain of L. giganteum form Louisiana for its maintenance and zoospore release.

Safety

Zoospores of the fungus appear harmless to vertebrates (Kerwin et al., 1990), most aquatic invertebrates (Nestrud & Anderson 1994) except Daphnia spp. and copepods (Couch, 1935), three cladoceran species and a chironomid species (Nestrud & Anderson, 1994).

Hyphomycetes

Most soil fungi used for the control of insect pests belong to the group hyphomycetes, such as Beauveria bassiana and Metarhizium anisopliae, formerly known as Entomophthora anisopliae . Some species have been developed as commercial products because of their ability to be mass produced. Most fungi in this group can cause natural outbreaks, to a wide range of insect hosts, on their own when environmental conditions are favorable. Soil provides protection against UV along with optimal conditions of temperature and moisture. Furthermore, the fungi may survive in the soil through recycling in insects or roots (Leger, 2008). Thus, they provide a long-term strategy for larvae and puparia control (Quesada-Moraga et al., 2006). B. bassiana infects both larvae and adults of a broad host range. Understanding the interactions between B. bassiana and other soil microorganisms is very important for the success of using of this fungus. Commertially available products based on B. bassiana are Mycotrol O (Emerald BioAgriculture), Naturalis Home and Garden (H&G), Naturalis L (Troy BioSciences, Inc.), and Biosect® (Kafr El Zayat - KZ Chemicals, Egypt). Khater (2003) used Biosect® to control larvae of both M. domestica and C.pepiens In vitro. LC50 an LC90 values were 29.2 x10^7 and 9.97 x108 spores/ ml for house flies and 1.2x107 and 4.17x109 spores/ ml, for mosquitoes, respectively. The total larval mortalities of mosquitoes were almost 100%, the few survived larvae died as pupae. At the lowest concentrations, Larval and pupal malformations were 30% and 45%, See fig (1,2,4,5) for symptoms of mycosis.

Entomophthorales

Entomophthorales is a group of fungi that tend to be much more host specific and can cause natural outbreaks in insect populations. Several different Entomophthora muscae sensu stricto genotypes were documented and each type was restricted to a single host species, indicating a very high degree of host specificity at or below the level of the subfamily (Jensen et al., 2001). Currently, no commercially-based product is available because of difficulties in mass production. All available literature deal with E. muscae as a pathogenic fungi of adult M. domestica, but Khater (2003) isolated it for the first time in Egypt, form Moshtohor, Toukh, Qlubia governorate, such strain has a unique ability to infect larvae of house flies for the first time also. The isolated fungus was replicated on Sabouraud Dextrose Agar (SDA) medium for hyphenation and on liquid medium for laboratory – scale Entomophthora' hyphal production for enhancement of zoosporogenesis. Khater (2003) also made a bioassay using early 3rd larval stages of house flies which were infected In vitro by adding different concentrations of E. muscae (from 25 to 16 x 102 spores/ml) to the breeding medium of larvae and followrd up till adult emergence. LC_{50} and LC_{90} values for larval mortalities were 2.19 x 10^2 and 20 x 10^2 spores/ ml, respectively. At lower concentrations, larval durations were elongated (7.2 days) than that of the control group (2.6 days). Larval abnormalities were larval pigmentation, black, red- pin point pigmentation, ulceration, maceration and larval pupal intermediates. Mortality and morphological changes of pupae (symptoms of mycosis) reached 100% at 8 x 102 spores /ml. Adult mortalities, deformity, and emergence reduction rates increased as the concentrations increased (79.3%, 60%, 98.2%, respectively, at 4x10^2 spores/ ml). Adult abnormalities were failure of adult eclosion, small adults, and deformed wings and legs (Fig. 1-3). In addition, this pathogen manipulates the behavior of infected host for his own dissemination as the apparently normal adults exhibited the characteristic position of Entomophthora spp. (Fig. 6) (Krasnoff et al., 1995; and Khater, 2003) as follows:

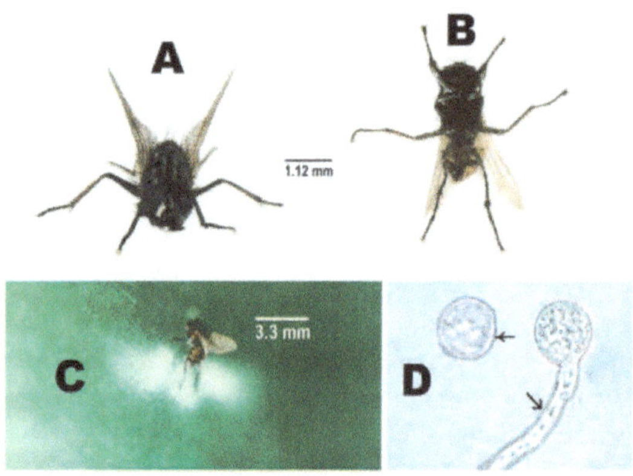

Fig. 6. Characters of adult house fly infected with Entomophthora spp.: A. Adult exhibited elevated abdomen, wings, and extended legs whereas the mouth parts were stoked downwards (anterior view), B. Adult with extended legs (ventral view), C. Conidial shower (white color), D. Hyphal body (zoosporangia) right arrow, spore (left arrow) of Entomophthora spp. Adapted form Khater (2003).

First, adults died in elevated positions, with the proboscis extended and attached to the substrate, legs spread, the abdomen angled away form the substrate, and the wings rose above the thorax. The conidial discharge reached 3-9 cm, depending on the size of the cadavers (Fig. 6). Conidia are forcibly ejected from conidiophores that emerge most profusely from the intersegmental membranes on the insect's abdomen, such posture enhance dispersal of conidia (Krasnoff et al., 1995; Khater, 2003). Similarly, an onion maggot fly, Delia antiqua (Meigen), infected by E.muscae Cohn (Fresenius) and dying at the top of a grass stem with its proboscis fastened to the substrate and its abdomen well raised by an atypical disposition of the legs appears to be an ideal launch pad for maximizing the effective area covered by the conidial shower (see Krasnoff et al., 1995 for more details). Second, the pathogen also benefits indirectly from the sexual attractiveness of female cadavers. Males attempting necrophilous copulations invariably become infected with the fungus (Moller, 1993). Third, both field observations (Roffey, 1968) and controlled experiments (Bellini et al., 1992; Khater 2003, unpublished data) indicate that host mortality usually occurs in the late afternoon. This may be a temporal manipulation serving to enhance the likelihood that conidia produced from the dead host will be released at a time when the possibility for germination is highest, after dark when dew has set and surface moisture has had a chance to accumulate. The presumed adaptive payoff of elevation and posture for the fungus enhances dispersal of propagules.

Safety

Concerning the safety of entomopathogenic fungi developed for commercial use in microbial control of insect pests. Most fungi showed no infectivity to man or other vertebrates. Safety tests with Nomuraea rileyi , Hirsutella thompsonii , Verticillium lecnii, and L. giganteum assured negative findings to different mammals and birds. On the other hand, B. bassiana has been reported to cause allergies in humans and is at least an opportunistic pathogen to man and other mammals. B. bassiana and M. anisopliae affect non-target invertebrates with various degrees; see El-Husseini (2006) for more details. Soil treatment with M. anisopliae and B. bassiana on Tapinoma nigerrimum colonies indicated that there were no significant differences in ant, as a non- target host, activity before and after fungal treatment (Garrido-Jurado et al., 2011). For more information about entomophathogenic fungi, see Hakjek et al. (2007), Lacey & Kaya (2007), Gilbert & Gill (2010), Singh et al. (2011), and van Emden & Service (2001). For using of fungi for controlling arthropods of medical and veterinary importance, see Steenberg et al. (2001), Samuels, et al. (2002), Khater (2003), Kirkland et al. (2004), Schotle et al. (2004), Hartelt et al. (2008), Zabalgogeazcoa et al. (2008), Mochi et al. (2010), and Stephen & Kurtböke (2011).

Nematodes

Entomopathogenic nematodes cause damage to soilborne insect pests. Nematodes from the families Steinernematidae and Heterorhabditidae have proven to be the most effective as biological control organisms (Lacy & Kaya, 2007) to control a wide range of insect pests including filth flies, German cockroaches, cat fleas, armyworms, carpenter worms, crown borers, cutworms, flea beetles, leaf miners, mole crickets, phorid flies, plume moths, root weevils, sciarid flies, stem borers, webworms, and white grubs (Smart, 1995).

Life Cycle

Generally speaking, the life cycle of most nematodes includes an egg stage, four juvenile stages, and an adult stage. The third juvenile stage, dauer, is the only infective and freeliving stage which is capable of surviving in the soil; its function is to locate, attack, and infect an insect host through its breathing holes, mouth, or anus, but some species are capable of penetrating thin areas of the insect's cuticle. After that, the nematodes release special bacteria into the insect. The toxins produced by the bacteria kill the insect after a few days. The bacteria multiply inside the body of the insect and the nematodes eat the bacteria. The nematodes mature, mate, and multiply inside the insect. Eventually, the insect's body becomes filled with nematodes. Infective stage nematodes then exit the insect body searching of other insects to infect. Once inside the body cavity of the host, the infective juveniles release bacteria that live symbiotically within the nematode's gut. The nematodebacterium relationship is highly specific: only Xenorhabdus spp. bacteria co-exist with steinermatids, and only Photorhabdus bacteria co-exist with heterorhabditids. Under optimal conditions, it takes 3–7 days for steinernemadtids and heterorhabditids to complete one life cycle inside a host from egg to egg. Emergence of infective juveniles from the host requires about 6–11days for steinernemadtids and 12–14 days for heterorhabditids (Kaya & Koppenhöfer, 1999). The ability of any biological control nematode to infect a particular insect can be affected by nematode and insect behavior, physical barriers, and immune responses.

Using Beneficial Nematodes

Over 30 species of beneficial nematodes have been identified. Seven species have been commercialized worldwide and seven are currently available in the United States: Steinernema carpocapsae, S. feltiae, S. glaseri, S. riobravis, Heterorhabditis bacteriophora, H. megidis, and H. marelatus. (Kaya & Koppenhöfer, 1999). There were two commercial nematode products available for termite control, Spear ® and Saf T-Shield ®. The use of the entomopathogenic nematodes S. feltiae and H. megidis in high numbers was shown to be effective against housefly larvae (Renn, 1995) and effective after formulation into a housefly bait, as effective as the spraying of a carbamate pesticide in pig units (Renn, 1998). Also the nematode Paraiotonchium muscadomesticae could infect housefly larvae, but mortality was low except at high nematode concentrations, but P. muscadomesticae infect housefly larvae and descendants of the nematodes invade and damage the ovaries of adult female flies and are deposited in the larval habitat when the flies attempt to oviposit. Infected adults lived about half as long as uninfected flies (Geden, 1997; Khater, unbuplished data); consequently, P. muscadomesticae reduce fly population indirectly. Nematodes were effective in the laboratory but persisted in manure only for 3-7 days. Moreover, mermithid nematodes which parasitize grasshoppers, cockroaches, and mosquitoes can play a part in natural suppression of pest insects. Nematodes must be raised in live insects and this is not a very efficient means of production.

Safety

Entomopathogenic nematodes are not harmful to humans, animals, plants, or earthworms. They can therefore be used as biological control organisms. Such beneficial nematodes have been released extensively in the field with negligible effects on nontarget insects and are regarded as remarkably safe to the environment. Finally, microbial control agents provide a more environmentally acceptable and sustainable form of insect pest management than chemical insecticides, but are most effective when underpinned by a detailed knowledge of specific host-pathogen interactions. For more information about successful application of entompathgenic nematodes, see Mill et al. (2000), Gaugler (2002), Crow (2006), Hajek et al. (2007), Lacey & Kaya (2007), and Gilbert & Gill (2010).

TRANSGENIC PESTICIDES

Genetically modified organisms (GMO) and crops (GMC) could be engineered to augment pest control. Transgenic organisms are genetically altered by artificial introduction of DNA from another organism and the artificial gene sequence is referred to as a transgene. Plants with such transgenes are called GMC, plant pesticides or plant incorporated protectants.

Genetically Modified Crops

By the mid 1980s, one company, Monsanto had committed to a research program designed to create crop protection products through the application of biotechnology (Glover, 2008). Charles (2001) has produced a very readable history of pesticide-related transgenic crops. Different endotoxins have different biological spectra, thus different genes are used in GMC, for protection from attack by various insects. For example, Monsanto has used cryIA(c) genes from Btk in cotton and tomatoes and cryIIIA genes from Btt in potatoes. Moreover, Novartis and Mycogen used cryIA(b) genes from Btk and Monsanto used cryIIA and cryIA(b) genes, both from Btk in corn. AgrEvo (Plant Genetic Systems) is using cry9C toxin genes from B thuringiensis subsp tolworthi in transgenic corn. GMC have been enhanced sometimes by use of stacked genes. i.e. more than one transgene is introduced into the same crop to achieve multiple desired characteristics and avoidance of resistance. Many experimental studies conducted to date indicate that transgenic plants have no adverse effects on nontarget organisms. In addition, there is no scientific evidence as yet that the commercial cultivation of GM crops has caused environmental impacts beyond the impacts that have been caused by conventional agricultural management practices. Nonetheless, studies are still on-going to assess the potential environmental impacts of GM crops (see Makonde et al., 2010). The effect of GM crops on natural enemies remains a controversial topic (Andow et al., 2006; Marvier et al., 2007; Lövei et al., 2009).

Genetically Modified Organisms

Molecular biology impact searching for new products through understanding the chemistry and mode of action of a natural product and the discovery of a new protein (and its gene) that may be used to transform a target crop or organism. For more information about proteases as insecticidal agents; see Harrison & Bonning (2010).

Recombinant Bacteria

It has been shown that plasmids with a molecular weight of 60–94 MDa play an essential role in the crystal toxin production. Cloning various toxin genes into host organism's increase the persistence or the insecticidal properties. The development of recombinant DNA techniques improves the efficacy of Bti through combining the most potent insecticidal proteins from Bti, Btj, and Bs into new bacterial strains that are ten-fold more toxic than wild type species of Bti and Bs used in current commercial formulations. New bacterial larvicides offer environmentally compatible options for use as components in integrated vector control programs aimed at reducing malaria, filariasis, and many important viral diseases of humans (Federici, 2010). Recombinant (new) bacterial larvicides are used as components in IPM programs because they are much more highly efficient than the wild type strains from which they were originally derived. Their costs are similar to that of new chemical insecticides, and they are much more environmentally compatible than most chemical insecticides. Therefore, recombinant bacterial larvicides will play an important role in controlling pests and vectors in the near future. For more information, see Berón & Salerno (2007), Lacey & Kaya (2007), Gilbert & Gill (2010), Federici (2010), and Singh et al. (2011).

Recombinant Baculoviruses

Most insect viruses take several days to kill their host insect. In the same time, the pest is still causing damage. Very high doses are often necessary for adequate control and early larval life stages are highly most susceptible to virus infection. For solving the previous problems, there has been steady interest in the potential to produce recombinant BVs to increase their speed of kill or reduce insect feeding (Inceoglu et al., 2001), The activity of BVs against their natural hosts may be enhanced by introduction of insect-specific toxins, such as toxin genes isolated from the scorpion or spiders, or by interference with insect physiology (Bonning & Hammock, 1996). Arthropod toxins usually attack insect sodium channels producing final effect similar to the chemical insecticides of the pyrethroid group. Though, the specific target in sodium channels is different, so there is a potential possibility to produce synergistic effect by biopesticide/chemical pesticide application (McCutchen et al., 1997). BV recombinants that produced occlusion bodies incorporating Bt toxin were constructed by making a fusion protein consisting of polyhedron and Bt toxin (Chang et al., 2003). This new biopesticide is highly pathogenic than wild type virus as it combines the advantages of the virus and the bacterial toxin. The changes to host physiology were done by introducing genes coding for some insect hormones or hormone-modifying enzymes into BV genome, or by deletion of the BVencoded ecdysteroid glucosyltransferase (egt) gene. Cloning juvenile hormone esterase gene into BV genome which over expressed decreases the concentration of the juvenile hormone which is a signal for a caterpillar to stop feeding and pupate (Inceoglu et al., 2001). The product of the egt gene interacts with larval moulting and indirectly affect the time of feeding of infected caterpillars. Furthermore, recombinant juvenile hormone esterase act as a biochemical anti- JH hormone agent and it affects ovarian development in the house cricket Acheta domesticu (Bonning et al., 1997). The deletion of the baculovirus encoded egt gene was used first by O'Reilly and Miller (1991). Such deletion from BV genome resulted in 30% faster killing of caterpillars. The egt gene is not essential for viral replication and can be replaced with an exogenous gene; the product of which may enhance the insecticidal activity of the recombinant virus (Sun et al., 2004). Genetically modified BVs are safe as they infect only their hosts. They are not pathogenic to bees and vertebrate species (Sun et al., 2004) as well as to the natural enemies of larvae such as parasitoids and predators (Boughton et al., 2003). Consequently, recombinant BVs lead to the expansion of BV use worldwide. For more details about recombinant BVs, see Szewczyk et al. (2009, 2011), Gilbert & Gill (2010), Ahmed et al. (2011), and Singh et al. (2011).

Transgenic Insects

Genetic modification of mosquitoes (which renders them genetically modified organisms, GMOs) offers opportunities for controlling malaria. Transgenic strains of mosquitoes have been developed to replace or suppress wild vector populations and reduce transmission and deliver public health gains. The transition of this approach from confined laboratory settings to open field trials in disease endemic countries (DECs) is a staged process that aims to maximize the likelihood of epidemiologic benefits while minimizing potential pitfalls during implementation. Unlike conventional approaches to vector control, application of GM mosquitoes will face contrasting expectations of multiple stakeholders, the management of which will prove critical to safeguard support and avoid antagonism, so that potential public health benefits can be fully evaluated. Inclusion of key stakeholders in decision making processes, transfer of problem-ownership to DECs, and increased support from the wider malaria research community are important prerequisites for this. It is argued that the many developments in this field require coordination by an international entity to serve as a guiding coalition to stimulate collaborative research and facilitate stakeholder involvement. Contemporary developments in the field of modern biotechnology, and in particular GM, requires competencies beyond the field of biology, and the future of transgenic mosquitoes will hinge on the ability to govern the process of their introduction in societies in which perceived risks may outweigh rational and responsible involvement (Knols et al., 2007). For more information about of transgenic insects and improving their ecological safety for field release, see Vreysen et al. (2007).

FUTURE TRENDS, NANOPARTICLES

The potential uses and benefits of nanotechnology are enormous. These include enhancement involving nanocapsules for vector and pest management and nanosensors for pest detection. Nanoparticles are 1-100 nm in diameter, whereas the size of a virus is roughly 100 nm. Such particles are agglomerated atom by atom. Nanotechnology deals with the targeted nanoparticles which exhibit different physical strength, chemical reactivity, electrical conductance, and magnetic properties. Nanoparticles are present in insect entire body parts. Insects are incredible nanotechnologists. The surfaces of many insect wings have evolved properties materials scientists only dream of for their creations (Watson et al., 2010). Nanoparticles help to produce new pesticides, insecticides and insect repellants (Owolade et al., 2008). Nanoencapsulation is a process through which a chemical (ex. an insecticide) is slowly but efficiently released to a particular host for insect pest control. Release mechanisms include diffusion, dissolution, biodegradation and osmotic pressure with specific pH (Vidhyalakshmi et al., 2009). Such process can also deliver DNA and other desired chemicals into plant tissues for protection of host plants against insect pests (Torney, 2009). Nanoparticles loaded with garlic essential oil are efficacious against Tribolium castaneum Herbst (Yang et al., 2009). Aluminosilicate filled nanotube can stick to plant surfaces while nano ingredients of nanotube have the ability to stick to the surface hair of insect pests and ultimately enters the body and influences certain physiological functions (Patil, 2009). Nanotechnology is used widely in agriculture and food (Joseph & Morrison, 2006). One of the world's largest agrochemical corporations, Syngenta, is using nanoemulsions in its pesticide products. Encapsulated product from Syngenta delivers a broad control spectrum on primary and secondary insect pests of cotton, rice, peanuts, and soybeans. Marketed under the name Karate® ZEON, a quick release microencapsulated product containing the active compound lambda-cyhalothrin (a synthetic insecticide based on the structure of natural pyrethrins) which breaks open on contact with leaves. The encapsulated product "gutbuster" only breaks open to release its contents when it comes into contact with alkaline environments, such as the stomach of certain insects. Furthermore, the new technology improve pesticide and fertilizer delivery systems which can take action to environmental changes, ex. they will release their cargo in a controlled manner (slowly or quickly) in response to different signals e.g. heat, moisture, ultrasound, magnetic fields, etc. Recently, nanotechnology is widely acceptable publicly because it is not yet linked to any toxicological and ecotoxicological risks. Research on nanoparticles and insect control should be directed toward production of faster and ecofriendly pesticides to deliver into the target host tissue through nanoencapsulation. This will control pests efficiently and accelerate the green revolution. For more in rank about usages of nanoparticles, see Joseph & Morrison (2006), Torney (2009), Yang et al. (2009), Bhattacharyya et al. (2008, 2010), Ahmed et al. (2011), and Hashim (2011).

CONCLUSION

IPM programmes have demonstrated that current levels of pesticide use in many circumstances are not necessary and, frequently, are even counter-productive. Excessive and otherwise inappropriate pesticide use is an unnecessary burden on farmers' health and income, on public health, and on the environment (Pretty, 2009). Biorational insecticides have emerged as an alternative or as supplemental forms of pest control. The use of biopesticides will help in preventing the discarding of thousands of tons of pesticides on the earth and provide the residue free food and a safe environment to live. In spite of intensive research on plant natural products and insect-plant chemical interactions over the past three decades, only two new types of botanical insecticides have been commercialized in the past 15 years, those based on neem seed extracts (azadirachtin), and those based on plant essential oils. There are some obstacles toward commercialization of new botanical products, such as the availability of sufficient quantities of plant material to produce the pesticides, standardization, refinement, and quality control of the products; regulatory approval, patents, difficulties in registration; as well as problems related to their volatility, poor water solubility, and aptitude for oxidation. These challenges should be overcome for botanicals to be of particular use for human and high value animals and crops. For more information about standardization, regulatory approval, and commercialization of botanical insecticides, see Tripathi et al. (2009), Isman (2006, 2010), Dubey (2011), and Isman et al. (2011). The insect growth regulator, nonsteroidal ecdysone agonist bisacylhydrazine (BSH) insecticides are generally faster acting than the JHA and CSI insecticides, thus, preventing crop damage by inhibition of feeding within 3–12 h after application. Although both JHA and CSI were discovered long before the BSH, the mode of action of BSH is the best understood at the molecular level. This allows cloning and expression of cDNAs encoding the ecdysone receptor complex, a heterodimer of two proteins: ecdysone receptor (EcR) and ultraspiracle (USP), from several insects, and the availability of stable and easy to synthesize BSH. Moreover, the molecular targets for BSH and reasons for the selective insect toxicity of bisacylhydrazine insecticides are also well known and understood. More investigations are required to explore several aspects about other IGRs, for instance, the molecular targets and basis of actions of about JHAs and CSI and if the JHAs use the same molecular target/site as the natural JHs do. Development of more efficient high-throughput assays for discovery of new and novel CSIs is allowed through understanding of the biosynthetic pathway for chitin synthesis and cuticle deposition, molecular characterization of the various enzymes involved, and the precise mode of action of CSIs, see Gilbert & Gill (2010) for more details. For designing better applications and protocols for pheromone application, understanding the mechanisms of mating disruption systems for different target species and dispensers is very curtail. It is better to suppress highly mobile insects at regional rather than local scales and use newer formulations, such as puffers, attract-and-kill, and sprayable formulations, that offer opportunities to reduce costs, increase program flexibility, and mix strategies. More reliable, economic, and widespread applications of pheromone can be achieved if a joint effort is made by people from different disciplines and organizations. Because of some difficulties with high populations of pests, pheromones should not be viewed as stand-alone strategies but integrated with other IPM options. Simply, biological control could be defined as using biota to reduce biota safely and economically. In the last 50 years, microbial insecticides illustrate remarkable developments associated with distinct good results under optimized laboratory conditions, followed many times by unsatisfactory results in the field applications. It is very important to produce reliable, effective, and safe entomo-pathogens for microbial control through the integration of research efforts about physiology, pathology, genetics, mass production, formulation, and application strategies, see Ravensberg (2011), for more information about a roadmap to the successful development and commercialization of microbial pest control products for control of arthropods. Production of biopesticides sharpens the action of genetic engineering for amalgamation of two or more effective lethal processes to finally tailor them into one agent/organism. As a result, the probability of development of resistant strains will decrease due to little morbidity and effective mortality. In spite of concerns about the potential environmental hazards of their long-term use, such opportunities will continue to persuade companies to seek new products and producers. GM microbes and transgenic crops are the new comer in IPM strategies and are gaining popularity because of their efficacy in eradicating pests. Nanotechnology is a promising field of research launches in the present decade a wide array of prospects and is anticipated to give major force in pest control. It is accepted by the public and not yet linked to any great concerns about health and the environment. Limited shelf-life, high specificity and variable field performance could be considered as advantages and disadvantages of biorational insecticides. As a result, proper identification of a target insect pest is essential and biorational products must be applied when the pest is in its most vulnerable life stage. Otherwise, applications may be ineffective, and applications of a conventional product may be necessary. Insect could also be controlled through environmental/ cultural controls as well as genetic control, see Vreysen et al. (2007), Gurr & Kvedaras (2010), Atkinson (2010), and van Emden & Service (2011) for more details. For more information abut area-wide control of insect pests from research to field implementation, see Vreysen et al. (2007). Achieving a zero pesticide strategy in tropical agroecosystems may be easier than in temperate zones, as in many instances farmers have not yet begun the generalized use of pesticides. This gives special opportunities for scientists and farmers to work on a systems approach to minimize pest impact before agroecosystems have been disturbed. For those areas already heavily impacted by the use of pesticides, such as cash crops, horticultural crops and livestock, adjustment to pesticide-free systems management will take some adaptation. Time is needed to establish or re-establish conditions in the system that are conducive to increased natural control such as habitat management and agronomic practices,.as well as to introduce farmers to new biological and physical control concepts and methods (Pretty, 2009). In general, biorational insecticides are much slower acting than those acting on neural target sites. The end user has, of course, been used to seeing insects die within a very short time following application of neuroactive insecticides. The change to the new insecticides has necessitated educating the distributors and the users on the mode of action and safety of the new products. There is a great potential to clean up agriculture form conventional insecticides. The road will be open to detoxify agriculture in case of endorsement of strong political will, consumer awareness, and market responses. Indisputably, the number and quality of biorational products will increase and the costs will fall. These progresses will assure an increasing place in the market for biopesticides for the foreseeable future.

ACKNOWLEDGEMENTS

The author would like to express her sincerest gratitude to Prof. Dr. Aza Moustafa, Professor of Insecticides at the Research Institute of Medical Entomology, Egypt) for her support and suggestions; Prof. Dr. Adel Shaheen, Head of the Department of Fish Diseases and management, Faculty of Veterinary Medicine, Benha University, Egypt and the father of aquatic mycotic diseases in Egypt, for his honest help and advise; Prof. Dr. Abdou Mahdi, Plant Pathology Department, Faculty of Agriculture, Benha University, Egypt, for his helpful advice about the mycotic part of this work; Prof. Dr. Mohamed Hafez, Plant Pathology Department, Faculty of Agriculture, Benha University, Egypt, for his valuable help with the viral part of this chapter as well as the quality of the figures; Dr. Nagwa Ahmed, Dr. Mohsen Negm El Din and and Lubna El- Akabawi, Department of Parasitology, Faculty of Veterinary Medicine, Benha University, Egypt, for their valuable advice

REFERENCES

1. Ahmad, I., Ahmad, F. & Pichtel, J. (2011). Microbes and Microbial Technology, Springer. Andow, D. A., Lövei, G. L. & Arpaia, S. (2006). Bt transgenic crops, natural enemies and implications for environmental risk assessment, Nature Biotechnology 24: 749-751.
2. Atkinson, PW. (2010). Vector Biology, Ecology and Control, Springer. DOI 10.1007/978-90-481- 2458-9
3. Awad, HH., El-Tayeb, TA., Abd El-Aziz, NM. & Abdelkader, MH. (2008). A semi-field study on the effect of novel hematoporphyrin formula on the control of Culex pipiens larvae, Journal of Agriculture & Social Sciences 4: 85–8
4. Barfod, KK. (2010). Occupational exposure to Bacillus thuringiensis based biopesticides, Development of an aerosol hazard identification model. University of Copenhagen, PhD Dissertation.
5. Baum, JA., Johnson, TB. & Carlton, BC. (1999). Bacillus thuringiensis natural and recombinant bioinsecticide products, in FR. Hall & JJ. Menn (eds), biopesticides use and delivery, Humana Press, Totowa, New Jersey, pp. 189–209.
6. Bellini, R., Mullens. B. A. & Jespersen. J.B. (1992). Infectivity of two members of the Entomophthora muscae complex (Zygomycetes: Entomophthorales) for Musca domestica (Diptera: Muscidae), Entomophaga 37: 1 l-19.
7. Ben Amor, T. & Jori, G. (2000). Sunlight-activated insecticides: historical background and mechanisms of phototoxic activity, Insect Biochemistry and Molecular Biology 30: 915– 925
8. Berón, CM., & Salerno, GL. (2007). Cloning and Characterization of a Novel Crystal Protein from a Native Bacillus thuringiensis Isolate Highly Active Against Aedes aegypti, Current Microbiology 54: 271–276.
9. Betarbet, R., Sherer, TB., MacKenzie, G., Garcia-Osuna, M., Panov, AV. & Greenamyre, JT. (2000). Chronic systematic pesticide exposure reproduces features of Parkinson's diseas, Nature Neuroscience 3:1301–1306. www.intechopen.com
10. Bhattacharyya, A., Bhaumik, A., Rani, PU., Mandals. & Epidi, TT. (2010). Nano-particles – A recent approach to insect pest control, African Journal of Biotechnology 9 (24): 3489- 3493.
11. Bhattacharyya, A., Gosh, M., Chinnaswamy, KP., Sen, P., Barik, B., Kundu, P. & Mandal, S. (2008). Nano-particle (allelochemicals) and Silkworm Physiology, in KP.
12. Chinnaswamy & Rao A Vijaya Bhaskar (eds), Recent Trends in Seribiotechnology, Bangalore. Indian. pp. 58-63.
13. Boeke, SJ., Boersma, MG., Alink, GM., van Loon, JJA., van Huis A., Dicke, M. & Rietjens, IMCM. (2004). Safety evaluation of neem (Azadirachta indica) derived pesticides, Journal of Ethnopharmacology 94: 25–41.
14. Bomford, MK. & Isman, MB. (1996). Desensitization of fifth instar Spodoptera litura (Lepidoptera: Noctuidae) to azadirachtin and neem, Entomologia Experimentalis et Applicata 81:307–13.

15. Bonning, BC. & Hammock, B D. (1996). Development of recombinant baculoviruses for insect control. Annual review of Entomology 41:191-210.

16. Bonning, BC., Loher, W., Bruce, D. & Hammock, BD. (1997). Recombinant Juvenile Hormone Esterase as a Biochemical Anti-Juvenile Hormone Agent: Effects on Ovarian Development in Acheta domesticus, Archives of Insect Biochemistry and Physiology 34:359–368.

17. Boughton, AJ., Obrycki, JJ. & Bonning, BC. (2003). Effects of a protease-expressing recombinant baculovirus on nontarget insect predators of Heliothis virescens, Biological Control, 28: 101 –110.

18. Bowen, Fl., Fisara, P., Junquera P., Keevers, DT., Mahoney, RH. & Schmid, HR. (1999). Longlasting prevention against blowfly strike using the insect growth regulator dicyclanil, Australian Veterinary Journal. 77 (7): 454–460.

19. Bowers, W.S. & Nishida, R. (1980). Juvocimences L potent juvenile hormone mimics from sweet basil, Science 209: 1030–1032.

20. Bray, DP., Bandi, KK., Brazil, RP., Oliveira, AG. & Hamilton, JG. (2009). Synthetic sex pheromone attracts the leishmaniasis vector Lutzomyia longipalpis (Diptera: Psychodidae) to traps in the field, Journal of Medical Entomology 46(3): 428-34.

21. Casida, JE. & Quistad, GB. (1995). Pyrethrum Flowers: Production, Chemistry, Toxicology and Uses. Oxford, UK: Oxford Univ.Press. 356 pp.

22. Chang, JH., Choi, JY., Jin, BR., Roh, JY., Olszewski, JA., Seo, SJ., et al., (2003). An improved baculovirus insecticide producing occlusion bodies that contain Bacillus thuringiensis insect toxin, Journal of Invertebrate Pathology 84:30-37.

23. Charles, D. (2001). Lords of the harvest: Biotech, big money, and the future of food. Perseus Publishing, Cambridge, MA, 348 pp.

24. Collins, D.A. (2006). A review of alternatives to organophosphorus compounds for the control of storage mites, Journal of Stored Products Research 42: 395–426

25. Copping, LG. & Menn, JJ. (2000). Biopesticides: a review of their action, applications and efficacy, Pest Management Science 56: 651- 76.

26. Couch, JN. (1935). A new saprofytic species of Coelomomyces, with notes on other forms, Mycologia 27: 376-387. www.intechopen.com Ecosmart Biorational Insecticides: Alternative Insect Control Strategies 51

27. Cristofari-Marquand, Ec., Kacel, M., Milhe, F., Magnan, A. & Lehucher-Michel, M. (2007). Asthma caused by peracetic acid-hydrogen peroxide mixture, Occupatinal Health 49 (2): 155–158.

28. Crow, WT. (2006). Using Nematodes to Control Insects: Overview and Frequently Asked Questions. ENY-050 (IN468), one of a series of the Entomology & Nematology Department, Florida Cooperative Extension Service, Institute of Food and Agricultural Sciences, University of Florida. http://edis.ifas.ufl.edu/ TOPIC_ GUIDE_Insect_Management_Guide.

29. De Groot, I.J.M. (1991). Protection of Stored Grains and Pulses, Agrodok 18 Agromisa Foundation, Wageningen, The Netherlands, 45 pp.

30. Dethier, V.G. (1982). Mechanisms of host plant recognition, Entomologia Experimentalis et Applicata 31: 49-56.

31. Dubey, N.K. (2011). Natural Products in Plant Pest Management, CAB International.

32. El-Husseini, M.M. (2006). Microbial Control of Insect Pests: is it an effective and environmentally safe alternative? Arab Journal of Plant Protection 24 (2): 162-169.

33. Enan, EE. (2005). Molecular and pharmacological analysis of an octopamine receptor from American cockroach and fruit fly in response to plant essential oils, Archives of Insect Biochemistry and Physiology 59: 161-171.

34. EPA. (1993). Flower and Vegetable Oils. Prevention, Pesticides, and Toxic Substances (7508W). EPA-738-F-93-027. www.epa.gov/oppsrrd1/REDs/factsheets/4097fact.pdf

35. Eto, M. (1990). Biochemical mechanism of insecticidal activities, in G. Haug & H. Hoffman (eds.) Chemistry of Plant Protection, Springer Verlag. 6: 65-107.

36. Federici, BA. (1981). Mosquito control by the Fungi Culicinomyces Coelomomyces and Coelomomyces. in HD. Burges. (ed.) Microbial Control of Pests and Plant Diseases 1970- 1980, London, Academic Press, 29: 555-572.

37. Federici, BA. (2010). Recombinant bacterial larvicides for control of important mosquito vectors of disease, in PW. Atkinson (ed.), Vector Biology, Ecology and Control, Springer, Part III, 163-176.

38. Fillinger, U. & Lindsay, SW. (2000). Suppression of exposure to malaria vectors by an order of magnitude using microbial larvicides in rural Kenya, Tropical Medicine and International Health 11:1629–1642.

39. Garrido-Jurado, I., Ruano, F., Campos M. & Quesada-Moraga, E. (2011). Effects of soil treatments with entomopathogenic fungi on soil dwelling non-target arthropods at a commercial olive orchard. Biological Control, in Press.

40. Gaugler, R. (2002). Entomopathogenic nematology, CABI Publishing

41. Geden, C.J. (1997). Evaluation of Paraiotonchium muscadomesticae (Nematoda: Tylenchida: Iotonchiidae), a potential biological control agent of the housefly (Diptera: Muscidae), Biological Control 10: 42-47.

42. Gilbert, LI. & Gill, SS. (2010). Insect Control: Biological and Synthetic Agents. Academic Press.

43. Gill, S.A., Reeser, R. & Raupp, M.J. (1998). Battling thrips, five pesticides put to the test, Grower Talks 62: 46-48.

44. Glover, D. (2008). The Corporate Shaping of GM Crops as a Technology for the Poor, Made by Monsanto, www.steps-centre.org/PDFs/GM%20Crops%20web%20final_small.pdf www.intechopen.com

45. Graf, JF. (1993).The role of insect growth regulators in arthropod control, Parasitology Today 9 (12): 471-474.

46. Gunasena, HPM. & Marambe, B. (1998). Neem in Sri Lanka: A monograph. University of Peradeniya, Sri Lanka.

47. Gurr, G.M. & Kvedaras, O.L. (2010). Synergizing biological control scope for sterile insect technique, induced plant defenses and cultural techniques to enhance natural enemy impact, Biological Control 52 (3): 198-207.

48. Hajek, AE., McManus, M.L.& Delalibera Junior, I. (2007). A review of introductions of pathogens and nematodes for classical biological control of insects and mites, Biological Control 41: 1-13.

49. Hallmon, CF., Schreiber, ET., Vo, T. & Bloomquist, MA. (2000). Field trials of three concentrations of Laginex as biological larvicide compared to Vectobac-12AS as a biocontrol agent for Culex quinquefasciatus. Journal of the American Mosquito Control Association 16: 5-8.

50. Harrison, L. & Bonning, BC. (2010). Proteases as Insecticidal Agents. Toxins 2: 935-953.

51. Hartelt, K., Wurst E., Collatz, J., Zimmermann, G., Kleespies, RG., Oehme, RM., Kimmig, P., Steidle, JLM. & Mackenstedt, U. (2008). Biological control of the tick Ixodes ricinus with entomopathogenic fungi and nematodes: Preliminary results from laboratory experiments, International Journal of Medical Microbiology 298 (1): 314-320.

52. Hashim, A. (2011). Advances in nanocomposite technology, InTech, Croatia.

53. Hohmann, CL., Flávia, A C., Silva, FAC. & de Novaes, TG. (2010). Selectivity of neem to Trichogramma pretiosum Riley and Trichogrammatoidea annulata De Santis (Hymenoptera: Trichogrammatidae), Neotropical Entomology 39(6):985-990.

54. Hollingworth, R.M., Johnstone, E.M. & Wright, N. (1984). In: P.S. Magee, G.K. Kohn and J.J. Menn (eds.), Pesticide Synthesis through Rational Approaches, ACS Symposium Series No. 255, American Chemical Society, Washington, DC (1984), pp. 103–125.

55. Hoover, K., Kishida, KT., DiGiorgio, L.A., Workman, J., Alaniz, S.A., Hammock, B.D. & Duffey, S.S. (1998). Inhibition of baculoviral disease by plant-mediated peroxidase activity and free radical generation. Journal of Chemical Ecology 24: 1949-2001.

56. H□fte, H. & Whiteley, HR. (1989). Insecticidal crystal proteins of Bacillus thuringiensis. Microbiology and Molecular Biology Reviews 53:242–255.

57. Inceoglu, AB., Kamita, SG., Hinton, AC., Huang, Q., Severson, TF.,Kang, KD. & Hammock, B. D. (2001). Recombinant baculoviruses for insect control. Pest Management, Science 57: 981 - 7.

58. Isman, MB. (2005). Problems and opportunities for the commercialization of botanical insecticides, in: C. Regnault-Roger, BJR. Philog` ene. & C. Vincent (eds.), Biopesticides of Plant Origin, pp. 283–91. Paris: Lavoisier.

59. Isman, M.B. (2006). Botanical insecticides, deterrents, and repellents in modern agriculture and an increasingly regulated world, Annual Review of Entomology 51:45–66.

60. Isman, MB. (2010). Botanical insecticides, deterrents, repellents and oils, in B.P. Singh (ed.), in: Industrial crops and uses. CAP International, pp. 433-445.

61. Isman, M.B. & Akhtar, Y. (2007). Plant natural products as a source for developing environmentally acceptable insecticides. in: I. Shaaya, R. Nauen, & A.R. Horowitz www.intechopen.com (eds), Insecticides Design Using Advanced Technologies. Springer, Berlin, Heidelberg, pp. 235–248.

62. Isman, MB., Miresmailli, S. & Machial, C. (2011). Commercial opportunities for pesticides based on plant essential oils in agriculture, industry and consumer products. Phytochemistry Reviews, 10:197–204.

63. Isman, MB., Matsuura, H., MacKinnon, S., Durst, T., Towers, GHN. & Arnason, JT. (1996). Phytochemistry of the Meliaceae. So many terpenoids, so few insecticides. In JT. Romeo, JA. Saunders & P. Barbosa (eds.) Phytochemical Diversity and Redundancy, New York: Plenum, pp. 155– 78.

64. Jaronski, S. & Axtell, RC. (1983). Persistence of the mosquito fungal pathogen Coelomomyces giganteum (Oomycetes; Lagenidiales) after introduction into natural habitats, Mosquito News 43; 332-337.

65. Jehle, JA., Blissard, G.W., Bonning B.C., Cory, J.S, Herniou, E.A, Rohrmann, G.F., Theilmann, D.A., Thiem, S.A. & Vlak,. J.M. (2006). On the classification and nomenclature of baculoviruses: a proposal for revision. Archives of Virology 151: 1257–1266.

66. Jensen, AB., Thomsen L. & Eilenberg J. (2001). Intraspecific Variation and Host Specificity of Entomophthora muscae sensu stricto Isolates Revealed by Random Amplified Polymorphic DNA, Universal Primed PCR, PCR-Restriction Fragment Length Polymorphism, and Conidial Morphology, Journal of Invertebrate Pathology 78 (4): 251-259.

67. Joseph, T. & Morrison, M. (2006). Nanotechnology in Agriculture and Food. A Nanoforum report. www.nanoforum.org .

68. Karmakar, P.R. & Chatterjee, B.P. (1994). Isolation and characterization of two IgE-reactive proteins from Azadirachta indica pollen, Molecular and Cellular Biochemistry 131: 87– 96.

69. Kaya, H. K. & Koppenhöfer, A.M. (1999). Biology and ecology of insecticidal nematodes, in S. Polavarapu (ed.) workshop proceedings: Optimal use of insecticidal nematodes in pest management, Rutgers University, p. 1–8.

70. Kerwin, JL. & Petersen, EE. (1997). Fungi: Oomycetes and Chytridiomycetes. In: Lacey LA. (ed.), Manual of techniques in insect pathology, 5-4: 251- 268. New York. Academic Press.

71. Kerwin, JL. & Washino, RK. (1987). Ground and aerial application of the asexual stage of Coelomomyces giganteum for control of mosquitoes associated with rice culture in the central valley of California, Journal of the American Mosquito Control Association 3: 59- 64.

72. Kerwin, JL., Dritz, DA. & Washino, RK. (1990).

Confirmation of the safety of Coelomomyces giganteum (Oomycetes: Lagenidiales) to mammals. Journal of Economic Entomology 83: 374-376.

73. Kerwin, JL., Dritz, DA. & Washino, RK. (1994). Pilot scale production and application in wildlife ponds of Coelomomyces giganteum (Oomycetes: Lagenidiales), Journal of the American Mosquito Control Association 10: 451-455.

74. Khater, H.F. & Khater, D.H. (2009). The insecticidal activity of four medicinal plants against the blowfly Lucilia sericata (Diptera: Calliphoridae), International Journal of Dermatology 48 (5): 492- 497.

75. Khater, H.F. & Ramadan, M.Y. (2007). The acaricidal effect of peracetic acid against Boophilus annulatus and Argas persicus, Acta Scientiae Veterinariae 35 (1): 29-40.

76. Khater, H.F. (2011). Biorational Insecticides for integrated pest management. Benha Veterinary Medical Journal, In press.

77. Khater, H.F., Hanafy, A.M., Abdel- Mageed, A.D., Ramadan, M.Y., & El- Madawy, R.S. (2011). The insecticidal effect of some Egyptian plant oils against Lucilia sericata (Diptera: Calliphoridae), International Journal of Dermatology 50 (2): 187–194.

78. Khater, H.F., Ramadan, M.Y. & El- Madawy, R.S. (2009).The lousicidal, ovicidal, and repellent efficacy of some essential oils against lice and flies infesting water buffaloes in Egypt, Veterinary Parasitology 164 (2-4): 257-266.

79. Khater, HF. & Shalaby, AA. (2008). Potential of biologically active plant oils for control mosquito larvae Culex pipiens (Diptera: Culicidae) from an Egyptian locality, Revista do Instituto de Medicina Tropical de Sao Paulo 50 (2): 107-112.

80. Khater, HF. (2003). Biocontrol of some insects. PhD thesis, Zagazig University: Benha Branch, Egypt.

81. Kirkland, BH., Cho, E. & Keyhani, NO. (2004). Differential susceptibility of Amblyomma maculatum and Amblyomma americanum (Acari:Ixodidea) to the entomopathogenic fungi Beauveria bassiana and Metarhizium anisopliae, Biological Control 31 (3): 414-421.

82. Kleeberg, H. (2006). Demands for plants protection products- Risk assessment botanicals and semiochemicals REBECA workshop, Brussels, 13–14 June; http://www.rebeca.net.de/downloads/Risk%20assessment%20 botanicals%20semi ochemicals%20Minutes.pdf.

83. Knols, BG., Bossin HC., Mukabana, WR., & Alan S. Robinson, AS. (2007). Transgenic mosquitoes and the fight against malaria: managing technology push in a turbulent GMO world, The American Journal of Tropical Medicine and Hygiene 77 (Suppl 6): 232– 242.

84. Krasnoff, S.B., Watson, D.W., Gibson, D.M. & Kwan, E.C. (1995). Behavioral Effects of the Entomopathogenic Fungus, Entomophthora muscae on its Host Musca domestica: Postural Changes in Dying Hosts and Gated Pattern of Mortality, Journal of Insect Physiology 41 (10): 895-903.

85. Kumar, P., Singh, HP. & Poehling, H. (2010). Effects of neem on adults of Eretmocerus warrae (Hym., Aphelinidae), a parasitoid of Bemisia tabaci (Hom., Aleyrodidae) in tropical horticulture systems, Journal of Plant Diseases and Protection 117 (6): 273–277.

86. Kumar, U. (2002). Neem as a potential biopesticide and soil conditioner, Agrobios Newsletter 1(6): 8–12.

87. Lacey, L.A. & Kaya, H.K. (2007). Field manual of techniques in invertebrate pathology. 2nd Ed. Springer. Dordrecht, The Netherlands

88. Leal, WS., Blomquist, GJ. & Vogt, RG. (2003). Proteins that make sense. Insect Pheromone Biochemistry and Molecular Biology. Elsevier, Amsterdam.

89. Leger, R.J.S. (2008). Studies on adaptations of Metarhizium anisopliae to life in the soil, Journal of Invertebrate Pathology 98: 271-276

90. Leighton, TE., Marks, D. & Leighton, F. (1981). Pesticides: Insecticides and fungicides are chitin synthesis inhibitors, Science 213: 905-907.

91. Lu, FC. (1995). A review of the acceptable daily intakes of pesticides assessed by the World Health Organization, Regulatory Toxicology and Pharmacology 21: 351-364.

92. Lukšienė Z., Kurilčik N., Juršėnas S., Radžiutė, S. & Būda V. (2007). Towards environmentally and human friendly insect pest control technologies:

93. Photosensitization of leafminer flies Liriomyza bryoniae, Journal of Photochemistry and Photobiology B: Biology 89: 15–21.

94. Lövei, GL., Andow, DA. & Arpaia, S. (2009). Transgenic Insecticidal Crops and Natural Enemies: A Detailed Review of Laboratory Studies, Environmental Entomology 38 (2): 293-306.

95. Maia, MF. & Moore, SJ. (2011). Plant-based insect repellents: a review of their efficacy, development and testing, Malaria Journal 10 (Suppl 1): S11.

96. Makonde, H.M., Lenga, F.K., Masiga, D., Mugo, S. & Boga, H.I. (2010). Effects of Bacillus thuringiensis CRY1A(c) d--endotoxin on growth, nodulation and productivity of beans [Phaseolus vulgaris (L.) and siratro (Macroptilium atropurpureum DC.)], African Journal of Biotechnology 9 (1): 017-024.

97. Mangan, RL. & Moreno, DS. (2001). PhotoActive Dye Insecticide Formulations: Adjuvants Increase Toxicity to Mexican Fruit Fly (Diptera: Tephritidae), Journal of Economic

98. Entomology 94 (1):150-156.

99. Marvier, M., C. McCreedy, J. Regetz, & P. Kareiva. (2007). A meta-analysis of effects of Bt cotton and maize on non-target invertebrates, Science 316: 1475- 1477.

100. McCutchen, B.F., Hoover, K., Preisler, H.K., Betana, M.D., Herrmann, R., Robertson, J.L. &

101. Hammock, B.D. (1997). Interaction of recombinant and wild-type baculoviruses with classical insecticides and pyrethroid-resistant tobacco budworm (Lepidoptera: Noctuidae), Journal of Economic Entomology 90: 1170 - 80.

102. McNeil J. (2010).Viruses as biological control agents

of insect pests, organic agriculture. E:\1 chapter book 2 microbial control\McNeil 2010 Viruses as biological control agents of insect pests - eXtension.mht

103. Mehlhorn, H. (2011). Nature Helps, Parasitology Research Monographs 1, Springer-Verlag Berlin Heidelberg.

104. Merrriam, TL. & Axtell, RC. (1982). Evaluation of the entomogenous fungi Culicinomyces clavosporus and Coelomomyces giganteum for control of the salt marsh mosquito, Aedes taeniorhynchus, Mosquito News 42: 594-602.

105. Metchnikoff, E. (1879). Zur Lehre über Insektenkrankheiten, Zoologische Anzeiger, 3: 44-47.

106. Miles, C., Blethen, C. & Gaugler, A. (2000). Using beneficial nematodes for crop insect pest control. Washington State University Cooperative Extension. PNW544

107. Mochi, DA., Monteiro, AC., Machado ACR. & Yoshida, L. (2010). Efficiency of entomopathogenic fungi in the control of eggs and larvae of the horn fly Haematobia irritans (Diptera: Muscidae), Veterinary Parasitology 167 (1): 62-66.

108. Moller, A.P. (1993). A fungus infecting domestic flies manipulates sexual behaviour of its host, Behavioral Ecology and Sociobiology 33: 403- 407.

109. Mordue Luntz, A.J. (2003). Arthropod semiochemicals: mosquitoes, midges and sealice, Biochemical Society Transactions 31 (1): 128-133.

110. Mordue, A.J. & Blackwell, A. (1993). Azadirachtin : an update, Journal of Insect Physiology 39: 903-924.

111. Morgan, ED. (2009). Azadirachtin, a scientific gold mine, Bioorganic & Medicinal Chemistry 17(12): 4096-4105.

112. Moscardi, F. (1999). Assessment of the application of baculoviruses for control of Lepidoptera, Annual Review of Entomology 44: 257–289.

113. Naumann, K. & Isman, M.B. (1996). Toxicity of neem (Azadirachta indica A. uss) seed extracts to larval honeybees and estimation of dangers from field applications, American Bee Journal 136:518–20

114. Nerio, LS., Olivero-Verbel, J. & Stashenko E. (2010). Repellent activity of essential oils: a review, Bioresource Technology 101(1): 372-378.

115. Nestrud, LB. & Anderson, RL. (1994). Aquatic safety of Coelomomyces giganteum: effects on freshwater fish and invertebrates, Journal of Invertebrate Pathology 64: 228-233.

116. O'Brien, D.J. and Fahey, G. (1991). Control of fly strike in sheep by means of a pour-on formulation of cyromazine, Veterinary Record 129: 351–353.

117. Orduz, S. & Axtell, RC. (1991). Compatibility of Bacillus thuringiensis var. israelensis and Bacillus sphaericus with the fungal pathogen Coelomomyces giganteum (Oomycetes:Lagenidiales), Journal of the American Mosquito Control Association 7: 188- 193.

118. Owolade, OF., Ogunleti, DO. & Adenekan, MO. (2008). Titanium Dioxide affects disease development and yield of edible cowpea, Electronic journal of Environmental Agriculture and Food chemistry 7(50): 2942-2947.

119. Palevitch, D. & Craker, L.E. (1994). Volatile oils as potential insecticides. Herb, Spice and Medicinal Plant Digest 12: 1–5.

120. Patil, SA. (2009). Economics of agri poverty: Nano-bio solutions. Indian Agricultural Research Institute, New Delhi, Indian.

121. Pinto, MC. & do Prado, AP. (2001). Resistance of Musca domestica L. Populations to Cyromazine (Insect Growth Regulator) in Brazil, Memórias do Instituto Oswaldo Cruz 96(5): 729-732.

122. Pitasawat, B., Choochote, W., Tuetun, B., Tippawangkosol, P., Kanjanapothi, D., Jitpakdi, A. & Riyong, D. (2003). Repellency of aromatic turmeric Curcuma aromatica under laboratory and field conditions, Journal of Vector Ecology 28 (2): 234-240.

123. Pretty, J. (2009). The Pesticide Detox, Towards a More Sustainable Agriculture. Earthscan, London.

124. Price, DN. & Berry, MS. (2006). Comparison of effects of octopamine and insecticidal essential oils on activity in the nerve cord, foregut, and dorsal unpaired median neurons of cockroaches, Journal of Insect Physiology 52: 309-319.

125. Priestley, C.M., Burgess, I.F. & Williamson, E.M., (2006). Lethality of essential oils constituents towards the human louse Pediculus humanus, and its eggs, Fitoterapia 77, 303–309.

126. Priestley, CM., Williamson, EM., Wafford, KA. & Sattelle, DB. (2003). Thymol, a constituent of thyme essential oil, is a positive allosteric modulator of human GABAA receptors and a homo-oligomeric GABA receptor from Drosophila melanogaster, Britich Journal of Pharmacology 140:1363–7278.

127. Quesada-Moraga, E., Ruiz-Garcia, A. & Santiago-Alvarez, C. (2006). Laboratory evaluation of entomopathogenic fungi Beauveria bassiana and Metarhizium anisopliae against puparia and adults of Ceratitis capitata (Diptera : Tephritidae), Journal of Economic Entomology 99: 1955-1966.

128. Ragaei, M. & Khater, H.F. (2004). Laser and photobiology applications: Photosensitizers as photopesticides, Phototherapy and phototoxicity in animals. Workshop on Laser Chemistry and Applications Materials and Biomedical Research. National Research Center, Cairo, October 2-5, 2004.

129. Ragaei, M., Abdel- Latef, H. & Khater, H.F. (2004) Using of Bacillus sphaericus as microbial control agent against mosquito larvae. 1s1 international scientific Conference of toxicology 18- 19 Dec. (2004) Faculty of Pharmacy, 6th October University.

130. Ravensberg, WJ. (2011). A Roadmap to the Successful Development and Commercialization of Microbial Pest Control Products for Control of Arthropods. Springer.

131. Regnault-Roger C. (2011). Risks of Biocontrol Agents Containing Compounds of Botanical Origin or Semiochemicals, in Ehlers R. (ed.), Regulation of Biological Control Agents, Springer, Part 2, pp. 215-242, DOI: 10.1007/978-90-481-3664-3_10

132. Renn, N. (1995). Mortality of immature houseflies (Musca domestica L.) in artificial diet and chicken manure after exposure to encapsulated entomopathogenic nematodes (Rhabditida: Steinernematidae, Heterorhabditidae), Biocontrol Science and Technology 5: 349-359.

133. Renn, N. (1998).The efficacy of entomopathogenic nematodes for controlling housefly infestations of intensive pig units, Medical and Veterinary Entomology 12: 46–51.

134. Retnakaran, A., Hiruma, K., Palli, S.R. & Riddiford, L.M. (1995). Molecular analysis of the mode of action of RH-5992, a lepidopteran-specific, non-steroidal ecdysteroid agonist, Insect Biochemistry and Molecular Biology 25: 109–117.

135. Retnakaran, A., Macdonald, A., Tomkins, W.L., Davis, C.N., Brownright, A.J., et al., 1997. Ultrastructural effects of a non-steroidal agonist, RH-5992, on the sixth instar larvae of spruce budworm, Choristoneura fumiferana, Journal of Insect Physiology 43: 55–68.

136. Roffey J. (1968). The occurrence of the fungus Entomophthora grylli Fresenius on locusts and grasshoppers in Thailand, Journal of Invertebrate Pathology 11: 2377 241.

137. Salako, EA., Anjorin, ST., Garba, CD., & Omolohunnu, EB. (2008). A review of neem biopesticide utilization and challenges in Central Northern Nigeria, African Journal of Biotechnology 7 (25): 4758-4764.

138. Samuels, R. I., Coracini, D.L.A., Martins dos Santos, C.A. & Gava, CTA. (2002). Infection of Blissus antillus (Hemiptera: Lygaeidae) eggs by the entomopathogenic fungi Metarhizium anisopliae and Beauveria bassiana, Biological Control 23: 269–273.

139. Schmutterer, H. (1990). Properties and potential of natural pesticides from the neem tree, Azadirachta indica, Annual Review of Entomology 35: 271–297.

140. Schmutterer, H. (1995). In: The neem tree Azadirachta indica (A. Juss) and other meliaceous plants: sources of unique natural products for integrated pest management, medicine, industry and other purposes, VCH Verlagsgesellschaft mbH, Weinheim, Germany, p. 696.

141. Scholte, EJ., Knols, BGJ., Samson, RA. & Takken, W. (2004). Entomopathogenic fungi for mosquito control: A review, Journal of Insect Science 4(19): 1-24

142. Schuster, D. J. & Stansly, P.A. (2006). Chemical Control: Biorational Insecticides, pp199-210. In J. L. Gillett, H. N. HansPetersen, N. C. Leppla and D. D. Thomas (Eds.) Grower›s IPM Guide for Florida Tomato and Pepper Production. University of Florida, Extension. 206 p.

143. Shalaby, A.A. & Khater, H.F. (2005). Toxicity of certain solvent extracts of Rosmarinus officinalis against Culex pipiens larvae, Journal of the Egyptian- German Society of Zoology (48E): 69- 80.

144. Sharma, RN. (1993). The utilization of essential oils and some common allelochemic constituent for non-insecticidal pest management strategies. In: Newer Trends Essent-51. New Delhi (India): Tata McGraw Hill.

145. Shathele, M.S. (2009). New isolate media of the mosquito pathogenic fungus Lagenidium giganteum (Oomycetes: Lagenidiales) for fungal maintenance and zoospores release. Journal of Entomology 6 (3):167-175.

146. Singh, A., Parmar, N. & Kuhad, RC. (2011). Bioaugmentation, Biostimulation and Biocontrol, Springer-Verlag Berlin Heidelberg.

147. Smart Jr., GC. (1995). Entomopathogenic Nematodes for the Biological Control of Insects, Supplement to the Journal of Nematology 27(4S): 529-534. 1995.

148. Staal, G.B. (1986). Anti-juvenile horone agents, Annual Review of Entomology 31:391-429.

149. Steenberg, T., Jespersen, J.B., Jensen, K.M.V., Nielsen, B.O. & Humber, R.A. (2001). Entomopathogenic Fungi in Flies Associated with Pastured Cattle in Denmark, Journal of Invertebrate Pathology 77: 186–197.

150. Stephen, K. & Kurtböke, DI. (2011). Screening of Oomycete Fungi for Their Potential Role in Reducing the Biting Midge (Diptera: Ceratopogonidae) Larval Populations in Hervey Bay, Queensland, Australia. International Journal of Environmental Research and Public Health 8: 1560-1574.

151. Sun, X., Wang, H., Sun, X., Chen, X., Peng, C., Pan, D., et al. (2004). Biological activity and field efficacy of a genetically modified Helicoverpa armigera SNPV expressing an insect-selective toxin from a chimeric promoter, Biological Control 29:124-137.

152. Sur B, Bihari V, Sharma A, Joshi AK. (2001). Studies on physiology, zoospore morphology and entomopathogenic potential of the aquatic oomycete: Coelomomyces giganteum.

153. Mycopathologia 154: 51-54.

154. Szewczyk, B, Rabalski, L., Krol E., Sihler W. & Lobo de Souza, M. (2009). Baculovirus biopesticides - a safe alternative to chemical protection of plants, Journal of Biopesticides 2 (2): 209-216.

155. Szewczyk, B., Lobo de Souza, M., Batista de Castro, ME., Moscardi, ML. & Moscardi, F. (2011). Baculovirus Biopesticides. In M. Stoytcheva (ed.), Pesticides - Formulations, Effects, Fate, InTech. p. 25-36.

156. Taylor, MZ. (2001). Recent Developments in Ectoparasiticides, The Veterinary Journal 161(3):253-268.

157. Toloza, A., Lucía, A., Zerba, E., Masuh, H. & Picollo, MI. (2008) Interspecific hybridization of

158. Eucalyptus as a potential tool to improve the bioactivity of essential oils against permethrin-resistant head lice from Argentina, Bioresour Technology 99:7341–7347.

159. Tomlin, C.D.S. (2000). The Pesticide Manual, 12th edn. British Crop Protection Council Publications.

160. Torney F (2009). Nanoparticle mediated plant transformation. Emerging technologies in plant science research. Interdepartmental Plant Physiology Major Fall Seminar Series. Phys. p. 696.

161. Tripathi, AK., Upadhyay, S., Bhuiyan, M. & Bhattacharya, P.R. (2009). A review on prospects of essential oils as biopesticide in insect-pest management, Journal of Pharmacognosy and Phytotherapy 1(5): 052-063.

162. Tunaz, H. (2004). Insect Growth Regulators for Insect Pest Control, Turkish Journal of Agriculture and Forestry 28: 377-387.

163. UNEP. (2000). The Montreal Protocol on Substances that Deplete the Ozone layer. United Nations Environment Programme, Nairobi, Kenya. P. 1-54 (www.unep.org/ozone). van Emden, H. F. & Service, M.W. (2011). Pest and Vector Control. Cambridge University Press, 2001. Cambridge Books Online. Cambridge University Press. 31 http://dx.doi.org/10.1017/CBO9780511616334

164. Vasconcelos S.D.V., Cory J.S., Wilson K.R., Sait S.M. & Hails R.S. (1996). The effect of baculovirus infection on the mobility of Mamestra brassicae L. (Lepidoptera: Noctuidae) larvae at different developmentalstages, Biological Control 7: 299–306.

165. Veal, L. (1996). The potential effectiveness of essential oils as a treatment for headlice, Complement ther nurs midwifery 2: 97–101.

166. Vidhyalakshmi, R., Bhakyaraj, R. & Subhasree, RS. (2009). Encapsulation the Future of Probiotics-A Review, Advances in Biological Research 3(3-4): 96-103.

167. Vreysen, M.J.B., Robinson, A.S. & Hendrichs, J. (2007). Area-wide control of insect pests, from research to field implementation. Springer, Dordrecht, The Netherlands.

168. Walker, KR. & Welter, SC. (2010). Potential for outbreaks of leafrollers (Lepidoptera: Tortricidae) in California apple orchards using mating disruption for codling moth suppression, Journal of Economic Entomology 94:80.

169. Wan, MT., Watts, RG., Isman, MB. & Strub, R. (1996). An evaluation of the acute toxicity to juvenile Pacific northwest salmon of azadirachtin, neemextract and neem-based products, Bulletin of Environmental Contamination and Toxicology 56:432–439.

170. Ware, GW. & Whitacre, DM. (2004). The Pesticide Book, 6th Ed. Meister Media Worldwide,

171. Willoughby, Ohio.

172. Watson, G.S., Watson, JA., Hu, S., Brown, CL., Cribb, BW. & Myhra, S. (2010). Micro and nanostructures found on insect wings - designs for minimising adhesion and friction, International Journal of Nanomanufacturing 5:112-128.

173. Welter, SC., Millar, CG., Cave, F., Van Steenwyk, RA. & Dunley, J. (2005). Pheromone mating disruption offers selective management options for key pests, California Agriculture 59(1):16-22.

174. Wirth, MC., Park, HW., Walton, WE., & Federici, BA. (2005). CytlA of Bacillus thuringiensis delays evolution of resistance to Cry11A in the mosquito Culex quinquefasciatus, Applied and Environmental Microbiology 71:185–189

175. Witzgall, P. (2001). Pheromones - future techniques for insect control? Pheromones for Insect Control in Orchards and Vineyard, IOBC wprs Bulletin 24 (2): 114-122.

176. Yang, FL., Li, XG., Zhu, F. & Lei, CL. (2009). Structural Characterization of Nanoparticles Loaded with Garlic Essential Oil and Their Insecticidal Activity against Tribolium castaneum (Herbst) (Coleoptera: Tenebrionidae), Journal of Agricultural and Food Chemistry 57(21): 10156-10162.

177. Zabalgogeazcoa. I., Oleaga A. & Pérez-Sánchez, R. (2008). Pathogenicity of endophytic entomopathogenic fungi to Ornithodoros erraticus and Ornithodoros moubata (Acari: Argasidae), Veterinary Parasitology 158 (4): 336-343.

178. Zhou, M.Z., Sun, H.C., Hu, Z.H.& Sun, X.L. (2004). SOD enhances infectivity of Helicoverpa armigera single nucleocapsid nucleopolyhedrosis against H.armigera larvae, Virologica Sinica, 18: 506-507.

CHAPTER 4 ALTERNATIVE WEED CONTROL METHODS

G.R. Mohammadi

INTRODUCTION

Weed interference is one of the most important limiting factors which decrease crop yields and consequently global food production. Weed represent about 0.1% of the world flora and in agroecosystems, weeds and crops have co-evolved together right from the prehistoric times as revealed by pollen analysis studies (Cousens and Mortimer 1995). Weed can suppress crop yield by competing for environmental resources like water, light and nutrients and production of allelopathic compounds. Therefore, weed management have been a major challenge for crop producers from the start of agriculture. At the earlier times, since no synthetic chemicals were known, weed control was achieved by some methods such as hand weeding, crop rotation, polyculture and other management practices that were low input but sustainable. With the discovery of synthetic herbicides in the early 1930s, there was a shift in control methods toward high input and target-oriented ones (Singh et al. 2003).

However, herbicide-reliant weed control methods can cause high costs for crop producers due to the consumption of fossil fuels (the non-renewable energy resources) (Lybecker et al. 1988). Moreover, ground and surface water pollution by these synthetic chemicals are causes for concern (Hallberg 1989). Fast-developing herbicide-resistant ecotypes of weeds due to increased herbicide application is another serious threat for agriculture production (Holt and LeBaron 1990). Therefore, there is an urgent need to develop alternative weed control methods for use in agroecosystems. Many studies have revealed that the alternative methods such as the use of allelopathy phenomenon, cover crops and living mulches, competitive crop cultivars, suitable nutrient management, etc. can be proposed as the low cost, effective and eco-friendly practices for sustainable weed management in cropping systems. In this chapter, the most important alternative weed control methods are discussed.

ALLELOPATHY

The term allelopathy was first introduced by Hans Molisch in 1937 and refers to chemical interactions among plants, including those mediated by microorganisms. Rice (1984) defined allelopathy as the effects of one plant (including microorganisms) on another plant through the release of a chemical compounds into the environment. Allelopathy can play a beneficial role in various cropping systems such as mixed cropping, multiple cropping, cover cropping, crop rotations, and minimum and no–tillage systems. The exploitation of allelopathy in agricultural practices as a tool for weed control has shown weed reduction, pathogen prevention and soil enrichment (Kohli et al., 1998).

Ways by Which Allelopathy Can Be Used To Control Weeds in Cropping Sysyems

In general, the use of allelopathy as a tool to control weeds can be achieved in different ways:

1. Use of crop cultivars with allelopathic properties.
2. Application of residues and straw of allelopathic crops as mulches or incorporated into the soil.

Citation: G.R. Mohammadi (2013). Alternative Weed Control Methods, Weed and Pest Control - Conventional and New Challenges, Dr. Sonia Soloneski (Ed.), InTech, DOI: 10.5772/54164.

3. Use of an allelopathic crop in a rotational sequence.
4. Application of allelochemicals or modified allelochemicals as herbicides (Kruse et al. 2000).
5. Modification of crops to enhance their allelopathic effects.

Use of Crop Cultivars with Allelopathic Properties

It's clear that the crop cultivars differ in their allelopathic ability and thus superior cultivars can be selected for weed management programs (Wu et al. 1999; Olofsdotter et al. 2002). Differences in allelopathic potential between genotypes has been investigated among accessions (genetical different lines or strains of a species) of barley, cucumber (*Cucumis sativus*), oats, soybean (*Glycine max*), sunflower, sorghum (*Sorghum bicolor*), rice and wheat (Copaja et al. 1999, Dilday et al. 1994, Narwal 1996, Miller 1996,Yoshida et al. 1993, Wu et al. 1998).

In a study on 3000 accessions of *Avena* spp. Fay and Duke (1977) found that four accessions apparently exuded up to three times as much scopoletin (a chemical identified as phytotoxic) as a standard oat cultivar. When one of the accessions were grown in sand culture with wild mustard (*Brassica kaber*), the growth of the mustard was significantly less than when it was grown with an accession that exuded a lower amount of scopoletin. In a field experiment, 1000 accessions of rice were screened for allelopathic activity against the two weedy species, barnyardgrass (*Echinochloa crus-galli* Beauv) and *Cyperus difformis*. Of these, 45 accessions showed allelopathic activity against one of the weeds and five accessions inhibited both species (Olofsdotter et al. 1997). Dilday et al. (2001) evaluated 12,000 rice accessions from 110 countries for allelopathy to ducksalad [*Heteranthera limosa*(S.w.) Willd.] and about 5000 have been assessed for allelopathy to redstem (*Ammannia coccinea*) and barnyardgrass. Results indicated that among them, 145 accessions were allelopathic to ducksalad and redstem and 94 accessions demonstrated apparent allelopathic activity to barnyardgrass.

Many weed species are most susceptible to allelochemicals in the seed and seedling stages. Therefore, the ideal allelopathic cultivar must therefore release allelochemicals in bioactive concentrations before the target weeds grow to old. Knowledge about both the critical developmental stage where the crop starts releasing allelochemicals and the critical sensitive stage of the target weeds is therefore essential (Inderjit and Olofsdotter 1998).

Application of Residues and Straw of Allelopathic Crops

Weed suppressive effects of crop residues have been explained by different mechanisms, including initial low nitrogen availability following cover crop incorporation (Dyck and Liebman 1994; Kumar et al. 2008; Samson 1991), mulch effects (Mohler 1996; Mohler and Callaway 1991; Mohler and Teasdale 1993), stimulation of pathogens or predators of weed seeds (Carmona and Landis 1999;Conklin et al. 2002; Davis and Liebman 2003; Gallandt et al. 2005; Kremer 1993), and allelopathy (Chou 1999; Weston 1996).

Allelopathic compounds released from crop residues during decomposition can reduce both emergence and growth of weeds. Allelochemicals can be released either through leaching, decomposition of residues, volatilization or root exudation (Chou 1999). In production systems with no-till or conservation tillage that leave nearly all crop residues on the soil surface, the release of allelochemicals from both the growing plants and during residue decomposition could be advantageous (Kruse et al. 2000).

Barnes and Putnam (1983) reported that rye residue used as mulch reduced total weed biomass by 63%. It was found that disappearance of rye allelochemicals was more closely related to weed suppression than to the disappearance of rye residues. Especially due to the massive production of biomass, rye has the potential to influence the growth of succeeding plant species through the release of allelochemicals from the residue (Barnes et al. 1985).

Wheat residues suppress weeds due to the physical effect and to the production of allelochemicals (Petho 1992). their allelopathic effects was positively correlated with the total phenolic content in the tissue of the wheat cultivars (Wu et al 1998). Hydroxamic acids have also been identified in shoot and root tissue of wheat.

The residues of barley have also been associated with phytotoxicity (Overland 1966, Lovett and Hoult 1995). Phytotoxic phenolic compounds, including ferulic, vanillic and phydroxybenzoic acids, have been identified in barley (Börner 1960). The two alkaloids, gramine and hordenine have been confirmed to play an important role in the phytotoxic ability of barley (Lovett and Hoult 1995,Overland 1966). In a study, allelopathic compounds released from residues of barley apparently inhibit the emergence of yellow foxtail (*Setaria glauca*) (Creamer et al. 1996).

In another study, the use of sorghum plant tissues as a mulch or incorporated into the soil led to the reduced weed growth in corn field (Mohammadi et al. 2009). This can be attributed to the allelopathic compounds released from the sorghum plant tissues.

Use of an Allelopathic Crop in a Rotational Sequence

The entrance of allelopathic crops into the crop rotations can effectively control weeds. In a study, under reduced or no-till condition a considerable reduction in the population of giant foxtail (*Setaria faberii* Herrm.) was occurred when allelopathic soybean-corn-wheat rotation was followed than in corn alone (Schreiber 1992). In a 5–year field study with sunflower (*Helianthus annuus* L.)–oat rotation, the weed density increase was significantly less in sunflower plots than in control plots (Leather, 1983 a, b; 1987). It was found that sunflower plants possess chemicals, which inhibit the growth of common weed species. Macias et al., (1999) reported some sesquiterpene lactones with germacranolide and guaianolide skeletons and heliannuol from different cultivars of sunflower.

In another study, the inclusion of alfalfa in the crop rotation sequence significantly decreased the interference of weeds in the next crops (Entz et al. 1995). Ominski et al. (1999) conducted a survey in 117 fields in Manitoba, Canada, and found that rotation with alfalfa can effectively reduce the interference of wild oat (*Avena fatua* L.), Canada thistle (*Cirsium arvense* L.), wild mustard (*Brassica kaber* L.) and catchweed bedstraw (*Galium aparine* L.) in the succeeding cereal crops. Therefore, it can be concluded that the inclusion of alfalfa in crop rotation can be an efficient tool in an integrated weed management program. However, climatic and economic conditions are important limiting factors which can notably influence the regional crop rotation scenarios.

Application of Allelochemicals or Modified Allelochemicals as Herbicides

A promising way to use allelopathy in weed control is using extracts of allelopathic plants as herbicides (Dayan, 2002; Singh et al., 2005). Because biosynthesized herbicides are easily biodegradable, they are believed to be much safer than synthesized herbicides (Rice, 1984, 1995; Dayan et al., 1999; Duke et al., 2000). Duke et al. (2000) discussed that natural compounds have several benefits over synthetic compounds. For example, natural compounds may have novel structure due to diversity of molecular structure. This diversity is because synthetic chemists have been biased toward certain types of chemistry. They have had almost no interest in water-soluble compounds. Unlike a high proportion of synthetic pesticides, natural compounds are mostly water-soluble and non-halogenated molecules. Natural products relatively have short half-life and therefore considered safe of environmental toxicology standpoint (Duke et al., 2002).

Although, allelochemicals have the potential to be explored as natural herbicides, but prior to using them as herbicides, the following questions should be considered (Bhowmik and Inderjit 2003):

1. At what minimum concentration does each compound have phytotoxic activity?
2. Whether the compound is accurately separated and correctly identified?
3. What is the residence time and fate of the compound in the soil environment?
4. Does the compound influence microbial ecology and physicochemical properties of the soil?
5. What is the mode of action of the compound?
6. Has the compound any adverse effect on desired crops?
7. Whether the compounds are safe from health standpoint?
8. Whether the large production of the compound at commercial scale is economical?

Plant chemicals associated with allelopathic activity have been reported in most cases to be secondary metabolites from shikimic acid, acetate, or terpenoid pathways (Rizvi and Rizvi 1992; Vokou 2007). Some of the natural products exploited as commercial herbicides are triketone, cinmethylin, bialaphos, glufosinate and dicamba. The compounds having potential herbicidal activity but not commercially used are sorgoleone, artemisinin and ailanthone (Bhowmik and Inderjit 2003).

Sorgoleone is an allelochemical of sorghum which constitutes more than 80% of root exudate composition (Nimbal et al., 1996a; Czarnota et al., 2003). This compound inhibited the evolution of O_2 during photosynthesis in potato (*Solanum tuberosum* L.) and in common groundsel (*Senecio vulgaris* L.) (Nimbal et al. 1996a). Nimbal et al. (1996b) carried out a study on sorgoleone using triazine-susceptible potato and redroot pigweed thylakoids. Sorgoleone was a competitive inhibitor of atrazine binding sites. Sorgoleone also inhibited the photosystem II electron transport reactions (Gonzalez et al., 1997).

However, sorghum shoots produce higher amounts of cynogenic glucosides whose phenolic breakdown products inhibit plant growth (Einhellig and Rasmussen, 1989; Weston et al., 1989; Se´ne et al., 2001). In a study, Mohammadi et al. (2009) reported that the spray of sorghum shoot extract (Sorgaab) reduced weed infestation in corn field.

Artemisinin, a sesquiterpenoid lactone is an allelochemical of annual wormwood (*Artemisia annua* L.). It has been shown to inhibit the growth of redroot pigweed, pitted morning glory (*Ipomoea lacunosa*L.), annual wormwood and common purslane (*Purtulaca oleracea* L.) (Duke et al., 1987). Duke et al. (1987) concluded that artemisinin is a selective phytotoxin with herbicidal activity similar to cinmethylin (Bhowmik, 1988).

Ailanthone an allelochemical of *Ailanthus altissima* L. exhibited a strong herbicidal activity when sprayed on soil before the seed germination. It, however, also had dramatic effects when sprayed onto seedlings after their emergence from soil (Bhowmik and Inderjit 2003).

However, most of allelochemicals indicate a poor performance under field conditions compared to laboratory conditions. Moreover, many allelochemicals exhibit rapid dissipation under natural situations and thus fail to give desired results (Singh et al. 2003). Therefore, further studies are needed to enhance performance and stability of allelochemicals under field conditions.

Modification of Crops to Enhance the Allelopathic Effect

Breeding of crops for allelopathic ability by using the methods like screening and biotechnology is another promising strategy for efficient weed control. Just as crop plants are bred for disease resistance, crop plants can be bred to be allelopathic to weeds common to specific regions (Rice, 1984,1995; Jensen et al., 2001; Wu et al., 2000, 2003; Olofsdotter et al., 2002; He et al., 2004). Allelopathic effect against a broad spectrum of weeds has been proposed as a valuable character of an allelopathic crop and the possibility of inserting resistance genes towards one or several weeds as part of a breeding strategy of a crop has been mentioned (Olofsdotter et al. 1997).

Genetic modification of crop plants to improve their allelopathic properties and enhancement of their weed-suppressing ability has been suggested as a possibility (Kruse et al 2000). Use of biotechnological transfer of allelopathic traits between cultivars of the same species or between species has also been proposed (Chou 1999, Macias 1995, Macias et al. 1998, Rice 1984).

Several researchers have suggested improvement of allelopathic properties of crop cultivars by traditional breeding or by genetic manipulation. For example, there has been significant progress in isolating rice allelochemicals (Rimando et al., 2001) and locating genes controlling allelopathic effects of rice (Jensen et al., 2001). These researchers identified quantitative trait loci (QTL) associated with the rice allelochemicals against barnyardgrass. This is an important step toward breeding allelopathic rice varieties. It was found that 35% of the total phenotypic variation of allelopathic activity of population was explained by four main effect QTLs situated on three chromosomes.

In wheat, the control of hydroxamic acid accumulation seems to be multigenic involving several chromosomes. Chromosomes of group 4 and 5B are apparently involved in the accumulation of hydroxamic acids (Niemeyer and Jerez 1997).

In barley, a gramine synthesis gene has been detected on chromosome 5 (Yoshida et al. 1997).Moharramipour et al. (1999) reported that one or two genes control the synthesis of gramine. DIBOA is a hydroxamic acid compound which has been found in wild *Hordeum* species by Barria et al. in 1992 and the production of DIBOA by cultivated barley could possibly be achieved by transferring genetic material from wild barley species (Gianoli and Niemeyer 1998).

Duke et al. (2000) suggested that biotechnology may eventually allow for the production of highly allelopathic crops through the use of transgenes to increase allelochemical production to levels that effectively manage weeds without herbicides or with reduced herbicides input. However, it has been stated by Wu et al. (1999), that even though genetic manipulation seems promising, it might be more feasible to select for crop cultivars with improved allelopathic properties using conventional breeding methods, because of the strict regulation and public concern about transgenic crops.

Aromatic Plants

Aromatic plants could play an important role in the establishment of sustainable agriculture because of their ability to produce essential oils that could be used in the development of biological pesticides (Isman 2000). Eessential oils are increasingly adopted in agriculture for their use as pesticides (Daferera et al. 2003; Isman 2000; Tuncw and Sahinkaya 1998).

Palmer amaranth (*Amaranthuspalmeri* L.) germination was inhibited by essential oils of certain aromatic plants including lemon basil (*Ocimum citriodorum* L.), oregano (*Origanum vulgare* L.), and sweet marjoram (*Origanum majorana* L.) (Dudai et al. 1999). Dhima et al. (2009) found that anise, sweet fennel, lacy phacelia, and coriander aqueous extracts inhibited by 100% germination, root length, and seedling fresh weight of barnyardgrass. In another study, Tworkoski (2002) tested 25 plant-derived essential oils for herbicidal activity and found that those from red thyme (*Thymus vulgaris* L.), summer savory (*Satureja hortensis* L.), cinnamon (*Cinnamomum zeylanicum* L.) and clove (*Syzygium aromaticum* L.) were most toxic and caused cell death due to rapid electrolyte leakage on the detached leaves of dandelion (*Taraxacum officinale* L.).

COVER CROPS AND LIVING MULCHES

The term cover crop refers to a plant which is grown in rotation during periods when main crops are not grown. Cover crops are usually killed (mechanically or chemically) before the planting of the main crop. However, living mulches are cover crops that are planted between the rows of a main crop and are maintained as a living ground cover during the growing season of the main crop. Although, living mulches are sometimes referred to as cover crops, they grow at least part of the time simultaneously with the main crop. Apart from the definitions, both cover crop and living mulch suppress weeds by the similar mechanisms.

In general, cover cropping systems have high potentials for weed management in agroecosystems. Cover cropping has long and short-term weed control effects (Barberi 2002) as a result of competition and/or allelopathy exerted by the crop (Randall et al. 1989; Boydston and Hang 1995). These effects can enhance the effectiveness of other non-chemical weed control means in view of an effective integrated approach (Creamer et al. 1996; Bond and Grundy 2001). Long-term weed control effects are due to the prevention of emergence and/or seedling suppression of species of different seasonality compared to the following crop, while short-term effects take place when emergence prevention and seedling suppression occur in species presenting the same seasonality of the following crop (Campiglia et al. 2009).

How can a Cover Crop or Living mulch affect Weed Growth?

The effects of a cover crop or living mulch are achieved by a rapid occupation of the open space between the rows of the main crop or generally, the niches that would normally be filled by weeds (Teasdale 1998). This prevents germination of weed seeds and reduces the growth and development of weed seedlings. Generally, the weed suppressing ability of these systems is thought to be based on alleopathic properties, physical impedance of germination and seedling growth, and competition for light, water, and nutrients (Teasdale, 1993; Teasdale and Mohler, 1993).

Germination of weed seeds may be inhibited by complete light interception (Phatak, 1992) by cover crops or by secretion of allelochemicals (White et al., 1989; Overland, 1966). A delay in emergence of weeds because of the presence of living mulches or cover crops can also adversely affect weed seed production. Moreover, the presence of living mulches or cover crops leads to greater seed mortality of weeds by favoring predators (Cromar et al. 1999).

Once established, cover crops and living mulches can use the light, water, and nutritional resources that would otherwise be available to weeds. This can result in the inhibition of weed seed germination and reduction in the growth and development of weed seedlings. Therefore, weeds attempting to establish along with a cover crop or living mulch would be in competition for resources and may not develop sufficiently. Moreover, physical impediment to weed seedlings is another mechanism by which these crops suppress weeds (Facelli and Pickett 1991; Teasdale 1996; Teasdale and Mohler 1993).

Since, most crop-living mulch systems are sufficiently supported by water and nutrients, it seems that light is the most important resource for competition between living mulches and weeds. In a study,Kruidhof et al. (2008) found that weed suppression is positively correlated to early light interception by the living mulch and is sustained by the strong negative correlation between cumulative light interception and weed biomass. Similarly, Steinmaus et al. (2008) reported that weed suppression was linked to light interception by the mulch cover for most weed species.

Allelopathy is another mechanism by which living mulches may suppress weeds (Fujii 1999). However, this is difficult to separate experimentally from mechanisms relating to competition for growth resources. Allelopathic compounds can be released into the soil by a variety of mechanisms that include decomposition of residues, root exudation, and volatilization (Weston 2005). According toWestra (2010) root exudation produces allelopathic compounds that are actively secreted directly into the soil rhizosphere by living root systems. The allelochemicals then move through the soil by diffusion and come into contact neighboring plants. This creates a radius effect, where proximity to the allelopathic species results in greater concentrations of the allelochemical, which, in turn, typically decreases the growth of neighboring plants.

Factors Influencing the Weed Suppressive Ability of a Cover Cropping System

The success of a cover cropping system to suppress weeds can be influenced by some factors including:

Cover Crop Species

The variability among cover crop species determines the importance and opportunities of species selection as a component in the design of a suitable weed management system (den Hollander et al. 2007). Cover crop species are significantly different in their ability to suppress weeds. In a study, among six leguminous species (Persian clover, *Trifolium resupinatum* L.; white clover, *T. repens* L.; berseem clover, *T. alexandrinum* L.; hairy vetch; alfalfa; and black alfalfa, *M. lupulina* L.), the lowest weed dry weights were obtained from the plots interseeded with hairy vetch as compared with the other species (Mohammadi 2009).

Morphological growth characteristics, such as early relative growth rate of leaf area and earliness of height development, have been identified to determine competition in intercropping systems (Kropff and van Laar 1993). Weed suppression benefits from a rapid soil cover, as this reduces the germination and establishment of weeds as well as the relative competitive ability of established weed seedlings (Ross and Lembi, 1985). The relative growth rate reflects the increase of characteristics like soil cover and dry matter accumulation during early development, when growth is still exponential. The relative growth rate of a plant species is thus affected by its light capturing ability, by the efficiency by which it converts light into biomass and by the fraction of newly produced biomass which is invested in leaves (den Hollander et al. 2007).

For weed competition and weed suppression, earliness has also been reported an important characteristic (De Haan et al., 1994). Particularly for competition for light, which is asymmetric (Weiner, 1986), obtaining a good starting position seems highly relevant. From that perspective the relative growth rate seems to be a more important characteristic than the maximum accumulated amount of biomass (den Hollander et al. 2007).

Apart from soil cover development, height is also an important characteristic, determining competition for light (Berkowitz, 1988). Akanvou et al. (2001) found that *Crotalaria juncea*, *Cajanus cajan* and*M. pruriens* can be considered as species with a higher competitive ability than *Calopogonium mucunoides*, *Stylosanthes hamata* and *A. histrix*. This was explained by the combination of high initial growth rates for height and leaf area development. Additionally, the high final height of *C. juncea* and*C. cajan* may confer higher competitiveness throughout the growing season.

However, differences in soil cover development do not only depend on species differences in morphology and physiology. The relative starting position, determined by, for instance, seed size, seeding rate and date of emergence, is another major factor in this respect. As small-seeded species may be more sensitive to conditions that might cause a poor establishment (den Hollander et al. 2007).

In general, ideal cover crops or living mulches for weed suppression should have the following characteristics:

1. Ability to provide a complete ground cover of dense vegetation.
2. Rapid establishment and growth that develops a canopy faster than weeds.
3 .Selectivity between suppression of weeds and the associated crop (Teasdale 2003).

Agricultural Practices

Among agricultural practices, both time and rate of living mulch or cover crop planting can be important factors determining the success of a these systems to suppress weeds. In a study,Mohammadi (2010) observed that increased hairy vetch planting rate from 0 to 50 kg ha^{-1} improved corn yield (by 11%) and reduced weed dry weight (by 50.9%). In another study, increasing berseem clover (as a living mulch) planting rate from 20 to 40 kg ha^{-1} reduced weed density and biomass by 20.2 and 10.1% respectively, in corn field (Mohammadi et al. 2012). It was hypothesized that as living mulch density is increased, canopy closure would occur more rapidly, decreasing the amount of photosynthetically active radiation (PAR) available beneath the canopy. This would result in a concomitant decrease in weed biomass until an optimum living mulch density is achieved, beyond which, no further decrease in weed biomass could be obtained (Collins et al. 2008).

Planting time of a cover crop or living mulch is also a very important factor which affects its weed suppressive ability. For example, when rye (*Secale* sp.) or small-grain living mulches were interseeded at or near planting of the main crop, they could provide higher levels of weed suppression (Rajalahti et al. 1999; Brainard and Bellinder 2004). Mohammadi et al. (2012) also reported that the lowest weed density and biomass were occurred in the treatment in which berseem clover (as a living mulch) was interseeded 15 days before corn planting as compared with the other interseeding times (simultaneous with or 15 days after corn planting). This may be related to the faster occupation of the open space between the rows of the main crop and consequently a more efficient use of the environmental resources by the living mulch which can ultimately lead to the reduced weed growth and development.

Generally, the biomass produced by a living mulch highly depends on its planting rate and time. Moreover, there is often a negative correlation between living mulch and weed biomasses (Akemo et al. 2000; Ross et al. 2001; Sheaffer et al. 2002). Similar findings were also reported by other researchers

(Meschede et al. 2007; Mohammadi 2010; Mohammadi et al. 2012).

However, an important concern on cover cropping systems is reduced main crop yield due to competition or allelopathic effects of living mulches or cover crops. Therefore, appropriate management of these systems is very critical to reduce harmful effects of cover crops or living mulches on main crop while allowing them to grow sufficiently to reap potential benefits. The selection of suitable cover crop and living mulch species is very important. Ideally, the main crop and the cover crop should differ to a high degree in the way they explore resources, thus avoiding competition between both species to at least some extent (Vandermeer, 1989). Generally, greater potential benefits might be expected from living mulches with a very different active growth period than the main crop.

Other attempts have also been made to reduce the unsuitable effects of the cover crops while maintaining their weed suppressing ability. Brandsaeter and Netland (1999) focused on temporal complementarity by separating periods of vigorous growth of the cover crop and the main crop, whileVrabel (1983) used chemical control of the cover crop to reduce yield losses. Brainard et al. (2004) evaluated different options, particularly, cover crop species, time of seeding, use of supplemental nitrogen and herbicide regulation. Ross et al. (2001) conducted mechanical control of the cover crop and combined this with a screening of different cover crops.

It can be concluded that, although, inclusion of cover crops and living mulches in cropping systems can be useful to reduce harmful effects of weeds, but an appropriate management program is very essential to obtain the best results.

(For more information on the weed suppressing role of living mulches and their management in cropping systems, please see Mohammadi (2012), Living mulch as a tool to control weeds in agroecosystems: A Review).

PLANTING ARRANGEMENT

An integrated weed management approach should employ multiple control strategies (Walker and Buchanan 1982), possibly including the development of weed suppressing cropping systems (Shrestha and Fidelibus 2005). Alteration of planting arrangement can be proposed as an efficient practice to suppress weeds in agroecosystems. This can be achieved by the change of planting density, row spacing, row orientation, etc.

Planting Density and Row Spacing

The practice of increasing crop plant density by using higher seeding rates associated with narrower row spacing can lead to earlier canopy closure, thus shading weeds in their early developmental stages (Vera et al. 2006). Sharma and Angiras (1996 a,b) and Angiras and Sharma (1996) found that reduced row spacing increased light interception by crops and reduced weed biomass, increasing crop yield. The studies conducted on barley (*Hordeum vulgare* L.) have shown that higher seeding rates using cultivars with differing competitive abilities enhanced crop competitiveness against wild oat (*Avena fatua* L.) (Harker et al. 2009; Watson et al. 2006; O'Donovan et al. 2000).

In general, increasing crop seeding rates can hasten and increase resource use, and thereby reduce the negative effect of weeds (Berkowitz 1988; Mohler 1996). Therefore, weed management and cereal and pulse crop yields were improved with higher than recommended seeding rates in the absence of herbicides (Barton et al. 1992; Kirkland 1993; Townley-Smith and Wright 1994; Khan et al. 1996;Ball et al. 1997; O'Donovan et al. 2000). In another study, tartary buckwheat (*Fagopyrum tataricum*) was effectively suppressed when canola (*B. rapa*) seeding rate was increased from 2 to 8 kg ha^{-1}(O'Donovan and Newman 1996).

Weed suppression by crops appears to be enhanced by size-asymmetric competition, in which the larger crop plants suppress the initially smaller weed plants (Schwinning and Weiner 1998; Weiner 1990). At high-density, size-asymmetric competition is stronger and starts earlier, whereas the crop still has a large size advantage. At relatively low crop densities, crop cover early in growing season is low, leaving a larger amount of resources available for the weeds, thus enabling them to establish and grow quickly (Kristensen et al. 2008).

Row spacing can also affect the crop competitive ability against weeds. In a study, rice grown in 30-cm rows had greater weed biomass and less grain yield than in15-cm and 10–20–10-cm rows and crops in the wider spacing (30-cm) were vulnerable to weed competition for the longest period (Chauhan and Johnson 2011). In another study, Mohammadi et al. (2012) reported that corn yield improved and weed biomass reduced in response to increasing plant density and decreasing row spacing.

Row spacing can also influence the critical period of weed control in crops. It is hypothesized that narrow row spacings may decrease the interval of critical weed competition periods (Chauhan and Johnson 2011). According to Chauhan and Johnson (2011) the critical weed-free periods for rice planted at the 30-cm rows were up to 8 days longer than the other two rows spacings (15-cm and 10–20–10-cm rows).

Moreover, several studies have documented the reduced competitive ability of short-stature cultivars (Harker et al., 2009; O'Donovan et al., 2000) and improvements in the competitive ability of shorter varieties could be derived from narrower row spacing (Drews et al., 2004).

In general, the higher weed densities typical in low-input and organic systems may make narrow row spacing and higher planting density particularly attractive.

Row Orientation

Light is an important determinant of crop productivity. Crops can be manipulated to increase shading of weeds by the crop canopy, to suppress weed growth, and to maximize crop yield (Borger et al. 2010). In general, cropping systems that reduce the quantity and quality of light in the weed canopy zone suppress weed growth and reduce competition (Borger et al. 2010; Crotser and Witt 2000; Rajcan et al. 2002; Sattin et al. 1994; Shrestha and Fidelibus 2005; Teasdale 1995). During early growth stages, there is interference between crop and weed plants because of reflected light. The reflection of far-red photons by the stem of one plant lowers the red to far red photon ratio of light experienced by the stems of neighboring plants. This modifies the light environment in the plant stem tissue, which results in an increased stem elongation rate. As plants age, the crop canopy closes, and mutual shading further increases the competition for photosynthetic light (Borger et al. 2010).

One possible way to reduce light interception by weeds and to increase light interception by the crop canopy is to manipulate the crop row orientation (Holt 1995). So that, orientating crop rows at a near right angle to the sunlight direction increases the shading of weeds between the rows. In a study, within wheat and barley crops oriented east–west, weed biomass was reduced by 51 and 37%, and grain yield increased by 24 and 26% (compared with crops oriented north–south). This reduction in weed biomass and increase in crop yield likely resulted from the increased light (photosynthetically active radiation) interception by crops oriented east–west (i.e., light interception by the crop canopy as opposed to the weed canopy was 28 and 18% greater in wheat and barley crops oriented east–west, compared with north–south crops) (Borger et al. 2010).

According to Alcorta et al. (2011), rows oriented east-west allowed less light penetration to the weed canopy zone than north-south rows throughout the growing season and weed species responded to low light levels by producing leaves with larger specific leaf area and leaf area ratios than those in the north-south rows. Moreover, the leaf, stem and root dry weight of the weed species in the east-west rows was reduced by 30% compared to the weed species in north-south rows.

According to the results of another study, during periods of peak PAR, the Red : Far Red (R : FR) ratio was more than three times greater under the grape canopy in north-south rows than in east-west rows, indicating that row orientation can affect both quantity and quality of light available to weeds (Shrestha and Fidelibus 2005).

However, the effect of row orientation can vary with latitude and with the seasonal tilt of the earth in relation to the sun. Near the equator, north–south (as opposed to east–west) orientation gives crops higher levels of light absorption for most of the year. At higher latitudes (up to 55°), absorption is highest in north–south crops in summer and east–west crops for the rest of the year. From 65°upwards, east–west orientation gives greatest light absorption all year (although the difference between orientations is minor) (Mutsaers 1980).

It can be concluded that manipulation of row orientation can be an ideal method to incorporate into an integrated weed-management program because it does not cost growers anything to implement, and it is environmentally friendly compared with chemical weed control tactics (Mohler 2001). However, the geographical and seasonal conditions should be considered.

Spatial Uniformity

According to Kristensen et al. (2008) increased crop density and spatial uniformity can play an important role in weed management and a strategy based on increased crop density and spatial uniformity can reduce or eliminate herbicide application in conventional cereal production. Crop spatial uniformity decreases competition within the crop population early in the growing season (Olsen and Weiner 2007) and maximizes the total shade cast by the crop by reducing self-shading (Weiner et al. 2001). In a study, In the presence of weeds, the highest yields were obtained with high crop density and high spatial uniformity (Kristensen et al. 2008).

However, the early size advantage of the crop is the theoretical basis for our prediction of positive effects of increased density and spatial uniformity on weed suppression (Weiner et al. 2001). Therefore, it can be concluded that increased crop density and uniformity will not lead to effective weed suppression when weeds have the initial size advantage (e.g., perennial weeds), or are able to catch up in size with the crop before competition becomes intense (Kristensen et al. 2008). Moreover, one might expect the effects of high crop density and spatial uniformity on weeds to be more pronounced at low soil nitrogen levels because weeds grow more slowly at low fertilization levels (Blackshaw et al. 2003).

COMPETITIVE CROP CULTIVARS

In a plant community, competition occurs when the environmental resources are limited. Competition for limited resources is the primary causes of crop loss from weeds. Crop cultivars that better compete with weeds or preempt resources from weeds may benefit an integrated weed management program (Jordan 1993; Lemerle et al. 1996; Lindquist and Kropff 1996).

The competitive ability of a plant has two components, the competitive effect—ability of an individual to suppress other individuals—and the competitive response—ability of an individual to avoid being suppressed—corresponding to different abilities of plants to acquire and use resources (Goldberg, 1990). Competitive effect is related to resource acquisition, with large or tall plants competitively depressing smaller ones (Gaudet and Keddy, 1988; Keddy and Shipley, 1989; Aarssen, 1992). Plants avoid being suppressed by acquiring resources faster (foraging strategy), shifting resource acquisition site or time relative to neighbors (escaping strategy), or conserving scarce resources (persistent strategy) (Navas and Moreau-Richard 2005).

The development of competitive crop cultivars is an important aspect of integrated weed management and can reduce reliance on herbicides (McDonald 2003). The ideal weed competitive cultivars are high-yielding under both weed-free and weedy conditions and have strong weed-suppressive ability. Weed-suppressive ability is the ability to suppress weed growth and reduce weed seed production and, hence, benefit weed management in the subsequent growing season (Jannink et al., 2000; Zhao et al., 2006).

In a general view, crop competitive ability can be divided into two practical perspectives. Crop tolerance is defined as the ability of the crop to endure competitive stress from the presence of weeds without substantial reduction in growth or yield. Weed suppressive ability is the ability of the crop to reduce weed growth and fecundity (So et al. 2009). Weed suppressive ability is determined by assessing weed biomass or weed seeds under weedy conditions. Stronger Weed suppressive ability is not always associated with higher yield under weedy conditions (Saito et al. 2010). However, suppressing weeds reduces weed seed production and benefits weed management in future grow- treating seasons while tolerating weeds only benefits the current growing season. Moreover, weed pressure from unsuppressed weeds increases the likelihood of crop yield loss, irrespective of the crop's tolerance (Jannink et al. 2000).

Ideally, a competitive cultivar should both tolerate weeds and suppress their growth (Jordan 1993). The tolerance of a crop cultivar to weeds is the ability of that cultivar to maintain high seed yields when weeds are present. The weed suppression ability of a crop cultivar is the ability of that cultivar to reduce weed growth and subsequent seed production (Spies et al. 2011). Callaway (1992) documents genetic variability for both perspectives in numerous crop species and many authors suggest breeding to improve the traits (Garrity et al., 1992; Callaway and Forcella, 1993; Kropff and van Laar, 1993;Wortmann, 1993; Liebman and Gallandt, 1997; Bussan et al., 1997).

In general, the traits offering weed competitive ability to crop cultivars can be divided into the several groups including:

Canopy and Morphological Traits

Canopy architecture influences many canopy processes including interactions between the crop and specific aspects of its environment (Daughtry et al., 1983; Welles and Norman, 1991). Canopy architecture is a function of leaf number, shape, distribution, orientation, and plant size, which collectively determine the vertical distribution of light within the crop canopy (Williams et al., 1968;Girardin and Tollenaar, 1994).

Sinoquet and Caldwell (1995) reported that light is the primary resource for which weeds will compete in an irrigated and high N-input crop production system. Total canopy leaf area index, height, rate of leaf area development and their distribution in the canopy, are the most important traits in competition for light (Sinoquet and Caldwell, 1995), which can be improved through cultural practices or/and by plant breeding (Lindquist and Mortensen, 1998). Identifications and improvements in traits driving light interception, such as height, leaf area index and canopy diameter can increase competitiveness of several crops (Bennett and Shaw 2000; Callaway 1992; Lindquist and Mortensen 1998; Lindquist et al. 1998). According to Watson et al. (2002) a species competitive strength is strongly determined by its share in leaf area when the canopy closes and interplant competition starts. In general, rapid canopy closure and a large, late-maturing canopy were positively associated with competitive ability (So et al. 2009).

In field pea cultivars, vine length and the leafy characteristic may be important genetic characteristics associated with competition (Spies et al. 2011). However, Wall and Townley-Smith (1996) believed that vine length was more important than the leafed or semi leafless trait. Several traits relate to competitive ability of dent corn, including plant height, shoot growth rate, canopy density (Lindquist and Mortensen 1998), leaf uprightness (Sankula et al. 2004), crop maturity, leaf area growth rate (Begna et al. 2001 a, b), canopy closure, and maximum leaf area index (Lindquist et al. 1998).

In an experiment, when wild-proso millet competed with sweet corn, hybrids with a large canopy were best equipped to tolerate the weed and suppress wild-proso millet growth and seed production, even for late-maturing hybrids that competed the longest period of time (So et al. 2009).

Plant height and tillering ability are also key characteristics for wed suppressive ability under specific growing environments. Their relative contributions to weed suppressive ability could be affected by crop establishment method, agro-ecosystems (upland or lowland) or weed species (Saito et al. 2010). According to Lemerle et al., (1996) greater tiller numbers, taller plants, elevated photosynthetically active radiation interception, and greater early season biomass accumulation were all found in the most competitive genotypes in a study of wheat genotypes from around the world.

Wang et al. (2006) also reported that an erect cowpea genotype is more competitive due to its taller stature, greater height growth rate, and higher position of maximal leaf area density, despite a lower photosynthetic rate and light use efficiency than the other cowpea genotypes. They concluded that erect growth habit may be generally more competitive with weeds compared to semi-erect or prostrate growth habit. In another study, the size of the flag leaf has been correlated with competitive ability in barley (Watson et al. 2002).

Phenological Traits

Weed species differ markedly in their development phenologies. An effective, 'broad spectrum' weed-suppressive cultivar will therefore need a strong competitive presence over the full duration of the season. The positive correlation between weed biomass and time to maturity of cultivars in organic fields indicates that weed growth was higher in cultivars with increased time to maturity. Thus, it may be desirable for organic producers to use early maturing cultivars to reduce weed biomass in the field (Jannink et al. 2001). In other words, faster time to maturity was found to be associated with reduced weed biomass. Huel and Hucl (1996) in evaluating 16 genotypes of spring wheat (*Triticum aestivum*L.), found a positive correlation between early maturity and competitive ability. In another study, earlier flowering cultivars of soybean were more successful at suppressing weed growth.

In general, earlier maturing cultivars might have higher relative growth rate. Across a broad spectrum of species, researchers have found a negative relationship between leaf longevity and plant relative growth rate (Reich et al., 1997). They posit that this relationship occurs because long-lived leaves require more structural carbon and protective secondary metabolites than short-lived leaves, such that the cost to the plant of developing photosynthetic capacity is greater (Jannink et al. 2001).

Moreover, earlier maturing cultivars might produce larger seeds if they diverted a greater proportion of photosynthate to their fewer reproductive structures (Kollman et al., 1979; Openshaw et al., 1979;Wallace et al., 1993). With a larger initial size they might achieve a higher absolute growth rate despite equal relative growth rate. This can lead to the higher competitive abilities of these cultivars.

However, this idea contrasts with that of some workers who have suggested that later maturity confers greater competitive ability against weeds because cultivars that remain vegetative grow to be taller (Hinson and Hanson, 1962; McWhorter and Hartwig, 1972; Monks and Oliver, 1988). Jannink et al. (2000) also found earlier-maturity soybean cultivars displayed greater initial growth and weed suppression, compared to later-maturity cultivars, but were less able to sustain weed suppression throughout the season due to senescence. It can be concluded that earlier maturing cultivars have a higher weed suppressive ability if they can sustain this ability throughout the growing season.

Growth Parameters

In plant ecology, relative growth rate (RGR) is considered to be one of the key characteristics of plants that is positively correlated with competitive ability (Grime 1977; Grace 1990). Holt and Ocrutt (1991) showed that the RGR is one of the most important plant growth parameters to increase the competitive ability of cotton against weeds. As a result of a high RGR, a crop will rapidly increase in size and can occupy a larger space, both below and above ground. Consequently, such a crop has the opportunity to acquire a larger share of limiting resources, such as light, nutrients, and water, than a weed.

In a study conducted by Mohammadi (2007) some plant growth parameters including leaf area index, specific leaf area, crop growth rate, relative growth rate and net assimilation rate were evaluated to identify which of them can enhance corn competitive ability against weeds. The variable selection using the stepwise multiple linear regression method revealed that, among the growth parameters under study, both the relative growth rate and the specific leaf area (SLA) were the best predictors of corn cultivar competitiveness.

Broad surveys across taxa have found SLA to predict relative growth rate (Hunt and Cornelissen, 1997; Reich et al., 1997) and implicate SLA in competitive ability (van der Werf et al., 1993). A high SLA contributed to increased light interception by crops (Jannink et al. 2000) and led to a reduction in the amount of light available to weeds. According to Johnson et al. (1998), a high SLA is one of the best predictors of cultivar competitiveness against weeds. Dingkuhn et al. (1999) also reported that more weed competitive cultivars have a high SLA, which leads to a high leaf area index.

Tollenaar et al. (1994) showed that differences among corn hybrids in competition against weeds can be attributed to differences among them in the leaf area index and the transmission of late-season photosynthetic photon flux density. In the case of interactions with weeds, more rapid early leaf area development, higher leaf area index and biomass accumulation can play important roles. Leaf area and plant height might affect significantly crop–weed interactions (Blackshaw, 1994; Lemerle et al., 1996). Cultivars with strong weed suppressive ability accumulated more biomass, produced more tillers and displayed higher leaf area index during the vegetative growth stage than those with weak weed suppressive ability (Saito et al. 2010).

Plant height is another growth parameter which can influence weed suppressive ability of a crop cultivar. The negative correlation between weed biomass and plant height in organic fields implies that weed biomass decreased as height increased, suggesting that height does help to suppress weeds. In several studies, plant height was associated with competitive ability in both conventional (Huel and Hucl, 1996; Lemerle et al., 1996; Hucl, 1998) and organic systems (Gooding et al., 1993). In soybean, plant height 6 to 7 wk after emergence showed moderately high heritability, strong genetic correlation to weed suppressive ability and was quick and simple to measure. These characteristics make it an ideal indirect selection criterion, particularly in a practical soybean breeding program where labor needs at the time to measure early height are not as high as in the spring or the fall (Jannink et al. 2000).

In rice, under severe weed competition, higher biomass accumulation at 42 days after sowing was associated with higher weedy yield. For adaptation to both moderate and severe weed pressure, cultivars should have high-yielding ability, high plant height at maturity, and large biomass accumulation at 42 days after sowing (Saito et al. 2010). Dingkuhn et al. (1999) also considered relative yield (the ratio of grain yield under weedy conditions and grain yield under weed-free conditions) as an indicator of weed competitiveness and Rodenburg et al. (2009) showed that longer duration and higher yield under weed-free conditions were associated with higher grain yields under weedy conditions.

Seed Traits

In annual plants, and thus in many agricultural contexts, seed size can be an important determinant of success during initial stages of competition simply because larger seeds lead to greater initial growth and therefore to greater capture of available resources at the expense of other competitors (Black, 1958; Ross and Harper, 1972). Vigorous seed, expressed as early emergence and root growth contribute to cultivar competitiveness against weeds. In sweet corn, early vigor and seedling growth rate are largely attributed to endosperm phenotype, specifically, the result of starch concentration of the endosperm (Azanza et al. 1996).

In general, seed vigour within a genotype has been attributed to seed size, protein, which is in turn related to ATP production and ultimately, mitochondrial quality and quantity. Seed vigour has been positively related to both seedling vigour and final yield and can be improved to enhance crop competitive ability (Watson et al. 2002).

NUTRIENT MANAGEMENT

A suitable nutrient management program can be an effective tool to control weeds in cropping systems. The competitive relationship between crop and weeds is highly dependent on supply and availability of nutrients (Evans et al. 2003; Di Tomaso 1995). Manipulation of soil fertility, whether using organic or inorganic amendments should be considered as an important component of long-term weed management programs and effective fertilizer management is an important component of integrated weed management systems (Blackshaw et al. 2007; DiTomaso 1995). Unfortunately, nutrients applied to soils are also available for weeds. In most farming systems, competition for N is the most important source of nutrient interference (DiTomaso 1995). Walker and Buchanan (1982) also found that of all nutrients, plant response to nitrogen (N) fertilizer is the most widely observed and the manipulation of soil N supply offers the most promise in the short term as a means by which crop–weed competitive outcomes can be influenced.

Therefore, it is important to develop fertilization strategies for crop production that enhance the competitive ability of the crop, minimize weed competition, and reduce the risk of nonpoint source pollution from nitrogen (Cathcart and Swanton 2003; DiTomaso 1995).

Aspects of Nutrient Management

Different aspects of nutrient management including fertilizer rate, timing and application method can be successfully manipulated to reduce weed interference in crops (Angonin et al. 1996; Blackshaw et al. 2004; Van Delden et al. 2002).

Application Rate and Timing

Weed emergence and growth in the field can be stimulated by fertilizer application rate and timing. Generally, weed growth may increase as the nitrogen application rate increases, resulting in the need for more frequent POST herbicide applications or cultivation (Sweeney et al. 2008). In sugar beet, weed emergence from sown seed increased as the N application rate at planting increased from 56 to 224 kg N ha^{-1} (Dotzenko et al. 1969).

The timing of fertilizer application in early planted crops, such as sugar beet and corn, may especially influence the germination, emergence, and competitiveness of weeds that might otherwise remain dormant early in the growing season. In a study, when nitrogen was broadcasted in April at the time of planting, weed germination and emergence were stimulated. In contrast, nitrogen application at the time of planting in May did not influence seed germination and weed emergence because of greater N availability because of mineralization at this time of year or because seed germination has been stimulated by other environmental cues (Sweeney et al. 2008). Results of both greenhouse (Alka¨mper et al. 1979) and field experiments (Davis and Liebman 2001; Dyck et al. 1995) indicate that for certain crop-weed combinations, delaying soil N availability can shift the competitive balance to favor crop growth. For example, the competitiveness of wild mustard (*Sinapis arvensis* L.), a winter annual, in sugar beet was favored by early compared with late nitrogen fertilization (Paolini et al. 1999). Generally, delaying nitrogen applications, applying slow-release nitrogen fertilizers or placing nitrogen below the weed seed germination zone could be potential strategies for reducing early season weed establishment in cropping systems (Sweeney et al. 2008).

In the case of phosphorus, early-season application is critically important for vigorous plant growth and development (Grant et al. 2001). Thus, fertilization strategies that restrict weed access to phosphorus fertilizer early in the growing season would appear to have merit (Blackshaw and Molnar 2009).

Application Method

Crop-weed interactions can also influence by fertilizer application method. Since, weeds often germinate at or near the soil surface, especially in zero-tillage systems (Yenish et al., 1992; Hoffman et al., 1998), therefore, in this situation the greatest benefits may be realized by physically placing nitrogen (N) in an area of the soil profile where crop seeds, but not weed seeds, are germinating (Blackshaw 2005). Subsurface-banded N was often better than surface-broadcast N fertilizer in terms of N uptake by wheat vs. weeds, weed biomass production and wheat yield (Blackshaw et al. 2005).

Petersen (2003) reported that weed N uptake and weed biomass were 50% lower with subsurface-banded compared with surface-broadcast liquid swine manure. Rasmussen (2002) similarly documented lower weed biomass and higher crop yield with injected than with surface-broadcast liquid swine manure. In another study, subsurface-banded N compared with broadcast N fertilizer reduces N uptake by weeds and decreases weed growth and biomass. Moreover, banded N fertilizer resulted in the lowest seed bank numbers of both grass and broadleaf weeds (Blackshaw 2005).

Phosphorus (P) fertilization practices could also have an impact on the extent of interference by weeds. Researchers indicated that banding P near lettuce rooting system as opposed to broadcast P could potentially reduce the damage of spiny amaranth (*Amaranthus spinosus*), smooth pigweed, and common purslane by enhancing the competitive ability of the crop (Santos et al. 1997; Shrefler et al. 1994). In wheat, seed-placed or mid row-banded P compared with surface-broadcast P fertilizer often resulted in higher yields when wheat was in the presence of competitive weeds (Blackshaw and Molnar 2009).

In general, weed P concentration and biomass production were often greatest with surface-broadcast P fertilizer, indicating that this common application method of P fertilizer should be discouraged. Alternative practices such as seed-placed or subsurface-banded P fertilizer were less advantageous to weeds. Weed seed bank was also affected by P application method. So that, seedbank evaluation at the end of the experiment indicated that the seed density of five of six weed species under study was reduced with seed-placed or subsurface-banded P compared with surface-broadcast P (Blackshaw and Molnar 2009).

However, the benefit of seed-placed or subsurface-banded P fertilizer will likely be greatest in soils with low background P levels and within zero-tillage production systems, where weed seeds are not distributed throughout the soil profile but rather concentrated near the soil surface (Blackshaw and Molnar 2009).

Organic and Biofertilizer

Organic manure may affect crop–weed competitive interactions differently than chemical nitrogen fertilizer (Davis and Liebman, 2001), probably due to speed of N release or form of N. In a study, the gradual N release from manure and compost over years appeared to benefit weeds more than spring wheat. Moreover, fresh and composted manure had the greatest seed bank of both grass and broadleaf species as compared with chemical fertilizer treatment (Blackshaw 2005). In another study,Mohammadi et al. (2012) found that phosphate biofertilizer had no significant effect on corn yield, whereas, weed biomass was notably increased when phosphate biofertilizer was applied.

It seems that, in most cases, weed infestation level and duration may enhance by the use of organic fertilizers. However, the other beneficial aspects of these fertilizers should not be ignored.

Critical Period of Weed Control in Response to Nutrient Management

The critical period of weed control (CPWC) is an important principal of an integrated weed management program. It is a period in the crop growth cycle during which weeds must be controlled to prevent yield losses (Knezevic et al. 2002). Weeds that are present before or emerge after this period do not cause significant yield loss. Studies on the critical period of weed control are important in making weed control recommendations because they indicate the optimum time for implementing and maintaining weed control and reduce cost of weed control practices (Hall et al. 1992; Van Acker et al. 1993).

According to Weaver et al. (1992) the manipulation of edaphic factors including the alteration of soil nutrient supply can influence the crop-weed interference relationships, especially in determining the critical time of weed removal (the start of the critical period). Evans et al. (2003) reported that the addition of nitrogen fertilizer delayed the beginning and hastened the end of the critical period of weed control in corn. Their study showed that the effect of nitrogen fertilization on early season crop growth provided a competitive advantage for corn relative to weeds. In another study, Mohammadi and Amiri (2011) found that the use of mono ammonium phosphate as a starter fertilizer slightly delayed the end of the CPWC in soybean (by 5 days), but this condition shortened the CPWC by 12 days because of the later beginning of the CPWC (by 17 days).

Weed Response to Fertilization

Some weed species are considered to be luxury consumers of nutrients (Qasem 1992; Teyker et al. 1991) and this might contribute to their ability to take up higher amounts of N at higher N fertilizer rates. Weeds not only reduce the amount of N available to crops but the growth of many weed species is enhanced by higher soil N levels (Blackshaw et al. 2003; Henson and Jordan 1982; Supasilapa et al. 1992). Thus, adding N fertilizer in cropping systems can potentially have the unintended consequence of increasing the growth and competitive ability of weeds more than that of the crop.

In a greenhouse study, Teyker et al. (1991) reported greater N uptake for redroot pigweed (*Amaranthus retroflexus* L.) than corn when the addition of N was elevated, suggesting that redroot pigweed interference in corn may be greater at higher levels of N. Others also have postulated that weeds may be more competitive when fertility is enhanced with N addition because of the superior uptake efficiency of many weed species (Di Tomaso 1995; Sibuga and Bandeen 1980).

At the conclusion of a 47-yr soil fertility study, carpetweed (*Mollugo verticillata* L.) and henbit (*Lamium amplexicaule* L.) densities were greatest on plots that had received annual applications of P fertilizer (Banks et al. 1976). In another study, downy brome (*Bromus tectorum* L.) densities were reported to be higher on soils with higher P levels (Belnap et al. 2003). Verma et al. (1999) similarly reported that weed growth and competitiveness with fenugreek (*Trigonella foenum-graecum* L.) increased at the higher soil P levels.

Weed germination and dormancy are also influenced by fertilizer application. For example, germination of common lambsquarters seed from mother plants that received 280 kg ha^{-1} of ammonium nitrate was greater than germination of seed from a mother plant where no N was applied, suggesting that N deficiency increased dormancy in seeds (Baskin and Baskin 1998; Fawcett and Slife 1978). According to Cairns and de Villiers (1986) the dormancy of several grass weed species was broken by ammonia, but the gas had no effect on the dormancy of dicotyledonous weed seed. Redroot pigweed (*Amaranthus retroflexus* L.) seed germination was also stimulated by 10 to 100 ppmv of ammonium nitrate or urea (Sardi and Beres 1996). Other researchers found that the chilling or light requirement for seed germination in some species can be replaced with N, particularly nitrate (Cohn et al. 1983; Egley and Duke 1985; Sexsmith and Pittman 1963; Steinbauer and Grigsby 1957).

Generally, fertilizer management strategies that favour crops over weeds deserve greater attention when weed infestations consist of species known to be highly responsive to higher soil nutrient (e.g. N) levels. In these situations farmers should consider the benefits of specific fertilizer timing and/or placement methods that would minimize weed interference (Blackshaw et al. 2004; Kirkland and Beckie 1998; Mesbah and Miller 1999).

BIOLOGICAL CONTROL

Biological control of weeds refers to the use of any kind of organism (micro or macro) to suppress weeds and reduce their harmful effects in agroecosystems. Plant pathogens are potentially valuable additions to the arsenal of weapons for use against weeds.

Weed Biological Control Approaches Using Pathogens

Biological control of weeds using pathogens can be considered from two broad approaches including classical biological control (CBC) and inundative biological control (IBC) (Mohan babu et al. 2003).

Classical Biological Control

This approach is fairly simple in its concept: discover effective and highly host-specific agents from the weed's native geographic range, confirm their safety and effectiveness by rigorous experimental evaluation, and introduce them into regions where the weed has been newly introduced and requires control (Charudattan and Dinoor 2000). Host specificity tests provide the information on which to base the risk assessment and, thereby play the central role in any CBC project (Mohan babu et al. 2003).

Classical biological control by means of pathogens has been used in several parts of the world to control exotic weeds (Bruckart and Hasan 1991; Watson 1991). One of the most successful examples of classical biological control of weeds is the introduction of a rust fungus, *Puccinia chondrillina*, into Australia to control rush skeleton weed (*Chondrilla juncea*). A plant of Mediterranean origin, it became a serious weed in Australian cereal crops. The fungus, also from the Mediterranean was introduced along with three insects, as a classical biocontrol agent. Following the introduction and establishment, the fungus disseminated rapidly and widely and controlled the most common biotype of the weed (Cullen, 1985).

Other successful examples of classical biocontrol programs include the use of a smut fungus, *Entyloma ageratinae*, imported from Jamaica to control Hamakua pamakani (*Ageratina riparia*, Asteraceae) in Hawaiian forests and rangelands (Trujillo, 1985) and three other rust fungi, *Puccinia carduorum*, imported from Turkey and released into northeastern United States to control musk thistle (*Carduus thoermeri*) (Baudoin et al., 1993), *Phragmidium violaceum* to control weedy species of *Rubus* in Chile (Oehrens, 1977) and Australia (Bruzzese, 1995) and *Uromycladium tepperianum*, to control an introduced invasive tree species, *Acacia saligna* in South Africa (Morris, 1997). The last fungus causes extensive gall formation on branches and twigs accompanied by a significant energy loss. Heavily infected trees are eventually killed (Charudattan and Dinoor 2000).

However, there is a potential problem in using biological control, namely, a shift in the weed population toward more resistant weed biotypes. Although, it also illustrates the possibility to counter the presence of natural resistance in weed populations by the introduction of new pathogen strains (Charudattan and Dinoor 2000).

Inundative Biological Control

The strategy of inundative biological control is to simulate natural epiphytotics of a selected pathogen within the population of the target weed species, early in the season and thus kill or at least significantly reduce the competitive ability of the weed and so prevent crop losses. This approach is typically used against endemic weeds, in which indigenous pathogens are mass-produced and applied as formulated products (bioherbicides) (Mohan babu et al. 2003). A bioherbicide is defined as a plant pathogen used as a weed-control agent through inundative and repeated applications of its inoculum (Charudattan and Dinoor 2000). The specificity of bioherbicides is considered as a positive attribute (Mohan babu et al. 2003).

Some examples of registered bioherbicides consisted of DeVine®, composed of a Florida isolate of*Phytophthora palmivora*, is used for the control of *Morrenia odorata* (stranglervine or milkweed vine) in citrus in Florida. Collego®, based on *Colletotrichum gloeosporioides* f. sp. *aeschynomene*, is used to control *Aeschynomene virginica* (northern jointvetch), a leguminous weed in rice and soybean crops in Arkansas, Mississippi and Louisiana. BioMal®, registered in Canada for the control of *Malva pusilla*(round-leaved mallow), containing *Colletotrichum gloeosporioides* f.sp. *malvae*, is presently unavailable for commercial use. The fourth bioherbicide, Dr. BioSedge®, based on the rust fungus*Puccinia canaliculata* and registered for the control of *Cyperus esculentus* (yellow nutsedge) in the United States, is also unavailable for commercial use. An isolate of *Xanthomonas campestris* pv. *poae*, a wilt-inducing bacterium, isolated in Japan from *Poa annua* (annual bluegrass or wintergrass), is registered in Japan as the bioherbicide CAMPERICO® to control annual bluegrass in golf courses (Charudattan and Dinoor 2000).

Bioherbicides can make a significant contribution to weed control in the future, once the well-documented constraints have been overcome, particularly through improved target selection, formulation and marketing. The over-riding concern in using plant pathogens for weed control is their potential threat to non-targets (Mohan babu et al. 2003) which needs a serious attention.

Microbial-Derived Herbicides

Microbial-derived herbicides, especially microorganism secondary metabolites are a new kind of microbial herbicide to control weeds which are always phytotoxins. They are very different in chemical structure and size. These bioactive components invade into the host plant, cause pathogenicity, destroy their structure and lead them to produce necrotic lesions or chlorotic halo (Li et al. 2003). They are less poisonous to most of mammalian systems, easily degraded and so far result in no biological disaster compared to chemical herbicides (Charudattan, 1991).

Phytotoxins used for microbial herbicides can be divided into three types: bacterial, fungal and actinomycete derived product. The pathogens which produce phytotoxins as a microbial herbicide must fit certain requirements: (1) be reproduced by biological technique, (2) grow fast after spraying or be capable of killing weeds within definite time, (3) suit industrial production and (4) be suitable for packaging, transport and use (Li et al. 2003).

In comparison with fungi, bacteria have some advantageous characteristics such as short-growth period, simple fermentation technique and easily controlled production process. In addition, bacteria can produce secondary metabolites unlike fungal spores, which need strict conditions for action as herbicides and their residues are easily degraded. Bacterial herbicides have a good prospect in application and exploitation (Li et al. 2003). Most of the bacteria with an ability to produce toxins are Gram-positive bacteria such as *Pseudomonas*, *Erwinia*, *Xanthomonas* but there are a few Gram-negative bacteria such as *Streptomyces*, *Corynebacterium fasciomonads* and some are non-fluorescent*pseudomonads* (Kremer et al., 1990).

Two phytotoxins from actinomycetes including herbicidines and herbimycins are higher-plant toxins and produced by *Streptomyces saganonensis*. The former is used to control grassy weeds in paddy field as a selective herbicide, the latter controls monocotyledonous and dicotyledonous weeds (Stephen and Lydon, 1987).

However, the role of biomicrobial herbicides in agriculture is still problematic and insignificant (Mohan babu et al. 2003). Although, the current emphasis on lowering use of chemical herbicides may increase the production and use of biological-based herbicides in the future.

In general, the use of alternative weed control strategies can prevent or reduce the chemical herbicide application. This can lead to less reliance on fossil fuels (the non-renewable energy resources), lower environmental degradations and consequently a higher degree of sustainability for agroecosystems.

REFERENCES

1. Aarssen, L.W., 1992. Causes and consequences of variation in competitive ability in plant communities. J. Veg. Sci. 3, 165–174.
2. Akanvou, R., L. Bastiaans, M. J. Kropff, J. Goudriaan and M. Becker. 2001. Characterization of Growth, Nitrogen Accumulation and Competitive Ability of Six Tropical Legumes for Potential Use in Intercropping

Systems. Journal of Agronomy and Crop Science 187, 111-120.

3. Akemo, M.C., Regnier, E.E. and Bennett, M.A. 2000. Weed suppression in spring-sown rye-pea cover crop mixes. Weed Technology 14, 545-549.
4. Alcorta M., Fidelibus M. W., Steenwerth K. L. and Shrestha A. 2011. Effect of Vineyard Row Orientation on Growth and Phenology of Glyphosate-Resistant and Glyphosate-Susceptible Horseweed (*Conyza canadensis*). Weed Sc. 59(1):55-60.
5. Alka¨mper, J., E. Pessips, and D. V. Long. 1979. Einfluss der Du°ngung auf die Entwickelung und Nahrstoffnahme verschiedener Unkra°uter in Mais. European Weed Research Society Symposium, Mainz, Germany. Paris: European Weed Research Society.
6. Angiras, N. and V. Sharma. 1996. Influence of row orientation, row spacing and weed-control methods on physiological performance of irrigated wheat (*Triticum aestivum*). Indian J. Agron. 41:41–47.
7. Angonin, C., J. P. Caussanel, and J. M. Meynard. 1996. Competition between winter wheat and *Veronica hederifolia*: influence of weed density and the amount and timing of nitrogen application. Weed Res. 36:175–187.
8. Azanza, F., A. B. Zur, and J. Juvik. 1996. Variation in sweet corn characteristics associated with stand establishment and eating quality. Euphytica 87:7–18.
9. Ball, D. A., Ogg, Jr., A. G. and Chevalier, P. M. 1997. The influence of seeding rate on weed control in small-red lentil (*Lens culinaris*). Weed Sci. 45: 296–300.
10. Banks, P. A., P. W. Santelmann, and B. B. Tucker. 1976. Influence of long-term soil fertility treatments on weed species in winter wheat. Agron. J. 68:825–827.
11. Barberi, P., 2002. Weed management in organic agriculture: are we addressing the right issues? Weed Res. 42, 177–193.
12. Barnes, J. P., Putnam, A. R. and Burke, B. A. 1985. Allelopathic activity of rye (*Secale cereale* L.). In The Science of allelopathy (ed. A. R. Putnam and C. S. Tang), pp. 271-286. Wiley, New York.
13. Barnes, J.P., Putnam, A.R., 1983. Rye residues contribute to weed suppression in no–tillage cropping systems. J. Chem. Ecol. 9, 1045–1057.
14. Barton, D. L., Thill, D. C. and Shafii, B. 1992. Integrated wild oat (*Avena fatua*) management affects spring barley (*Hordeum vulgare*) yield and economics. Weed Technol. 6: 129–135.
15. Baskin, C. C. and J. M. Baskin. 1998. Seeds: Ecology, Biogeography, and Evolution of Dormancy and Germination. New York: Academic.
16. Baudoin, A.B.A.M., Abad, R.G., Kok, L.T., Bruckart, W.L., 1993. Field evaluation of *Puccinia carduorum* for biological control of musk thistle. Biol. Control 3, 53-60.
17. Begna, S. H., R. I. Hamilton, L. M. Dwyer, D. W. Stewart, D. Cloutier, L. Assemat, K. Foroutan-Pour, and D. L. Smith. 2001a. Weed biomass production response to plant spacing and corn (*Zea mays*) hybrids differing in canopy architecture. Weed Technol. 15:647–653.
18. Begna, S. H., D. L. Smith, R. I. Hamilton, L. M. Dwyer, and D. W. Stewart. 2001b. Corn genotypic variation effects on seedling emergence and leaf appearance of short-season areas. J. Agro. Crop Sci. 186:267–271.
19. Belnap, J., S. K. Sherrod, and M. E. Miller. 2003. Effects of soil amendments on germination and emergence of downy brome (*Bromus tectorum*) and *Hilaria jamesii*. Weed Sci. 51:371–378.
20. Bennett, A. C. and D. R. Shaw. 2000. Effect of Glycine max cultivar and weed control on weed seed characteristics. Weed Sci. 48:431–435.
21. Berkowitz, A. R. 1988. Competition for resources in weed-crop mixtures. Pages 89–119 *in* M. A. Altieri and M. Liebman, eds. Weed management in agroecosystems: Ecological approaches. CRC Press, Inc., Boca Raton, FL.
22. Bhowmik P. C. and Inderjit. 2003. Challenges and opportunities in implementing allelopathy for natural weed management. Crop Prot 22: 661-671.
23. Bhowmik, P.C., 1988. Cinmethylin for weed control in soybeans, *Glycine max*. Weed Sci. 36, 678–682.
24. Black, J.N., 1958. Competition between plants of different initial seed sizes in swards of subterranean clover (*Trifolium subterraneum* L.) with particular reference to leaf area and light microclimate. Aust J Agric Res 9: 299–318.
25. Blackshaw R. E. 2005. Nitrogen fertilizer, manure, and compost effects on weed growth and competition with spring wheat. Agron. J. 97:1612–1621.
26. Blackshaw R. E. and Molnar L. J. 2009. Phosphorus fertilizer application method affects weed growth and competition with wheat. Weed Sci. 57:311-318.
27. Blackshaw R. E., Molnar L. J. and Larney F. J. 2005. Fertilizer, manure and compost effects on weed growth and competition with winter wheat in western Canada. Crop Protection 24: 971–980.
28. Blackshaw, R. E., L. J. Molnar, and H. H. Janzen. 2004. Nitrogen fertilizer timing and application method affect weed growth and competition with spring wheat. Weed Sci. 52: 614–622.
29. Blackshaw, R. E., R. L. Anderson, and D. Lemerle. 2007. Cultural weed management. Pages 35–47 in M. K. Upadhyaya and R. E. Blackshaw, eds. Non-Chemical Weed Management: Principles, Concepts and Technology. Oxfordfordshire, UK: CABI.
30. Blackshaw, R.E., 1994. Differential competitive ability of winter wheat cultivars against Downy brome. Agron. J. 86, 649–654.
31. Blackshaw, R.E., Brandt, R.N., Janzen, H.H., Entz, T., Grant, C.A., Derksen, D.A., 2003. Differential response of weed species to added nitrogen. Weed Sci. 51: 532–539.
32. Bond W. and Grundy A. C. 2001. Non-chemical weed management in organic farming systems. Weed Res., 41, 383-405.
33. Borger C. P. D., Hashem A. and Pathan S. 2010. Manipulating crop row orientation to suppress weeds and increase crop yield. Weed Sci. 58:174-178.

34. Börner. 1960. Liberation of organic substances from higher plants and their role in the soil sickness problem. The Botanical Review 26, 396-424.

35. Boydston, R.A., Hang, A., 1995. Rapeseed (*Brassica napus*) green manure crop suppresses weeds in potato (*Solanum tuberosum*). Weed Technol. 9, 669–675.

36. Brainard, D.C. and Bellinder, R.R. 2004. Weed suppression in a broccoli-winter rye intercropping system. Weed Science 52, 281-290.

37. Brainard, D.C., Bellinder, R.R., Miller, A.J., 2004. Cultivation and interseeding for weed control in transplanted cabbage. Weed Technol. 18, 704–710.

38. Brandsaeter, L.O., Netland, J., 1999. Winter annual legumes for use as cover crops in row crops in northern regions: I. Field experiments. Crop Sci. 39, 1369–1379.

39. Bruckart, W.L., Hasan, S., 1991. Options with plant pathogens intended for classical control of range and pasture weeds. In: TeBeest, D.O. (Ed.), Microbial Control of Weeds. Chapman and Hall, New York, pp. 69-79.

40. Bruzzese, E., 1995. Present status of biological control of European blackberry (*Rubus fruticosus* aggregate in Australia. In: Delfosse, E.S., Scott, R.R. (Eds.), Proceedings of the Eighth International Symposium on Biological Control of Weeds. DSIR/CSIRO. Melbourne, Australia, pp. 297-299.

41. Bussan, A.J., O.C. Burnside, J.H. Orf, E.A. Ristau, and K.J. Puettmann. 1997. Field evaluation of soybean (*Glycine max*) genotypes for weed competitiveness. Weed Sci. 45:31–37.

42. Cairns, A.L.P. and O. T. de Villiers. 1986. Breaking dormancy of *Avena fatua* L. seed by treatment with ammonia. Weed Res. 26:191–198.

43. Callaway, M.B. 1992. A compendium of crop varietal tolerance to weeds. Am. J. Alternative Agric. 7:169–180.

44. Callaway, M.B., and F. Forcella. 1993. Crop tolerance to weeds. p. 100–131. *In* C.A. Francis and M.B. Callaway (ed.) Crop improvement for sustainable agriculture. Univ. Nebraska Press, Lincoln.

45. Campiglia E., Paolini R., Colla G. and Mancinelli R. 2009. The effects of cover cropping on yield and weed control of potato in a transitional system. Field Crops Research 112, 16–23.

46. Carmona, D. M. and D. A. Landis. 1999. Influence of refuge habitats and cover crops on seasonal-density of ground beetles (Coleoptera: Carabidae) in field crops. Environ. Entomol. 28:1145–1153.

47. Cathcart, R. J. and C. J. Swanton. 2003. Nitrogen management will influence threshold values of green foxtail (*Setaria viridis*) in corn. Weed Sci. 51:975–986.

48. Charudattan, 1991. The mycoherbicide approach with plant pathogens. In: TeBeest, D.O. (Ed.), Microbial Control of Weeds. Chapman and Hall, New York, pp. 24–57.

49. Charudattan, R., Dinoor, A., 2000. Biological control of weeds using plant pathogens: accomplishments and limitations. Crop Prot. 19, 691–695.

50. Chauhan B. S. and Johnson D. E. 2011. Row spacing and weed control timing affect yield of aerobic rice. Field Crops Res. 121:226-231.

51. Chou, C. H. 1999. Roles of allelopathy in plant biodiversity and sustainable agriculture. Critical Rev. Plant Sci. 18:609–636.

52. Cohn, M. A., D. L. Butera, and J. A. Hughs. 1983. Seed dormancy in red rice, III: response to nitrite, nitrate, and ammonium ions. Plant Physiol 73:381–384.

53. Collins A. S., Chase C. A., Stall W. M. and Hutchinson C. M. 2008. Optimum Densities of Three Leguminous Cover Crops for Suppression of Smooth Pigweed (*Amaranthus hybridus*). Weed Science 56, 753–761.

54. Conklin, A. E., M. S. Erich, M. Liebman, D. Lambert, E. R. Gallandt, and W. A. Halteman. 2002. Effects of red clover (*Trifolium pratense*) green manure and compost soil amendments on wild mustard (*Brassica kaber*) growth and incidence of disease. Plant Soil 238:245–256.

55. Copaja, S. V., Nicol, D. and Wratten, S. D. 1999. Accumulation of hydroxamic acids during wheat germination. Phytochemistry 50, 17-24.

56. Cousens, R., and Mortimer, M. 1995. Dynamics of Weed Populations. Cambridge University Press, Cambridge.

57. Creamer, N. G., Bennett, M. A., Stinner, B. R., Cardina, J. and Regnier, E. E. 1996. Mechanisms of weed suppression in cover crop-based production systems. Hort science 31, 410-413.

58. Cromar, H. E., Murphy, S. D., and Swanton, C. J. 1999. Influence of tillage and crop residue on post dispersal predation of weed seeds. Weed Science, 47, 184-194.

59. Crotser, M. P. and W. W. Witt. 2000. Effect of *Glycine max* canopy characteristics, *G. max* interference, and weed free period on *Solanum ptycanthum* growth. Weed Sci. 48:20–26.

60. Cullen, J.M., 1985. Bringing the cost benefit analysis of biological control of *Chondrilla juncea* up to date. In: Delfosse, E.S. (Ed.), Proceedings of the VI International Symposium on Biological Control of Weeds. Agriculture Canada, Ottawa, pp. 145-152.

61. Czarnota, M.A., A.M. Rimando, and L.A. Weston. 2003. Evaluation of root exudates of seven sorghum accessions. J. Chem. Ecol. 29: 2073–2083.

62. Daferera, D. J., B. N. Ziogas, and M. G. Polissiou. 2003. The effectiveness of plant essential oils on the growth of *Botrytis cinerea*, *Fusarium sp.* and*Clevibacter michiganensis* subsp. michiganensis. Crop Prot. 22:39–44.

63. Daughtry, C.S.T., Gallo, K.P., Bauer, M.E., 1983. Spectral estimates of solar radiation intercepted by corn canopies. Agron. J. 72, 527–531.

64. Davis, A. S. and M. Liebman. 2001. Nitrogen source influences wild mustard growth and competitive effect on sweet corn. Weed Sci. 49:558–566.

65. Davis, A. S. and M. Liebman. 2003. Cropping system effects on *Setaria faberi* seed bank dynamics. Aspects Appl. Biol. 69:83–91.

66. Dayan, F.E. 2002. Natural pesticides. p. 521–525. In

D. Pimentel (ed.) Encyclopedia of pest management. Marcel Dekker, Inc., New York.

67. Dayan, F.E., J.G. Romagni, M. Tellez, A. Rimando, and S. Duke. 1999. Managing weeds with bio-synthesized products. Pestic. Outl. 10: 185–188.

68. De Haan R. L., D. L. Wyse, N. J. Ehlke, B. D. Maxwell, and D. H. Putnam. 1994. Simulation of spring-seeded smoother plants for weed control in corn (*Zea mays*). Weed Sci. 42,35–43.

69. den Hollander N. G., Bastiaans L. and Kropff M. J. 2007. Clover as a cover crop for weed suppression in an intercropping design I. Characteristics of several clover species. Europ. J. Agronomy 26: 92–103.

70. Dhima, K. V., I. B. Vasilakoglou, Th. D. Gatsis, E. Panou-Philotheou, and I. G. Eleftherohorinos. 2009. Effects of aromatic plants incorporated as green manure on weed and maize development. Field Crops Res. 110:235–241.

71. Di Tomaso JM 1995 Approaches for improving crop competitiveness through the manipulation of fertilization strategies. Weed Sci 43: 491-497.

72. Dilday, R. H., Lin, J. and Yan, W. 1994. Identification of allelopathy in the Usda-Ars rice germplasm collection. Australian Journal of Experimental Agriculture 34, 907-910.

73. Dilday, R.H., Mattice, J.D., Moldenhauer, K., Yan, AW., 2001. Allelopathic potential in rice germplasm against ducksalad, redstem and barnyard grass. J. Crop Prod. 4, 287–301.

74. Dingkuhn M., Johnson D.E., Sow A. and Audebert A.Y. 1999. Relationships between upland rice canopy characteristics and weed competitiveness. Field Crops Res. 61, 79–95.

75. Dotzenko, A. D., M. Ozkan, and K. R. Storer. 1969. Influence of crop sequence, nitrogen fertilizer and herbicides on weed seed populations in sugar beet fields. Agron. J. 61:34–37.

76. Drews S., Juroszek P., Neuhoff D., Kopke U. 2004. Optimizing shading ability of winter wheat stands as a method of weed control. J. Plant Dis. Prot.:545–552.

77. Dudai, N., A. Poljakoff-Mayber, A. M. Mayer, E. Putievsky, and H. R. Lerner. 1999. Essential oils as allelochemicals and their potential use as bioherbicides. J. Chem. Ecol. 25:1079–1089.

78. Duke, S.O., Dayan, F.E., Ramagni, J.G., Rimando, A.M., 2000. Natural products as sources of herbicides: current status and future trends. Weed Res. 40: 99–111.

79. Duke, S.O., Dayan, F.E., Rimando, R.M., Schrader, K.K., Aliotta, G., Oliva, A., Romagni, J.G., 2002. Chemicals from nature for weed management. Weed Sci. 50, 138–151.

80. Duke, S.O., Vaughn, K.C., Croom, E.M., Elsohly, H.N., 1987. Artemisinin, a constituent of annual wormwood (*Artemisia annua*) is a selective phytotoxin. Weed Sci. 35, 499–505.

81. Dyck, E. and M. Liebman. 1994. Soil fertility management as a factor in weed control: the effect of crimson clover residue, synthetic nitrogen fertilizer, and their interaction on emergence and early growth of lambsquarters and sweet corn. Plant Soil 167:227–237.

82. Dyck, E., M. Liebman, and M. S. Erich. 1995. Crop-weed interference as influenced by a leguminous or synthetic fertilizer nitrogen source: I. Double cropping experiments with crimson clover, sweet corn, and lambsquarters. Agric. Ecosyst. Environ. 56:93–108.

83. Egley, G. H. and S. O. Duke. 1985. Physiology of weed seed dormancy and germination. Pages 27–64 in Weed Physiology: Reproduction and Ecophysiology, Volume 1. Boca Raton, FL: CRC.

84. Einhellig, F.A., and J.A. Rasmussen. 1989. Prior cropping with grain sorghum inhibits weeds. J. Chem. Ecol. 15:951–960.

85. Entz, M.H., W.J. Bullied, and F. Katepa-Mupondwa. 1995. Rotational benefits of forage crops in Canadian prairie cropping systems. J. Prod. Agric. 8:521-529.

86. Evans S.P., Knezevic S.Z., Shapiro C., Lindquist J.L. 2003. Nitrogen level affects critical period for weed control in corn. Weed Sci 51: 408-417.

87. Facelli, J. M., and S. T. A. Pickett. 1991. Plant litter, It's dynamics and effects on plant community structure. Bot. Rev. 57,1–32.

88. Fawcett, R. S. and F. W. Slife. 1978. Effects of field applications of nitrate on weed seed germination and dormancy. Weed Sci. 26:594–596.

89. Fay, P. K. and Duke, W. B. 1977. An assessment of allelopathic potential in *Avena* germplasm. Weed Science 25, 224-228.

90. Fujii, Y. 1999. Allelopathy of hairy vetch and *Macuna*; their application for sustainable agriculture. pp.289-300. *In* C.H. Chou *et al. Biodiversity and Allelopathy from Organisms to Ecosystems in the Pacific.* Academia Sinica, Taipei.

91. Gallandt, E. R., T. Molloy, R. P. Lynch, and F. A. Drummond. 2005. Effect of cover-cropping systems on invertebrate seed predation. Weed Sci. 53:69–76.

92. Garrity, D.P., M. Movillon, and K. Moody. 1992. Differential weed suppression ability in upland rice cultivars. Agron. J. 84:586–591.

93. Gaudet, C.L., Keddy, P.A., 1988. A comparative approach to predicting competitive ability from plant traits. Nature 334, 242–243.

94. Gianoli, E. and Niemayer, H. M. 1998. DIBOA in wild Poaceae: Source of resistance to the Russian wheat aphid (*Diuraphis noxia*) and the greenbug (*Schizaphis graminum*). Euphytica 102, 317-321.

95. Girardin, Ph. and Tollenaar, M. 1994. Effects of intraspecific interference on maize leaf azimuth. Crop Sci. 34: 151–155.

96. Goldberg, D.E., 1990. Components of resource competition in plant communities. In: Grace, J.B., Tilman, D. (Eds.), Perspectives on Plant Competition. Academic Press, San Diego, pp. 27–49.

97. Gonzalez, V.M., Kazimir, J., Nimbal, C., Weston, L.A., Cheniae, G.M., 1997. Inhibition of a photosystem II electron transfer reaction by the natural product sorgoleone. J. Agric. Food Chem. 45, 1415–1421.

98. Gooding, M.J., A.J. Thompson, and W.P. Davies. 1993. Interception of photosynthetically active radiation, competitive ability, and yield of organically grown wheat varieties. Aspects Appl. Biol. 34:355–362.

99. Grace J.B. 1990. On the relationship between plant traits and competitive ability. In: Perspective on Plant Competition (ed. by Grace J.B. and Tilman D.). Academic Press, San Diego, CA, 51–65.

100. Grant, C. A., D. N. Flaten, D. J. Tomasiewicz, and S. C. Sheppard. 2001. The importance of early season phosphorus nutrition. Can. J. Plant Sci. 81:211–224.

101. Grime J.P. 1977. Evidence for the existence of three primary strategies in plants and its relevance to ecological and evolutionary theory. *Am. Nat.*111, 1169–1194.

102. Hall M. R., Swanton C. J. and Anderson G. W. 1992. The critical period of weed control in grain corn (*Zea mays*). Weed Sci 40: 441-447.

103. Hallberg G. R. 1989. Pesticide pollution of ground water in the humid United States. Agric. Ecosyst. Environ. 26, 299–367.

104. Harker, K.N., J.T. O'Donovan, R.B. Irvine, T.K. Turkington, and G.W. Clayton. 2009. Integrating cropping systems with cultural techniques augments wild oat (*Avena fatua*) management in barley. Weed Sci. 57:326–337.

105. He, H.Q., L.H. Shen, J. Xiong, X.L. Jia, W.X. Lin, and H. Wu. 2004. Conditional genetic effect of allelopathy in rice (*Oryza sativa* L.) under different environmental conditions. Plant Grow. Reg. 44: 211–218.

106. Henson, J. F. and Jordan, L. S. 1982. Wild oat (*Avena fatua*) competition with wheat (*Triticum aestivum* and *T. turgidum* Durum) for nitrate. Weed Sci., 30: 297-300.

107. Hinson,K., andW.D.Hanson. 1962. Competition studies on soybeans. Crop Sci. 2:117–123.

108. Hoffman, M.L., M.D.K. Owen, and D.D. Buhler. 1998. Effects of crop and weed management on density and vertical distribution of weed seeds in soil. Agron. J. 90:793–799.

109. Holt J. S. and LeBaron H. N. 1990. Significance and distribution of herbicide resistance. Weed Technol. 4, 141-149.

110. Holt J.S. and Ocrutt D.R. 1991. Functional relationships of growth and competitiveness in perennial weeds and cotton (*Gossypium hirsutum*). Weed Sci. 39, 575–584.

111. Holt, J. S. 1995. Plant responses to light: a potential tool for weed management. Weed Sci. 43:474–482.

112. Hucl, P. 1998. Response to weed control by four spring wheat genotypes differing in competitive ability. Can. J. Plant Sci. 78:171–173.

113. Huel, D.G. and P. Hucl, 1996. Genotypic variation for competitive ability in spring wheat. Plant Breed 115: 325–329.

114. Hunt, R., and J.H.C. Cornelissen. 1997. Components of relative growth rate and their interrelations in 59 plant species. New Phytol. 135:395–417.

115. Inderjit and Olofsdotter M. 1998. Using and improving laboratory bioassays in rice allelopathy research. In Allelopathy in Rice. Proceedings of the workshop on Allelopathy in Rice, 25-27 Nov. 1996 (ed. M. Olofsdotter), pp. 45-55. IRRI, Manilla.

116. Isman, B. M. 2000. Plant essential oils for pest and disease management. Crop Prot. 19:603–608.

117. Jannink, J. L., N. R. Jordan, and J. H. Orf. 2001. Feasibility of section for high weed suppressive ability in soybean: absence of tradeoffs between rapid initial growth and sustained later growth. Euphytica 120:291–300.

118. Jannink, J.L., Orf, J.H., Jordan, N.R., Shaw, R.G., 2000. Index selection for weed suppressive ability in soybean. Crop Sci. 40, 1087–1094.

119. Jensen, L.B., Courtois, B., Shen, L., Li, Z., Olofsdotter, M., Mouleon, R.D. 2001. Locating genes controlling allelopathic effect against barnyardgrass in upland rice. Agron. J. 93: 21–26.

120. Johnson D.E., Ding Kuhn M., Jones M.P. and Mahamane M.C. 1998. The influence of rice plant type on the effect of weed competition on *Oryza sativa* and *Oryza glaberrima*.Weed Res. 38: 207–216.

121. Jordan N. 1993. Prospects for weed control through interference. Ecol. Appl. 3, 84–91.

122. Keddy, P.A., Shipley, B., 1989. Competitive hierarchies in herbaceous plant communities. Oikos 54: 234–241.

123. Khan, M., Donald, W. W. and Prato, T. 1996. Spring wheat (*Triticum aestivum*) management can substitute for diclofop for foxtail (*Setaria* spp.) control. Weed Sci. 44: 362–372.

124. Kirkland, K. J. 1993. Weed management in spring barley (*Hordeum vulgare*) in the absence of herbicides. J. Sustain. Agric. 3: 95–104.

125. Kirkland, K. J. and H. J. Beckie. 1998. Contribution of nitrogen fertilizer placement to weed management in spring wheat (*Triticum aestivum*). Weed Technol. 12:507–514.

126. Knezevic S. Z., Evans S. P., Blankenship E. E., Van Acker R. C. and Lindquist J. L. 2002. Critical period for weed control: the concept and data analysis. Weed Sci 50: 773-786.

127. Kohli, R.K., Batish, D., Singh, H.P., 1998. Allelopathy and its implications in agroecosystems. J. Crop Prod. 1, 169–202.

128. Kollman, G.E., J.G. Streeter, D.L. Jeffers and R.B. Curry, 1979. Accumulation and distribution of mineral nutrients, carbohydrate, and dry matter in soybean plants as influenced by reproductive sink size. Crop Sci 19: 729–734.

129. Kremer, R. J. 1993. Management of weed seed banks with microorganisms. Ecol. Appl. 3:42–52.

130. Kremer, R.J., Begonia, M.F.T., Stanley, L., Lanham, E.T., 1990. Characterization of rhizobacteria associated with weed seedlings. Appl. Environ. Microbiol. 56, 1649–1655.

131. Kristensen L., Olsen J. and Weiner J. 2008. Crop density, sowing pattern, and nitrogen fertilization effects on weed suppression and yield in spring wheat. Weed Sci. 56:97-102

132. Kropff, M.J., and H.H. van Laar. 1993. Modeling crop-weed interactions. CAB International, Wallingford, UK.

133. Kruidhof, H. M., Bastiaans L. and Kropff M. J. 2008. Ecological weed management by cover cropping, effects on weed growth in autumn and weed establishment in spring. Weed Res., 48, 492-502.

134. Kruse, M., Strandberg, M. and Strandberg, B. 2000. Ecological effects of allelopathic plants – A Review. National Environmental Research Institute, Silkeborg, Denmark. 66 pp. – NERI Technical Report No. 315.

135. Kumar, V., D. C. Brainard, and R. R. Bellinder. 2008. Suppression of Powell amaranth (*Amaranthus powellii*), shepherd's-purse (*Capsella bursa-pastoris*), and corn chamomile (*Anthemis arvensis*) by buckwheat residues: role of nitrogen and fungal pathogens. Weed Sci. 56:271–280.

136. Leather, G.R., 1983a. Sunflowers (*Helianthus annuus*) are allelopathic to weeds. Weed Sci. 31, 37–42.

137. Leather, G.R., 1983b. Weed control using allelopathic crop plants. J. Chem. Ecol. 9, 983–990.

138. Leather, G.R., 1987. Weed control using allelopathic sunflowers and herbicide. Plant Soil 98, 17–23.

139. Lemerle, D., Verbeek, B., Cousens, R.D., Coombes, N.E., 1996. The potential for selecting wheat varieties strongly competitive against weeds. Weed Res. 36, 505–513.

140. Li, Y., Sun, Z., Zhuang, X., Xu, L., Chen, S., Li, M., 2003. Research progress on microbial herbicides. Crop Prot. 22, 247–252.

141. Liebman, M., and E. Gallandt. 1997. Many little hammers: ecological approaches for management of crop-weed interactions. p. 291–343. *In* L.E. Jackson (ed.) Ecology in Agriculture. Academic Press, San Diego.

142. Lindquist, J. L. and D. A. Mortensen. 1998. Tolerance and velvetleaf (*Abutilon theophrasti*) suppressive ability of two old and two modern corn (*Zea mays*) hybrids. Weed Sci. 46:569–574.

143. Lindquist, J. L. and M. J. Kropff. 1996. Applications of an ecophysiological model for irrigated rice (*Oryza sativa*)—*Echinochloa* competition. Weed Sci 44:52–56.

144. Lindquist, J. L., D. A. Mortensen, and B. E. Johnson. 1998. Mechanisms of corn tolerance and velvetleaf suppressive ability. Agron. J. 90:787–792.

145. Lovett, J. V. and Hoult, A. H. C. 1995. Allelopathy and self-defence in barley. In Allelopathy. Organisms, Processes, and Applications. (ed. Inderjit, K. M. M. Dakshini and F. A. E. (eds.).), pp. 170-183. American Chemical Society, Washington, DC.

146. Lybecker D. W., Schweizer E. E. and King R. P. 1988. Economic analysis of four weed management systems. Weed Sci. 36, 846–849.

147. Macias, F. A. 1995. Allelopathy in the search for natural herbicides models. In Allelopathy. Organisms, Processes and Applications (ed. K. M. M. Inderjit and E. F.A.), pp. 310-329. American Chemical Society.

148. Macias, F. A., Simonet, A. M., Galindo, J. C. G., Pacheco, P. C. and Sanchez, J. A. 1998. Bioactive polar triterpenoids from Melilotus messanensis. Phytochemistry 49, 709-717.

149. Macias, F.A., Varela, R.M., Torres, A., Molinillo, J.M.G., 1999. Potential of cultivar sunflowers (*Helianthus annuus* L.) as a source of natural herbicide template. In: Inderjit, Dakshini, K.M.M., Foy, C.L. (Eds.), Principles and Practices in Plant Ecology: Allelochemical Interactions. CRC Press, Boca Raton, FL, pp. 531–550.

150. McDonald, G. K. 2003. Competitiveness against grass weeds in field pea genotypes. Weed Res. 43: 48–58.

151. McWhorter, C.G., and E.E. Hartwig. 1972. Competition of johnsongrass and cocklebur with six soybean varieties. Weed Sci. 20: 56–59.

152. Mesbah, A. O. and S. D. Miller. 1999. Fertilizer placement affects jointed goatgrass (*Aegilops cylindrica*) competition in winter wheat (*Triticum aestivum*). Weed Technol. 13: 374–377.

153. Meschede, D. K., Ferreira, A. B. and Ribeiro Junior, C. C. 2007. Evaluation of weed suppression using different crop covers under Brazilian cerrado soil conditions. Planta Daninha, 25, 465-471.

154. Miller, D. A. 1996. Allelopathy in forage crop systems. Agronomy Journal 88: 854-859.

155. Mohammadi G. R. 2007. Growth parameters enhancing the competitive ability of corn (*Zea mays* L.) against weeds. Weed Biology and Management. 7: 232-236.

156. Mohammadi G. R. and F. Amiri. 2011. Critical period of weed control in soybean (*Glycine max*) as influenced by starter fertilizer. Australian Journal of Crop Science. 5: 1350-1355.

157. Mohammadi G. R. 2012. Living mulch as a tool to control weeds in agroecosystems: A Review. In: Price, A. J. (Ed.), Weed Control. InTech Press, Croatia. pp. 75-100.

158. Mohammadi G. R., Kahrizi, D. and Rashidi, N. 2009. Evaluation of sorghum (*Sorghum bicolor* (L.) Moench) ability to control weeds in corn (*Zea mays* L.). Korean Journal of Weed Science. 29: 104-111.

159. Mohammadi, G. R. 2009. The effects of legumes as living mulches on weed control and plant traits of corn (*Zea mays* L.). Korean Journal of Weed Science 29: 222-228.

160. Mohammadi, G. R. 2010. Weed control in corn (*Zea mays* L.) by hairy vetch (*Vicia villosa* L.) interseeded at different rates and times. Weed Biology and Management 10: 25-32.

161. Mohammadi, G. R., M. E. Ghobadi and S. Sheikheh Poor. 2012. Phosphate biofertilizer, row spacing and plant density effects on corn (*Zea mays* L.) yield and weed growth. American Journal of Plant Sciences. 3: 425-429.

162. Mohammadi G. R., Mozafari S., Ghobadi M. E. and Najaphy A. 2012. Weed suppressive ability and productivity of berseem clover interseeded at different rates and times in corn field. The Philippine Agricultural Scientist 95:146-152.

163. Mohan Babu R., Sajeena A., Seetharaman K., Vidhyasekaran P., Rangasamy P., Som Prakash M, Senthil

Raja A. and Biji K. R. 2003. Advances in ioherbicides development—an overview. Crop Protection 22: 253–260.

164. Moharramipour, S., Takeda, K., Sato, K., Yoshida, H. and Tsumuki, H. 1999. Inheritance of gramine in barley. Euphytica 106, 181-185.

165. Mohler, C. L. 1996. Ecological bases for the cultural control of annual weeds. J. Prod. Agric. 9: 468–474.

166. Mohler, C. L. 2001. Enhancing the competitive ability of crops. Pages 269–321 in M. Liebman, C. L. Mohler, and C. P. Staver, eds. Ecological Management of Agricultural Weeds. Cambridge, UK: Cambridge University Press.

167. Mohler, C. L. and J. R. Teasdale. 1993. Response of weed emergence to rate of *Vicia villosa* Roth and *Secale cereale* L. residue. Weed Res. 33:487–499.

168. Mohler, C. L. and M. B. Callaway. 1991. Effects of tillage and mulch on emergence and survival of weeds in sweet corn. J. Appl. Ecol. 29:21–34.

169. Monks, D.W., and L.R. Oliver. 1988. Interactions between soybean (*Glycine max*) cultivars and selected weeds. Weed Sci. 36:770–774.

170. Morris, M.J., 1997. Impact of the gall-forming rust fungus *Uromycladium tepperianum* on the invasive tree *Acacia saligna* in South Africa. Biol. Control 10, 75-82.

171. Mutsaers, H.J.W. 1980. The effect of row orientation, date and latitude on light absorption by row crops. J. Agric. Sci. 95:381–386.

172. Narwal, S. S. 1996. Potentials and prospects of allelopathy mediated weed control for sustainable agriculture. In Allelopathy in Pest Management for Sustainable Agriculture. Procceding of the International Conference on Allelopathy, vol. ll (ed. S. S. Narwal and P. Tauro), pp. 23-65. Scientific Publishers, Jodhpur.

173. Navas Marie-Laure and Moreau-Richard J. 2005. Can traits predict the competitive response of herbaceous Mediterranean species? Acta Oecologica. 27: 107-114.

174. Niemeyer, H. M. and Jerez, J. M. 1997. Chromosomal location of genes for hydroxamic acids accumulation in *Triticum aestivum* L (wheat) using wheat aneuploids and wheat substitution lines. Heredity 79, 10-14.

175. Nimbal, C.I., J.F. Pedersen, C.N. Yerkes, L.A.Weston, and S.C.Weller. 1996a. Phytotoxicity and distribution of sorgoleone in grain sorghum germplasm. J. Agric. Food Chem. 44:1343–1347.

176. Nimbal, C.I., Yerkes, C.N., Weston, L.A., Weller, S.C., 1996b. Herbicidal activity and site of action of the natural product sorgoleone. Pest Chem. Physiol. 54, 73–83.

177. O'Donovan, J. T. and Newman, J. C. 1996. Manipulation of canola (*Brassica rapa*) plant density and herbicide rate for economical and sustainable weed management. Proc. of the 2nd International Weed Control Congress, 25–28 June 1996, Copenhagen, Denmark. Department of Weed Control and Pesticide Ecology, Flakkebjerg, DK-4200 Slagelse, Denmark. pp. 969–974.

178. O'Donovan, J. T., Harker, K. N., Clayton, G. W. and Hall, L. M. 2000. Wild oat (*Avena fatua*) interference in barley (*Hordeum vulgare*) is influenced by barley variety and seeding rate. Weed Technol. 14: 624–629.

179. Oehrens, E., 1977. Biological control of the blackberry through the introduction of rust, *Phragmidium violaceum*, in Chile. FAO Plant Protect. Bull. 25, 26-28.

180. Olofsdotter, M., L.B. Jensen, and B. Courtois. 2002. Improving crop competitive ability using allelopathy—an example from rice. Plant Breed. 121:1–9.

181. Olofsdotter, M., Navarez, D. and Rebulanan, M. 1997. Rice allelopathy - Where are we and how far can we get? In The 1997 Brighton Crop protection Conference, vol. l, pp. 99-104, Brighton.

182. Olsen, J. and J. Weiner. 2007. The influence of *Triticum aestivum* density, sowing pattern and nitrogen fertilization on leaf area index and its spatial variation. Basic Appl. Ecol. 8:252–257.

183. Ominski, P.D., M.H. Entz, and N. Kenkel. 1999. Weed suppression by *Medicago sativa* in subsequent cereal crops: A comparative survey. Weed Sci. 47:282–290.

184. Openshaw, S.J., H.H. Hadley and C.E. Brokoski, 1979. Effects of pod removal upon seeds of nodulations and non-nodulating soybean lines. Crop Sci 19: 289–290.

185. Overland, L. 1966. The role of allelopathic substances in the «smother crop» barley. Amer. J. Bot. 53, 423-432.

186. Paolini, R., M. Principi, R. J. Froud-Williams, S. Del Puglia, and E. Binacardi. 1999. Competition between sugarbeet and *Sinapis arvensis* and*Chenopodium album*, as affected by timing of nitrogen fertilization. Weed Res. 39:425–440.

187. Petersen, J. 2003. Weed-spring barley competition for applied nitrogen in pig slurry. Weed Res. 43:33–39.

188. Petho, M. 1992. Occurrence and physiological role of benzoxazinones and their derivates. lV. isolation of hydroxamic acids from wheat and rye root secretions. Acta Agronomica Hungarica 41, 167-175.

189. Phatak, S. C. 1992. An integrated sustainable vegetable production system. Hort. Sci. 27, 738–741.

190. Qasem, J. R. 1992. Root growth, development and nutrient uptake of tomato (*Lycopersicon esculentum*) and *Chenopodium album*. Weed Res. 33:35–42.

191. Rajalahti, R. M., Bellinder, R. R. and Hoffman, M. P. 1999. Time of hilling and interseeding affects weed control and potato yield. Weed Science 47, 215-225.

192. Rajcan, I., M. AghaAlikhani, C. J. Swanton, and M. Tollenaar. 2002. Development of redroot pigweed is influenced by light spectral quality and quantity. Crop Sci. 42:1930–1936.

193. Randall, H.W.,Worsham, D., Blum,W., 1989. Allelopathic potential of legume debris and aqueous extracts. Weed Sci. 37, 674–679.

194. Rasmussen, K. 2002. Influence of liquid manure application method on weed control in spring cereals. Weed Res. 42:287–298.

195. Reich, P.B., M.B. Walters and D.S. Ellsworth, 1997. From tropics to tundra: Global convergence in plant functioning. Proc Natl Acad Sci (USA) 94: 13730–13734.

196. Rice, E.L. 1984. Allelopathy. 2nd ed. Academic Press, New York.

197. Rice, E.L. 1995. Biological control of weeds and plant diseases. Univ. Oklahoma Press, Norman.

198. Rimando, A.M., Olofsdotter, M., Dayan, F.E., Duke, S.O., 2001. Searching for rice allelochemicals: an example of bioassay-guided isolation. Agron. J. 93, 16–20.

199. Rizvi, S. H. and V. Rizvi. 1992. Allelopathy—Basic and Applied Aspects. 1st ed. London: Chapman and Hall. 480 p.

200. Rodenburg, J., Saito, K., Kakai, R.G., Toure, A., Mariko, M., Kiepe, P., 2009. Weed competitiveness of the lowland rice varieties of NERICA in the southern Guinea Savanna. Field Crops Res. 114, 411–418.

201. Ross, M.A. and J.L. Harper, 1972. Occupation of biological space during seedling establishment. J Ecol 60: 77–88.

202. Ross, S. M., King, J. R., lzaurralde, R. C. and O'Donovan, J. T. 2001. Weed suppression by seven clover species. Agronomy Journal 93, 820-827.

203. Ross, M.A. and Lembi, C.A., 1985. Applied Weed Science. Burgees Pub. Co.

204. Saito K., Azoma K. and Rodenburg J. 2010. Plant characteristics associated with weed competitiveness of rice under upland and lowland conditions in West Africa. Field Crops Res., 116: 308-317.

205. Samson, R. A. 1991. The weed suppressing effects of cover crops. Pages 11–22 in Fifth Annual REAP Conference, Macdonald College, Ste.-Anne-de-Bellevue, Quebec.

206. Sankula, S., M. J. VanGessel, and R. R. Mulford. 2004. Corn leaf architecture as a tool for weed management in two production systems. Weed Sci. 52:1026–1033.

207. Santos, B. M., J. A. Dusky, W. M. Stall, D. G. Shilling, and T. A. Bewick. 1997. Influence of smooth pigweed and common purslane densities on lettuce yields as affected by phosphorus fertility. Proc. Fla. State Hortic. Soc. 110:315–317.

208. Sardi, K. and I. Beres. 1996. Effects of fertilizer salts on the germination of corn, winter wheat, and their common weed species. Commun. Soil Sci. Plant Anal. 27:1227–1235.

209. Sattin, M., M. C. Zuin, and I. Sartorato. 1994. Light quality beneath field grown maize, soybean, and wheat canopies—red:far red variations. Physiol. Plant. 91:322–328.

210. Schreiber, M.M. 1992. Influence of tillage, crop rotation, and weed management on giant foxtail (*Setaria faberi*) population dynamics and corn yield. Weed Sci. 40:645–653.

211. Schwinning, S. and J. Weiner. 1998. Mechanisms determining the degree of size asymmetry in competition among plants. Oecologia 113:447–455.

212. Se´ne, M., T. Dore´, and C. Gallet. 2001. Relationship between biomass and phenolic production in grain sorghum grown under different conditions. Agron. J. 93:49–54.

213. Sexsmith, J. J. and U. J. Pittman. 1963. Effect of nitrogen fertilizers on germination and stand of wild oats. Weeds 11:99–101.

214. Sharma, V. and N. N. Angiras. 1996a. Effect of row orientations, row spacings and weed-control methods on light interception, canopy temperature and productivity of wheat (*Triticum aestivum*). Indian J. Agron. 41:390–396.

215. Sharma, V. and N. N. Angiras. 1996b. Light interception, weed growth and productivity of irrigated wheat as influenced by crop geometry and weed control methods. Indian J. Plant Physiol. 1:157–162.

216. Sheaffer, C. C., Gunsolus, J. L., Grimsbo Jewett, J. and Lee, S. H. 2002. Annual *Medicago* as a smother crop in soybean. Journal of Agronomy and Crop Science 188, 408-416.

217. Shrefler, J. W., J. A. Dusky, D. G. Shilling, B. J. Brecke, and C. A. Sanchez. 1994. Effects of phosphorus fertility on competition between lettuce (*Lactuca sativa*) and spiny amaranth (*Amaranthus spinosus*). Weed Sci. 42:556–560.

218. Shrestha, A. and M. Fidelibus. 2005. Grapevine row orientation affects light environment, growth, and development of black nightshade (*Solanum nigrum*). Weed Sci. 53: 802–812.

219. Sibuga, K. P. and J. D. Bandeen. 1980. Effects of various densities of green foxtail (*Setaria viridis* L. Beav.) and lambsquarters (*Chenopodium album*) on N uptake and yields of corn. E. Afric. Agric. For. J. 45:214–221.

220. Singh H. P., Batish D. R. and Kohli R.K. 2003. Allelopathic interactions and allelochemicals: New possibilities for sustainable weed management. Critical Reviews in Plant Sciences. 22: 239-311.

221. Singh, H.P., D.R. Batish, S. Kaur, N. Setia, and R.K. Kohli. 2005. Effects of 2-benzoxazolinone on the germination, early growth and morphogenetic response of mung bean (*Phaseolus aureus*). Ann. Appl. Biol. 147:267–274.

222. Sinoquet, H., Caldwell, R.M. 1995. Estimation of light capture and partitioning in intercropping systems. P.79-80. In H. Sinoquet, P. Cruz (Eds.) Ecophysiology of tropical intercropping. Inst. Natl. de la Recherche Agronomique (INRA), Paris.

223. So, Y. F., M. M. Williams, II, J. K. Pataky, and A. S. Davis. 2009. Principal canopy factors of sweet corn and relationships to competitive ability with wildproso millet (*Panicum miliaceum*). Weed Sci. 57:296–303.

224. Spies, J.M., T.D. Warkentin and S.J. Shirtliffe. 2011. Variation in field pea (*Pisum sativum*) cultivars for basal branching and weed competition. Weed Sci. 59:218-223.

225. Steinbauer, G. P. and B. Grigsby. 1957. Interaction of temperature, light, and moistening agent in the germination of weed seeds. Weeds 5:175–182.

226. Steinmaus, S., Elmore, C. L., Smith, R. J., Donaldson, D., Weber, E. A., Roncoroni, J. A. and Teasdale, J. R. 1993. Reduced-herbicide weed management systems for no-tillage corn (*Zea mays*) in a hairy vetch (*Vicia villosa*) cover crop. Weed Technol. 7, 879–883.

227. Stephen, O.D., Lydon, J., 1987. Herbicides from natural compounds. Weed Technol. 1 (2), 122–128.

228. Supasilapa, S., Steer, B.T., Milroy, S.P., 1992. Competition between lupin (*Lupinus angustifolia* L.) and great brome (*Bromus diandrus* Roth.): development of leaf area, light interception and yields. Aust. J. Exp. Agric. 32, 71–81.

229. Sweeney A. E., Renner K. A., Laboski C. and Davis A. 2008. Effect of fertilizer nitrogen on weed emergence and growth. Weed Sci 56:714-721.

230. Teasdale, J. R. 1995. Influence of narrow row/high population corn (*Zea mays*) on weed control and light transmission. Weed Technol. 9:113–118.

231. Teasdale, J. R. 1996. Contribution of cover crops to weed management in sustainable agricultural systems. J. Prod. Agric. 9, 475–479.

232. Teasdale, J.R., 1998. Influence of corn (*Zea mays*) population and row spacing on corn and velvetleaf (*Abutilon theophrasti*) yield. Weed Sci. 46, 447–453.

233. Teasdale, J. R. 2003. Principles and practices of using cover crops in weed management systems. In, R. Labrada (ed), Weed Management for Developing Countries. Food and Agriculture Organization of the United Nations (FAO), Rome, Italy. pp. 169-178.

234. Teasdale, J. R. and C. L. Mohler. 1993. Light transmittance, soil temperature, and soil moisture under residue of hairy vetch and rye. Agron. J. 85, 673-680.

235. Teyker, R. H., H. D. Hoelzer, and R. A. Liebl. 1991. Maize and pigweed response to nitrogen supply and form. Plant Soil 135:287–292.

236. Tollenaar M., Nissanka S.P., Aguilera A.,Weise S.F. and Swanton C.J. 1994. Effect of weed interference and soil nitrogen on four maize hybrids.*Agron. J.* 86, 596–601.

237. Townley-Smith, L. and Wright, A. T. 1994. Field pea cultivar and weed response to crop seed rate in western Canada. Can. J. Plant Sci. 74: 387–393.

238. Trujillo, E.E., 1985. Biological control of Hamakua pa-makani with *Cercosporella* sp. in Hawaii. In: Delfosse, E.S. (Ed.), Proceedings of the VI International Symposium on Biological Control of Weeds. Agriculture Canada, Ottawa, pp. 661-671.

239. Tuncw, I. and S. Sahinkaya. 1998. Sensitivity of two greenhouse pests to vapours essential oils. Entomol. Exp. Appl. 86:183–187.

240. Tworkoski, T., 2002. Herbicide effects of essential oils. Weed Science, 50: 425-431.

241. Van Acker R. C., Swanton C. J. and Weise S. F. 1993. The critical period of weed control in soybeans [*Glycine max* (L.) Merr.]. Weed Sci 41: 194-200.

242. Van Delden, A., L. A. Lotz, L. Bastiaans, A. C. Franke, H. G. Smid, R.M.W. Groeneveld, and M. J. Kropff. 2002. The influence of nitrogen supply on the ability of wheat and potato to suppress Stellaria media growth and reproduction. Weed Res. 42:429–445.

243. Vandermeer, J., 1989. The Ecology of Intercropping. Cambridge University Press.

244. van der Werf, A., M.V. Nuenen, A.J. Visser, and H. Lambers. 1993. Contribution of physiological and morphological plant traits to a species' competitive ability at high and low nitrogen supply. Oecologia 94:434–440.

245. Vera C.L., Woods S.M. and Raney J.P. 2006. Seeding rate and row spacing effect on weed competition, yield and quality of hemp in the Parkland region of Saskatchewan. Can. J. Plant Sci. 86: 911-915.

246. Verma, R., H. R. Agarwal, and V. Nepalia. 1999. Effect of weed control and phosphorus on crop-weed competition in fenugreek (*Trigonella foenumgraecum*). Indian J. Weed Sci. 31: 265–266.

247. Vokou, D. 2007. Allelochemicals, allelopathy and essential oils: a field in search of definitions and structure. Allelop. J. 19: 119–134.

248. Vrabel, T. E. 1983. Effects of suppressed white clover on sweet corn yield and nitrogen availability in a living mulch cropping system. Ph.D Thesis (Diss. abstr. DA 8321911), Cornell Univ., Ithaca, NY.

249. Walker, R. H. and G. A. Buchanan. 1982. Crop manipulation in integrated weed management systems. Weed Sci. 30 (Suppl. 1): 17–24.

250. Wall, D. A. and L. Townley-Smith. 1996. Wild Mustard (*Sinapsis arvensis*) responses to field pea (*Pisum sativum*) cultivar and seeding rate. Can. J. Plant Sci. 76: 907–914.

251. Wallace, D.H., K.S. Yourstone, P.N. Masaya and R.W. Zobel. 1993. Photoperiod gene control over partitioning between reproductive vs. vegetative growth. Theor Appl Genet 86: 6–16

252. Wang G, Ehlers JD, Marchi ECS *and* McGiffen ME, Jr. 2006. Competitive ability of cowpea (*Vigna unguiculata*) genotypes with different growth habit. Weed Science 54: 775–782.

253. Watson, A.K., 1991. The classical approach with plant pathogens. In: TeBeest, D.O. (Ed.), Microbial Control of Weeds. Chapman and Hall, New York, pp. 3-23.

254. Watson, P. R., Derksen, D. A., Van Acker, R. C., and Blrvine, M. C., 2002. The contribution of seed, seedling, and mature plant traits to barley cultivar competitiveness against weeds. Proceedings of the 2002 National Meeting – Canadian Weed Science Society. 49-57.

255. Watson, P.R., D.A. Derksen, and R.C. Van Acker. 2006. The ability of 29 barley cultivars to compete and withstand competition. Weed Sci. 54:783–792.

256. Weaver S. E., Kropff M. J. and Groeneveld R. M. W. 1992. Use of ecophysiological models for crop-weed interference: the critical period of weed interference. Weed Sci 40: 302-307.

257. Weiner, J. 1990. Asymmetric competition in plant populations. Trends Ecol. Evol. 5:360–364.

258. Weiner, J., 1986. How competition for light and nutrients affects size variability in ipomoea-tricolor populations. Ecology 67, 1425–1427.

259. Weiner, J., H. W. Griepentrog, and L. Kristensen. 2001. Suppression of weeds by spring wheat (*Triticum aestivum*) increases with crop density and spatial uniformity. J. Appl. Ecol. 38:784–790.

260. Welles, J.M., Norman, J.M., 1991. Instrument for indirect measurement of canopy architecture. Agron. J. 83, 818–825.

261. Weston, L. A. 1996. Utilization of allelopathy for weed management in agroecosystems. Agron. J. 88:860–866.

262. Weston, L. A. 2005. History and current trends in the use of allelopathy for weed management. Hort. Technology 15, 529–534.

263. Weston, L.A., R. Harmon, and S. Mueller. 1989. Allelopathic potential of sorghum–sudangrass hybrid (sudex). J. Chem. Ecol. 15:1855–1865.

264. Westra, E. P. 2010. Can Allelopathy be incorporated into agriculture for weed suppression? http,//www.colostate.edu/Depts/Entomology/courses/en570/papers_2010/westra.pdf.

265. White R. H., Worsham A. D. and Blum U. 1989. Allelopathic potential of legume debris and aqeous extracts. Weed Sci. 37, 674–679.

266. Williams, W.A., Loomis, R.S., Duncan, W.G., Dovert, A., Nunez, F., 1968. Canopy architecture at various population densities and the growth of grain of corn. Crop Sci. 8: 303–308.

267. Wortmann, C.S. 1993. Contribution of bean morphological characteristics to weed suppression. Agron. J. 85:840–843.

268. Wu, H., J. Pratley, D. Lemerle, and T. Haig. 2000. Evaluation of seedling allelopathy in 453 wheat (*Triticum aestivum*) accessions against annual ryegrass (*Lolium rigidum*) by equal-compartment-agar method. Aust. J. Agric. Res. 51:937–944.

269. Wu, H., J. Pratley,W. Ma, and T. Haig. 2003. Quantitative trait loci and molecular markers associated with wheat allelopathy. Theor. Appl. Genet. 107:1477–1481.

270. Wu, H., pratley, J., Lemerle, D. and Haig, T. 1999. Crop cultivars with allelopathic capability. Weed Research. 39: 171-180.

271. Wu, H., Pratley, J., Lemerle, D., Haig, T. and Verbeek, B. 1998. Differential allelopathic potential among wheat accessions to annual ryegrass. In. Proceedings 9th Australian Agronomy Conference, Wagga Wagga, Australia, 567-571 .

272. Yenish, J.P., J.C. Doll, and D.G. Buhler. 1992. Effects of tillage on vertical distribution and viability of weed seed in soil. Weed Sci. 40:429–433.

273. Yoshida, H., Iida, T., Sato, K., Moharramipour, S. and Tsumuki, H. 1997. Mapping a gene for gramine synthesis. Barley genetics newsletter 27, 22-24.

274. Yoshida, H., Tsumuki, H., Kanehisa, K. and Corucuera, L. J. 1993. Release of gramine from the surface of barley leaves. Phytochemistry 34, 1011-1013.

275. Zhao, D.L., Atlin, G.N., Bastiaans, L., Spiertz, J.H.J., 2006. Cultivar weed-competitiveness in aerobic rice: heritability, correlated traits, and the potential for indirect selection in weed-free environments. Crop Sci. 46: 372–380.

CHAPTER 5

HEALTH AND INSECTICIDES

Nadeem Sheikh

INTRODUCTION

Pyrethroids are synthetic chemicals similar to pyrethrins in the pyrethrum extract which is obtained from Chrysanthemum plant. Historically, pyrethroids have been classified into two classes that differ in their chemical structure and symptoms of exposure: Type I pyrethroids include allethrin, tetramethrin, d-phenothrin, permethrin, and bioresmethrin. Type II pyrethroids include cypermethrin, cyphenothrin, deltamethrin, cyfluthrin, fenvalerate , (Klaassen et al, 1996; Ray, 1991). Pyrethroid kills the insects that eat or come in contact with it by quickly affecting the insect's central nervous system (Tomlin, 1994; Costa, 1997). Cypermethrin is one of the most widely used type II pyrethroid insecticide, first synthesized in 1974 (WHO, 1989; Patel et al, 2006).

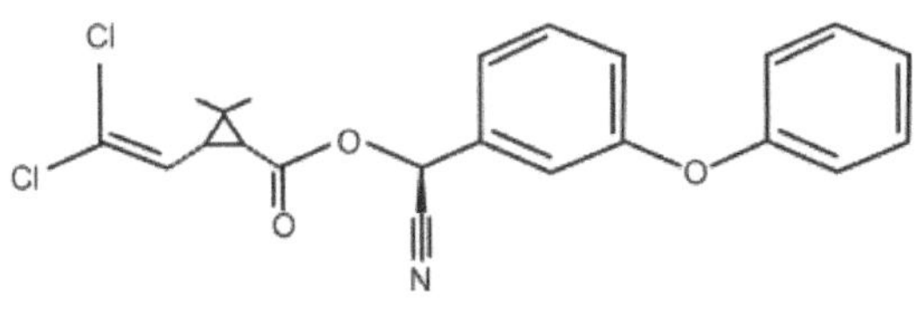

Figure 1. Structural Formula of Cypermethrin; Molecular formula: $C_{22}H_{19}C_{12}NO_3$; Molecular weight: 416.3.

It is considered to be environmentally safe and widely used in agriculture and veterinary medicine (Shukla et al, 2002). It is classified by the World Health Organization (WHO) as "moderately hazardous" (WHO Recommended Classification of Pesticides by Hazard 1994-95, WHO, Geneva). It is a fast-acting neurotoxin and is known to cause free radical-mediated tissue damage (Patel et al, 2006). Like other pyrethroids, cypermethrin kills the insects by interacting with the sodium channels in nerve cells through which sodium enters the cell in order to transmit a nerve signal. These channels can remain open for up to seconds instead of the normal period of a few milliseconds, after a signal has been transmitted. Cypermethrin also interferes with other receptors in the nervous system. The effect is that of long-lasting trains of repetitive impulses in sense organs. (Rodriguez et al, 2008; Tomlin, 1994; Costa, 1997).

Cypermethrin may become an air pollutant and its toxic effects in humans include abnormal facial sensations, coughing, dizziness, tingling, burning, itching, headache, nausea, anorexia and fatigue, vomiting and increased stomach secretion. It is also a skin and eye irritant (Klaassen et al, 1996; Sitting, 1991). Patients with severe exposure to cypermethrin may suffer from muscular twitching, coma and convulsive attacks. Mice on exposure to cypermethrin display symptoms including writhing, convolutions, salivation, etc (Lawrence et al, 1982). Chronic symptoms after exposure to cypermethrin include brain and locomotory disorders, polyneuropathy and immuno-suppression, and resemble the multiple chemical sensitivity syndromes (Müller-Mohnssen, 1995). Cypermethrin is classified by the US EPA as a weak category C oncogen, a possible human carcinogen with evidences of carcinogenesis in animals but no evidence of carcinogenicity in humans (US EPA, 1989; US EPA, 1997). It possesses complete carcinogenic and cocarcinogenic potential (tumor initiating and promoting potential) in both the sexes of Swiss albino mice and male as well as female mice develop benign

Citation: Nadeem Sheikh (2012). Health and Insecticides, Insecticides - Advances in Integrated Pest Management, Dr. Farzana Perveen (Ed.), InTech, DOI: 10.5772/29499.

tumors in skin upon exposure to cypermethrin (Shukla et al, 2002). Cypermethrin is genotoxic in mouse spleen and bone marrow cells where it induces the chromosomal aberration and sister chromatid exchange (Amer et al, 1993; Giri et al, 2003). It induces systemic genotoxicity in mammals by causing DNA damage in vital organs like brain, liver, kidney, apart from that in the hematopoietic system (Patel et al, 2006). It possesses mutagenic activity inducing dominant lethal mutations in male germ cells of mice (Shukla et al, 2002). It induces chromosomal aberrations and single stranded breaks in DNA in the cultured human lymphocytes. Moreover, it also affects the cell cycle causing a decrease in the proliferative rate index (Puig et al, 1989). Lungs are an important entry point for airborne contaminants (e.g., toxic gases, particulates, aerosols, volatile organic solvents etc). Toxicants present in the breathing zone may be absorbed in the nasopharyngeal, tracheobronchial, or pulmonary exchange surfaces of the lung, depending upon the physical and chemical properties of the toxicant. But the lung alveoli and the terminal bronchioles are one of the most effective surfaces in the body for absorption and are responsible for most of the resultant toxicity that occurs during respiratory exposure. Materials that remain within the respiratory system may produce local toxicity that may take form of lung cancer (Richards, 2008).

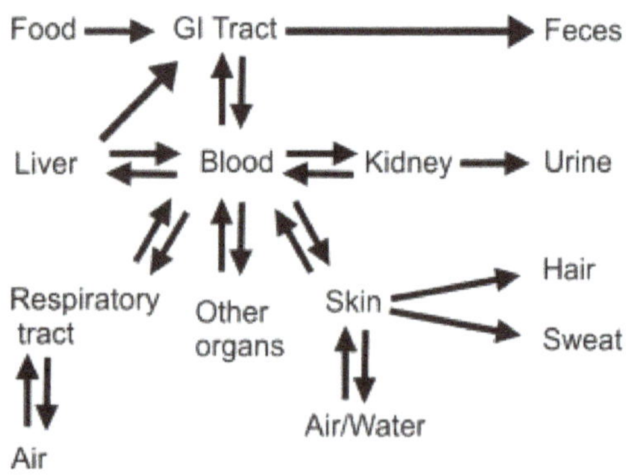

Figure 2. Absorption & distribution of chemicals in animals (Apostoli, 2002).

Toxicants may also pass into the systemic circulation from the lungs and affect the other body parts especially the liver because the hepatocytes directly receive the chemicals from the blood (Richards, 2008). The process of oncogenesis is a progression of events that lead to the formation of a tumor. Of the known carcinogenic agents (viruses, ultraviolet and ionizing radiations, and chemicals), chemicals appear to be of major importance in the induction of human cancers (Richards, 2008; Miller & Miller, 1981). Chemical carcinogenesis is a multistage process which involves initiation, promotion & progression (Simons, 1995; Pitot, 2001; Luch, 2005; Richards, 2008; Miller & Miller, 1981; Lyman, 1992). Initiation requires an irreversible change in the cellular genome within the portions of genome that are involved in regulating process of cellular growth and differentiation, whereas promotion moves initiated cells further along their transformation, commonly associated with prolonged and reversible exposure. Tumor progression results in genotypic and phenotypic changes associated with tumor growth, invasion, and metastasis (Richards, 2008; Miller & Miller, 1981; Lyman, 1992).

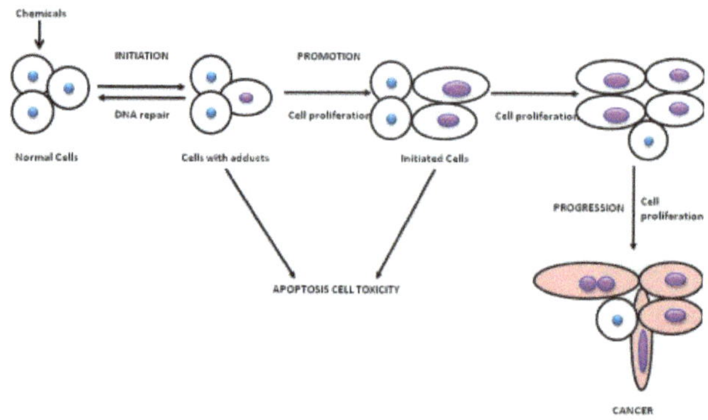

Figure 3. Malignant cell characteristics (Adapted Oliveira et al. An Acad Bras Cienc (2007), vol. 79(4): 595).

The process of oncogenesis requires several changes in the normal properties of a cell. Typically the malignant cell properties include acquisition of self-sufficiency in growth signals and loss of sensitivity to anti-growth signals (leading to uncontrolled growth), loss of capacity for apoptosis, acquisition of sustained angiogenesis, acquisition of ability to invade neighbouring tissues, acquisition of ability to build metastasis at distant sites and loss of capacity to repair genetic errors (Hanahan and Weinberg, 2000). Lung cancer is a disease of uncontrolled cell growth in tissues of the lung. This growth may lead to metastasis, invasion of adjacent tissue and infiltration beyond the lungs. The vast majority of primary lung cancers are carcinomas of the lung, derived from epithelial cells (Ashley, 1980). The major causes of lung cancer are chemical carcinogens such as those in tobacco smoke and air pollutants, ionizing radiation as radon gas, and viral infection (Lombard, 2006; Kotin & Falk,1959; Stock & Campbell,1955).

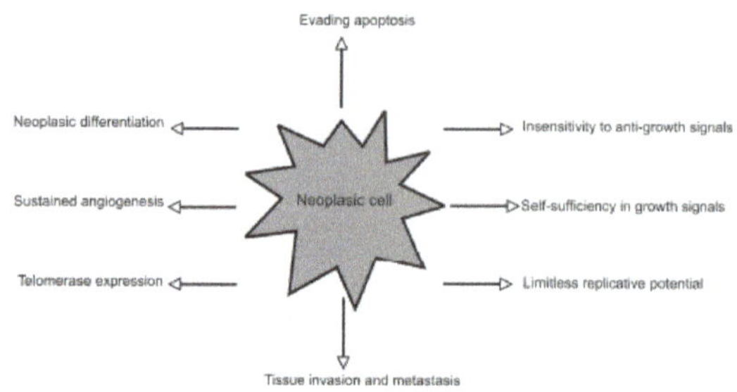

Figure 4. Chemical carcinogenesis stages and the occurrences involved in each one. (Adapted Oliveira et al. An Acad Bras Cienc (2007), Figure 2, , vol. 79(4): 597).

There are many histopathologic types of lung cancer and World Health Organization (WHO) classification, recommended originally in 1977 is usually accepted as the definitive classification of this cancer. Major types of lung cancer are adenocarcinoma, squamous cell carcinoma and large cell carcinoma collectively known as non-small-cell lung cancer (NSCLC) accounting for 75-80% of all lung cancers and small-cell lung cancer (SCLC) comprises the remaining 20-25% of lung cancer cases (Colby et al, 1995). Adenocarcinoma is the most common form of lung cancer originating in the peripheral glandular tissues of lung (Guillan et al, 1967). These tumors predominantly consists of mucin-secreting or non-mucinous columnar cells arranged as regular or irregular tubular and papillary elements supported by a well or poorly developed fibrous stroma (Ashley & Davies, 1967).

Squamous cell carcinoma also called epidermoid carcinoma usually appears as central tumors consisting mainly of clearly defined flattened or polygonal cells of pavement type arranged in keratinizing whorls, irregular masses and narrow or wide anatomizing or branching columns dispersed haphazardly through a fibrous stroma (Smith & Dexter, 1963). Large cell or bronchiolar carcinoma is the least common of all NSCLCs. It is composed of large cells with prominent nucleoli, and no mucin production or intercellular bridging is identified. A variant of large cell carcinoma has been identified; it contains neuroendocrine features and is called large cell neuroendocrine carcinoma (Zhiyong et al, 2006). Small cell lung carcinomas (SCLC) typically are centrally located, arising in peribronchial locations. These are richly cellular neoplasms consisting of more or less regular, darkly staining cells with characteristic nuclei, which because of their shape have been linked to oat grains. These nuclei are set in a cytoplasm so flimsy that it is barely discernible and common appearance is that of nuclei closely knit and lying in an indistinct web- like cytoplasm with cell membrane rarely clearly defined (Chaudhuri, 1973; Wurschmidt et al, 1996). Cancer of the liver refers to the uncontrolled and abnormal growth of cells in the liver. Most cases of the liver carcinomas are not primary but are the result from the metastases of other tumors as that of GI tract and lung cancer. The most frequent, malignant, primary liver cancer is hepatocellular carcinoma (HCC) an aggressive hepatic neoplasm which usually occurs in combination with cirrhosis (Gall, 1960). It has been suggested that HCC arises as a very well differentiated cancer and proliferates with a stepwise process of dedifferentiation of mature cells. Many HCCS seem to arise from dysplastic nodules (DNs) on the basis of evidence of the DNs containing HCC foci, frequent association of DNs in the vicinity of HCC, and clinical progression from DN to HCC (Kojiro and Roskams, 2005). Microscopically, hepatocellular carcinoma has four cytological and architectural patterns: fibrolamellar, pseudoglandular (adenoid), pleomorphic (giant cells) and clear cells. In well differentiated forms, tumor cells resemble hepatocytes, for trabeculae, cords and nests, and may contain bile pigment in cytoplasm. In poorly differentiated forms, malignant epithelial cells are discohesive, pleomorphic, anaplastic and giant. The tumor has a scant stroma and central necrosis because of the poor vascularization (Hou & McFadzean, 1956; Takayasu et al, 2004). Cholangiocarcinoma constitutes 10-22% of all the primary epithelial cancer of the liver and occurs most often in non-cirrhotic tissues (Menias et al, 2008; Patton & Horn, 1964; Steiner, 1957).It is a malignancy of biliary duct system that may originate in the liver and extrahepatic bile ducts, which terminate at the ampulla of Vater (Lake, 1993). A carcinoma arising from the intrahepatic bile ducts is typically a well differentiated adenocarcinoma that exhibit glandular or acinar structures; intracytoplasmic mucin is almost always observed. Characteristically, cells are cuboidal or low columnar and resemble biliary epithelium. A dense fibrous stroma is characteristic and may dominate the histologic architecture. It tends to invade lymphatics, blood vessels, perineural and periductal spaces, and portal tracts. A dense fibrous stroma is characteristic and may dominate the histologic architecture. It tends to invade lymphatics, blood vessels, perineural and periductal spaces, and portal tracts (Ashley, 1980). Cypermethrin and other pyrethroids have been used worldwide in the fields, gardens and homes for the last few decades. Although these insecticides have benefits but they also have side effects especially when they are used in increasing amount. Continuous exposure to these insecticides may cause severe disorders especially the mutations in the genome leading to cancer. The histopathological changes in the lung and liver tissues of cypermethrin exposed mice have shown significant changes. Exposure of the mice to cypermethrin through inhalation induces significant time dependent changes in the histopathology of lung as well as liver tissue. Inhalation is a major route of exposure to air born pollutants such as the pesticides (Emmendoerffer et al., 2000; Richards, 2008). Lungs and liver are the organs which are at highest risk to the environmental pollutants especially the air born chemicals (Dixon et al., 2008). Cypermethrin has shown the carcinogenic effect in the lung tissues of mice. Exposure to cypermethrin caused a gradual distortion of normal structure of alveoli in the lung. Cypermethrin induces hyperplasia and necrosis among the alveolar cells. There was also inflammation of lung tissue leading to pulmonary edema and alveolitis. It also induced pulmonary fibrosis due to which size of alveolar sac was reduced and alveolar walls became thicker. Epidemiologic data showed an increase in the number of cancer cases in persons involved in agricultural production using pesticides (Kornuta et al., 1996). Toxicological studies suggest that cypermethrin and other pyrethroids have carcinogenic effects (Shukla et al., 2002). There is evidence that pyrethroids induce benign tumors in the lungs of mice (Ishmael and Lithfield, 1988). Inhalation of pyrethroids may cause alveolitis, pulmonary edema, and damage to lung cells (Tian, 1993). Lung cancer develops through a series of progressive pathological changes occurring in the respiratory epithelium (Vineis and HusgafvelPursiainen, 2005) Pyrethroids are known to be genotoxic and may interact with DNA and damage its structure. These induce chromosomal aberrations and single strand breaks in DNA (Kornuta et al., 1996; Puig et al., 1989). Pyrethroids may lead to the molecular alterations such as loss of heterozygosity; gene mutations and aberrant gene promoter methylation which are potentially promising molecular biomarkers of lung carcinogenesis (Vineis and HusgafvelPursiainen, 2005). It has been suggested that pyrethroids induce the oxidative stress in lungs (Kale et al., 1999a). Oxidative stress is caused by an imbalance in the production of reactive oxygen. A particularly destructive aspect of oxidative stress is the production of reactive oxygen species, which includes free radicals and peroxides which cause damage to the cells (Schafer and Buettner, 2001; Rahman and MacNee, 2001; Evans and Cooke, 2004). DNA damage is caused by oxygen-derived species as free radicals resulting from oxidative stress (Abdollahi et al., 2004). Oxidative stress may cause pathological changes in the pulmonary epithelium. It may lead to the inflammation of lung tissue. There may be hyperplasia or proliferation of alveolar cells (MacNee, 2001; Erdogan et al., 2006). Intense stresses may cause necrosis of the alveolar cells (Lennon et al., 1991). Toxicological studies suggest that pyrethroids may induce inflammation of the lung tissues causing the oxidative stress (Emmendoerffer et al., 2000) and collagen deposition leading to the pulmonary fibrosis and edema (Erdogan et al., 2006). Idiopathic pulmonary fibrosis (IPF) is found to be associated with lung cancer as the epidemiological studies show greater incidences of lung cancer among the patients having IPF (Park et al., 2001). cypermethrin has also been seen to induce the liver injury. Exposure to cypermethrin damages the normal architecture of liver lobules. The number of the hepatocytes was found to be reduced with distorted polygonal shapes and widened sinusoids. Cypermethrin also induces liver fibrosis and necrosis. The magnitude of these findings was time dependent being more prominent in the tissues exposed for greater time (unpublished data). Pyrethroids are known to have hepatotoxic potential in mammals as they cause histopathological damage of liver through inhalation exposure (Tuzmen et al., 2008; Mani et al., 2004; Okuno et al., 1986; Sakr, 1999). Cypermethrin induces distortion of the normal polygonal shape of the hepatocytes as they become irregular and exhibit cloudy swelling. Histopathological changes induced by the cypermethrin include the liver vacuolar degeneration, enlargement of the sinusoids between hypertrophied hepatocytes, degeneration in hepatic cords and hepatocytes, vacuole formations in hepatocytes, pleomorphism in nucleus, congestion (Yavasoglu et al., 2006; Ksheerasagar and Kaliwal, 2006), extensive vacuolated pycnosis and necrosis (Singh and Singh, 2008; Abou-Zaid and El-Balshy, 1995; Sakr and Hanafy, 2002). Evidences have shown that hepatotoxic chemicals as cypermethrin may induce liver fibrosis (Singh and Singh, 2008). As in the case of lungs, liver damage and fibrosis is also induced by oxidative stress (Abdollahi et al., 2004). Exposure to cypermethrin introduces significant oxidative stress in the hepatic tissues due to which release of free radicals occurs in the liver. These free radicals cause destruction of the normal hepatic tissues (Giray et al., 2001; Kale et al., 1999b).

The inhalation exposure of mice to cypermethrin resulted in the development of skin tumor epidermal in origin as reported that cypermethrin and other pyrethroids possess carcinogenic (tumor initiating) and cocarcinogenic (tumor promoting) potential (Shukla et al., 2002; Kornuta et al., 1996). The carcinogenic property of cypermethrin observed may be attributed to its ability to interact with DNA and damage its structure (Kornuta et al., 1996). Such interactions are critical for the initiation of cells to transform into neoplastic cells. Cypermethrin may also induce the frequency of well established markers of genotoxicity such as chromosomal aberrations and micronuclei formation (Suralles et al., 1995). Commercial formulations of cypermethrin are reported to cause in vivo induction of sister chromatid exchange in mouse bone marrow cells (Chauhan et al., 1997). Taken together these reports we can conclude that cypermethrin causes hazardous effects in non target organisms through inhalation exposure. It has potential to induce carcinogenesis and fibrosis of the lungs and liver. Further more severe exposure of cypermethrin may cause the development of skin tumor due to mutations in the genome. However, further studies on carcinogenicity and acute as well as chronic toxicity of cypermethrin and other pyrethroids are required. In the agriculture sector of Pakistan about 70% of the population lives in villages and most of them are involved in agriculture directly or indirectly. The farmers use pyrethroids to save their crops from pest attacks. The tendency of farmers to use pyrethroids is increasing which is alarming. Also there are many malpractices, i.e., the spray men do not follow the necessary precautions. So it is strongly recommended that these workers should be educated about the harmful and toxic effects of the synthetic pesticides. Moreover, they should know the importance of the protective measures during spraying. Rural workers and public health authorities must become aware of the importance of protective equipment, periodic health examinations and reduced environmental pollution in order to lessen occupational risk of field workers and promote improved conditions of life for the population at large scale.

REFERENCES

1. Abdollahi, M., Ranjbar, A., Shadnia, S., Nikfa, S. And Rezaei, A., 2004. Pesticides and oxidative stress: a review. Med. Sci. Monit., 10(6): 141-147.
2. Abou-Zaid, F.A. And El-Balshy, R.M., 1995. Toxic effect of Ezalo, a synthetic pyrethroid on liver, kidney and lung of newly born mice. J. Union Arab Biol., 4: 235-264.
3. Amer, S.M., Ibrahim, A.A.E.S., El-Sherbeny, K.M., 1993. Induction of chromosomal aberrations and sister chromatid exchange in vivo and in vitro by the insecticide cypermethrin. J. Appl. Toxicol., 13(5): 341-345.
4. Apostoli, P., 2002. Element in environmental and occupational medicine. J. Chromatogr. B., 778: 63-97.
5. Ashley, 1980. Evan's Histological Appearances of Tumours. 3rd Ed. Churchill Livingstone Inc., New York. pp. 633-653.
6. Ashley, D.J.B. And Davies, H.D., 1967. Cancer of the lung. Histology and biological behavior. Cancer, 20(2): 165-174.
7. Chaudhuri, M.R., 1973. Primary pulmonary cavitating carcinomas. Thorax, 28(3): 354-366.
8. Chauhan, L.K., Agarwal, D.K. And Sundararaman V., 1997. In vivo induction of sister chromatid exchange in mouse bone marrow following oral exposure to commercial formulations of alpha cyano pyrethroids. Toxicol. Lett., 93:153–157.
9. Colby, T.V., Koss, M.N. And Travis, W.D. (Eds), 1995. Atlas of Tumor Pathology. Tumors of the Lower Respiratory Tract. Armed Forces Institute of Pathology, Washington, DC.
10. Costa, L.G., 1997. Basic Toxicology of Pesticides. In: Human Health Effects of Pesticides. Occupational Medicine. State of the Art Reviews. Keifer, M. C. (Ed.), (Vol.12 No.2). Hanley and Belfus. Inc., Philadelphia. pp. 251-268.
11. Dixon, D., Herbert, R.A., Kissling, G.E., Brix, A.E., Miller, R.A. And Maronpot, R.R., 2008. Summary of chemically induced pulmonary lesions in the National Toxicology Program (NTP) toxicology and carcinogenesis studies. Toxicol.Pathol., 36(3): 428-439.
12. Erdogan, S., Zeren, E.H., Emre, M., Aydin, O. And Gumurdulu, D., 2006. Pulmonary effects of deltamethrin inhalation: an experimental study in rats. Ecotoxicol. Environ. Safety, 63(2): 318-323.
13. Evans, M.D. And Cooke, M.S., 2004. Factors contributing to the outcome of oxidative damage to nucleic acids. BioEssays., 26(5): 533-542.
14. Gall E. A., Primary and metastatic carcinoma of the liver. Relationship to hepatic cirrhosis. Arch Pathol.1960Aug;70:226-32.

15. Giray, B., Gürbay, A. And Hincal, F., 2001. Cypermethrin-induced oxidative stress in rat brain and liver is prevented by Vitamin E or allopurinol. Toxicol. Lett., 118(3): 139-146.

16. Giri, S., Giri, A., Sharma, G.D. And Prasad, S.B., 2003. Induction of sister chromatid exchanges by cypermethrin and carbosulfan in bone marrow cells of mice in vivo. Mutagenesis, 18(1): 53-58.

17. Guillan, R.A., Zelman, S. And Alonso, R.A., 1967. Adenocarcinoma of the lungs. An analysis of 24 cases in men. Am. J. Clin. Pathol., 47(5): 580-584.

18. Hanahan, D. And Weinberg, R.A., 2000. The hallmarks of cancer. Cell., 100: 57-70.

19. Hou, P. C., and McFadzean, A. J. S., J. Path. Bact., 1956, 72, 411

20. Ishmael, J. And Lithfield, M.H. 1988. Chronic toxicity and carcinogenic evaluation of permethrin in rats and mice. Fundam. Appl. Toxicol., 11(2):308-322.

21. Kale, M., Rathore, N., John, S. And Bhatnagar, D., 1999a. Lipid Peroxidation and Antioxidant Enzymes in Rat Tissues in Pyrethroid Toxicity: Possible Involvement of Reactive Oxygen Species. . J. Nutr. Environ. Med., 9: 37-46.

22. Kale, M., Rathore, N., John, S., Bhatnagar, D., Nayyar, S.S. And Kothari, V., 1999b. The protective effect of vitamin E in pyrethroid-induced oxidative stress in rat tissues. J. Nutr. Environ. Med., 9(4): 281-287.

23. Klaassen, C.D., Amdur, M.O., And Doull, J. (Eds.)., 1996. Casarett and Doull's Toxicology. The Basic Science of Poisons. 5th Ed. McGraw-Hill Companies, Inc. Toronto.

24. Kojiro M and Roskams T. Early hepatocellular carcinoma and dysplastic nodules. Semin Liver Dis.2005;25(2):133-42.

25. Kornuta, N., Bagley, E. And Nedopitanskaya, N., 1996. Genotoxic effects of pesticides. J. Environ. Pathol. Toxicol. Oncol., 15(2-4): 75-78.

26. Kotin, P., And Falk, H.L., 1959. The role and action of environmental agents in the pathogenesis of lung cancer. I. Air pollution. Cancer, 12: 147-163.

27. Ksheerasagar, R.L. And Kaliwal, B.B., 2006. Histological and biochemical changes in the liver of albino mice on exposure to insecticide, carbosulfan. Caspian J. Env. Sci., 4(1): 67-70.

28. Lake J.R.Changing indications for liver transplantation. Gastroenterol Clin North Am.1993 Jun;22(2):213-29

29. Lawrence, L.J. And Casida, J.E., 1982. Pyrethroid toxicology.mouse intracerebral structure toxicity relationships. Pestic. Biochem. Physiol., 18(1): 9-14.

30. Lennon, S.V., Martin, S.J. And Cotter, T.G., 1991. Dose-dependent induction of apoptosis in human tumour cell lines by widely diverging stimuli. Cell. Prolif., 24(2): 203–214.

31. Lombard, H.L., 2006. An epidemiological study in lung cancer. Cancer, 18(10): 1301-1309.

32. Luch, A., 2005. Nature and nurture-lessons from chemical carcinogenesis. Nat. Rev. Cancer, 5(2): 113-125.

33. Lyman, G.H., 1992. Risk factors for cancer. Prim. Care, 19(3): 465-79. Macnee, W., 2001. Oxidative stress and lung inflammation in airways disease. Eur. J. Pharmacol., 429(1-3): 195-207.

34. Mani, U., Prasad, A.K., Sureshkumar, V., Kumar, P., Lal, K., Maji, B.K. And Dutta, K.K., 2004. Hepatotoxic alterations induced by subchronic exposure of rats to formulated fenvalerate (20% EC) by nose only inhalation. Biomed. Environ.Sci., 17(3): 309-314.

35. Menias CO, Surabhi VR, Prasad SR, Wang HL, Narra VR, Chintapalli KN. Mimics of cholangiocarcinoma: spectrum of disease. Radiographics. 2008 Jul-Aug;28(4):1115-29

36. Miller, E.C., Miller, J.A., 1981. Mechanisms of chemical carcinogenesis. Cancer, 47(5) Suppl): 1055-1064.

37. Müller-Mohnssen, H., 1999. Chronic sequelae and irreversible injuries following acute pyrethroid intoxication. Toxicol. Lett., 107(1-3): 161-176.

38. Okuno, Y., Ito, S., Seki, T., Hiromori, T., Murakami, A.L., Kadota, T. And Miyamoto, I., 1986. Fenvalerate induced granulomatous changes in rats and mice. J. Toxicol. Sci., 11:53-66.

39. Oliveira Paula A., Colaço Aura, Chaves Raquel, Guedes-Pinto Henrique, De-La-Cruz P. Luis F., Lopes Carlos. Chemical carcinogenesis. An. Acad. Bras. Ciênc.79(4): 593-616.

40. Park, J., Kim, D.S., Shim, T.S., Lim, C.M., Koh, Y., Lee, S.D., Kim, W.S., Kim, W.D., Lee, J.S. And Song, K.S., 2001. Lung cancer in patients with idiopathic pulmonary fibrosis. Eur. Respir. J., 17: 1216–1219.

41. Patel, S., Pandey, A.K., Bajpayee, M., Parmar, D. And Dhawan, A., 2006. Cypermethrininduced DNA damage in organs and tissues of the mouse: Evidence from the comet assay. Mutat. Res.–Genetic Toxicol. Environ. Mutagen., 607(2): 176- 183.

42. Patton R. B,Horn. R. C. Jr. Primary Liver Carcinoma. Autopsy Study Of 60 Cases. Cancer.1964 Jun;17:757-68.

43. Pitot, H.C., 2001. Animal models of neoplastic development. Dev. Biol. (Basel), 106: 53-57.

44. Puig, M., Carbonell, E., Xamena, N., Creus, A. And Marcos, R., 1989. Analysis of cytogenetic damage induced in cultured human lymphocytes by the pyrethroid insecticides cypermethrin and fenvalerate. Mutagenesis, 4(1): 72-74.

45. Rahman, I. And Macnee, W., 2001. Oxidative stress and regulation of glutathione in lung inflammation. Eur. Respir. J., 16(3): 534-554.

46. Ray, D.E., 1991. Pesticides derived from plants and other organisms. In: Handbook of Pesticide Toxicology. Hayes, Jr. W. J. and Laws, E. R. (Eds.), Vol. 2. Academic Press, Toronto. pp. 585-593.

47. Richards, I.S., 2008. Principles and Practices of Toxicology in Public Health. 1st Ed. Jones and Bartlett Publishers, USA. pp.115-190.

48. Rodriguez, H., Tamayo, C., Inostroza, J., Soto, C., Bustos-Obregón, E. and Paniagua, R., 2008. Cypermethrin

effects on the adult mice seminal glands. Ecotoxicol. Environ. Saf., [Epub ahead of print].

49. Sakr, S.A. And Hanafy, S.M., 2002. Histological alteration in the liver and kidney of toads (Bufo regularis) intoxicated with a pyrethroid insecticide. OnLine Journal of Biological Sciences, 2(4): 208-211.

50. Sakr, S.A., 1999. Pyrethroid inhalation induced hepatotoxicity in albino rats. Oxford Res. Forum J., 1: 27-31.

51. Schafer, F.Q. And Buettner, G.R., 2001. Redox environment of the cell as viewed through the redox state of the glutathione disulfide/glutathione couple. Free Radic. Biol. Med., 30(11): 1191–1212.

52. Shukla, Y., Yadav, A. And Arora A., 2002. Carcinogenic and cocarcinogenic potential of cypermethrin on mouse skin. Cancer Lett., 182(1): 33-41.

53. Simons, J.W., 1995. Genetic, epigenetic, dysgenetic, and non-genetic mechanisms in tumorigenesis. Crit. Rev. Oncog., 6: 261-273.

54. Singh, P.B. And Singh, V., 2008. Cypermethrin induced histological changes in gonadotrophic cells, liver, gonads, plasma levels of estradiol-17β and 11-ketotestosterone, and sperm motility in Heteropneustes fossilis (Bloch). Chemosphere, 72(3): 422-431.

55. Sitting, M., 1991. Handbook of toxic and hazardous chemicals and carcinogens. 3rd Ed. Noyes Publications, New Jersey.

56. Smith, J.F. And Dexter, D., 1963. Papillary neoplasms of the bronchus of low-grade malignancy. Thorax, 18: 340-349.

57. Steiner Pe . Carcinoma of the liver in the United States. Acta Unio Int Contra Cancrum. 1957;13(4-5):628-45.

58. Stocks, P. And Campbell, J. M., 1955. Lung cancer death rates among non-smokers and pipe and cigarette smokers; an evaluation in relation to air pollution by benzpyrene and other substances. Br. Med. J., 2: 923-929.

59. Suralles, J., Xamena, N., Creus, A., Catalan, J., Norppa, H. And Marcos, R., 1995. Induction of micronuclei by five pyrethroid insecticides in whole blood and isolated human lymphocyte cultures. Mutat. Res., 341: 169–184.

60. Takayasu K, Muramatsu Y, Mizuguchi Y, Moriyama N, Ojima H. Imaging of early hepatocellular carcinoma and adenomatous hyperplasia (dysplastic nodules) with dynamic ct and a combination of CT and angiography: experience with resected liver specimens. Intervirology. 2004;47(3-5):199-208.

61. Tian, H.L., 1993. Biochemical and toxicological studies of fenvalerate on the lung. Zhonghua Yu Fang Yi Xue Za Zhi., 27(5): 290-293.

62. Tomlin, C. (Ed.), 1994. A World Compendium. The Pesticide Manual. Incorporating the agrochemical handbook. 10th Ed. Crop Protection Publications, Bungay Suffolk, U.K.

63. Tuzmen, N., Candan, N., Kaya, E. And Demiryas, N., 2008. Biochemical effects of chlorpyrifos and deltamethrin on altered antioxidative defense mechanisms and lipid peroxidation in rat liver. Cell Biochem. Funct., 26(1): 119-124.

64. United States Environmental Protection Agency, Office of Pesticides and Toxic Substances. Office of Pesticide Programs. 1989. Pesticide fact sheet. Cypermethrin. No. 199. Washington, D.C. (Jan. 3.)

65. United States Environmental Protection Agency, 1997. Office of Pesticide Programs reference dose tracking report. Washington, D.C.

66. Vineis, P. And Husgafvel-Pursiainen, K., 2005. Air pollution and cancer: biomarker studies in human populations. Carcinogenesis, 26(11): 1846-1855.

67. World Health Organization, 1989. Environmental Health Criteria. Cypermethrin. Vol. 82. United Nations Environmental Programme, the International Labour Organization, and the World Health Organization. Geneva.

68. World Health Organization. WHO Recommended Classification of Pesticides by Hazard and guidelines to classification 1994–95, WHO, Geneva , Switzerland.

69. Wurschmidt, F., Bunemann, H., Heilmann, H.P., 1996. Small cell lung cancer with and without superior vena cava syndrome: A multivariate analysis of prognostic factors in 408 cases. Lung Cancer, 14(1): 160-160.

70. Yavasoglu, A., Sayim, F., Uyanikgil, Y., Turgut, M. And Karabayyavasoglu, N.U., 2006. The pyrethroid cypermethrin-induced biochemical and histological alterations in rat liver. J. Health Sci., 52(6): 774-780.

71. Zhiyong, L., Jianlin, W., Dong, Y. And Lizhi, Z., 2006. Large cell carcinoma of lung: analysis of CT signs and review of the literature. Chin. Ger. J. Clin. Oncol., 5(5): 309-311.

CHAPTER 6

COMPARATIVE STUDY OF INTEGRATED PEST MANAGEMENT AND FARMERS PRACTICES ON SUSTAINABLE ENVIRONMENT IN THE RICE ECOSYSTEM

Mohammad Zahangeer Alam, Md. Manjurul Haque, Md. Sirajul Islam, Emran Hossain, Sabiha Binta Hasan, Shahela Binte Hasan, and Md. Sakhawat Hossain

INTRODUCTION

Rice is a staple food for more than fifty percent of the world's population [1]. For decades rice crops have directly or indirectly played a key role in the livelihood of several billion people. In 2010, 154 million hectares of rice were cultivated worldwide, of which 137 million hectares were in Asia, with 48 million hectares harvested in Southeast Asia [2]. In Bangladesh, eighty percent of agricultural land is dedicated to rice cultivation [3]. Of the rice growing regions in Bangladesh, Barisal is one of the most important, due in part to the diversity of rice under cultivation. For this reason, as well as the region's consistently high yields, Barisal is known as the "Crop House of Bangladesh," although for several decades farmers have been plagued by a number of pest related issues [3].

Integrated pest management (IPM) is an adaptation to insect behavior and life cycles [4–6]. A farmer's agricultural practices and knowledge of pest species in a given agroecosystem are essential for developing a successful IPM plan [7]. An ecosystem level understanding of pest life cycles provides the basis for successful design and implementation of an IPM strategy. A farmer's practices and a well-designed IPM plan should be closely linked in rice ecosystems [8]. A successful IPM plan accounts for the protection of beneficial insects, secondary pest outbreaks, spread of disease, contamination of air, water, and soil resources, and pest resurgence [9, 10]. The wide scale adoption of IPM in rice agroecosystem could provide a tremendous net benefit to farmers in a number of countries. The innovative IPM program utilized in this study is from the "Farmer Field School Model" [8] and is applicable to rice agroecosystems throughout the world. It protects against pesticide-induced pest resurgence, which is the primary focus of the National IPM program of any country in the world [8].

Two hundred sixty-six insect species have been identified in rice ecosystems in Bangladesh. These insects form a complex food web within the rice ecosystem. In the rice agroecosystem 42 insect species are considered to be pests [11, 12]. Pest insect species can be categorized as minor and major pests. These pests cause severe damage to rice crops at different growth stages. The degree of damage is dependent on the growing season and surrounding environment [13].

One species, the yellow stem borer, Scirpophaga incertulas Walker, is considered to be a major rice pest [14]. According to DAE in Bangladesh [12] the stem borer can be eradicated through several methods: light trapping, hand picking eggs from rice leaves, sweeping with a net, perching, burning rice stubble, cultivation of insect resistant varieties, using bioagents, and applying Economic Threshold Level (ETL) based insecticides in Bangladesh. Currently, most rice farmers rely on insecticides rather than utilizing IPM strategies [15].

Integrated pest management (IPM) is an effective and environmentally friendly approach to pest management [16]. Sustainable and effective agricultural pest management techniques are important to the continued development of

Citation: Mohammad Zahangeer Alam, Md. Manjurul Haque, Md. Sirajul Islam, et al., "Comparative Study of Integrated Pest Management and Farmers Practices on Sustainable Environment in the Rice Ecosystem," International Journal of Zoology, vol. 2016, Article ID 7286040, 12 pages, 2016. doi:10.1155/2016/7286040.

alternative pest management strategies [17]. IPM techniques protect the natural enemies of pest insects and aid in the restoration of ecosystem activities [18]. This approach takes advantages of all appropriate pest management options, including the judicious use of pesticides. Organic food production employs similar methods to IPM but precludes the use of pesticides that are produced from synthetic chemicals [19]. Notwithstanding this, at a community level worldwide, chemical pesticides have played a vital role in providing an abundant and inexpensive food source [20]. Despite this, the persistent use and overuse of chemicals in non-IPM treated farming systems have resulted in a number of adverse environmental problems [21]. For local farmers, continuing education and training play a vital role in advancing their knowledge of IPM practices and the adverse impacts agrochemicals can have on the rice ecosystem. Certain IPM practices have been shown to significantly increase yields while also increasing environmental and economic benefits. In the long run, IPM can be a better pest management strategy than chemical control alone [19]. IPM and farmers practices both are important for the protection of sustainable environment with rice yield. We conclude that the effect of IPM on the yield components would be further researched substantively in this region.

MATERIALS AND METHODS

Geographical Position

The field experiment was conducted in the Barisal region of Bangladesh in 2010. Barisal is located in southern Bangladesh and encompasses an area of approximately 3,000 square kilometers. Barisal is bounded by Madaripur, Shariatpur, Chandpur, and Lakshmipur districts to the north; Patuakhali, Barguna, and Jhalokati districts to the south; Bhola and Lakshmipur districts to the east; Jhalokati, Pirojpur, and Gopalganj districts to the west. Our field site was located at 22.7000°N 90.3667°E and covers a total area of 324.41 km^2 [22].

Farmer's Land Selection

To determine an appropriate field site a preliminary meeting was organized with farmers and personnel of the DAE in 2010 at their office in Barisal Sadar of Bangladesh. At this meeting, an observation was made of farmer's ongoing practices within their rice ecosystems. The experiment site was selected based on uniformity of cropland, the presence of major rice insects, cultivated rice varieties, and transplanting season. Yellow stem borer was considered as the major rice pest for this study. To maintain a degree of food security the Bangladesh Rice Research Institute (BRRI) releases numerous rice cultivars. Among these, BRRI dhan 28 and BRRI dhan 29 are popular among local farmers in the Barisal region due in part to yield performance during the Boro season (November to February). BRRI dhan 29 seed was supplied to the farmers for seedling preparation. Agronomic practices were applied according to guidelines established by the Bangladesh Rice Research Institute (BRRI) (Table 2). These practices include land preparation, fertilizer application, irrigation, and hand weeding. All documented practices in each plot were similar to the farmer's preexisting practices (Table 2). Insects' egg mass collection, perching, sweeping, and ETL based insecticides application are the most recognized components of an IPM strategy in rice agroecosystems (Table 1).

Table 1: Details of treatments in the rice agroecosystem in 2010-11.

Treatments	Description of each treatment
T_1	Collection of egg masses, sweeping, and perching: egg mass collection and sweeping both were employed at 7- and 15-day intervals, respectively; sweeping was over before flowering of rice plants; bamboo perching was put in the rice field 10 × 10 m distance as nest for keeping birds in the rice fields as biological agent
T_2	Collection of egg masses, sweeping, perching, and ETL based insecticides application: collection of egg masses, sweeping, and perching activities were similar to treatment one; the rate of insecticides (Cartap) application was 30 mL @ 10 L water and sprayed two times for the total duration of rice crops
T_3	Prophylactic insecticides application: only insecticides were applied for three times of the total duration of rice plant and dose was 40 mL @ 10 L water
T_4	Farmers' practices: farmers' practices were what they understand to apply of insect's management in the rice ecosystem; they did not follow egg mass collection and sweeping in their rice fields; they were mostly dependent on insecticide (Cartap) application for control of rice pests; the insecticide application rate was about 45 mL @ 10 L water and application frequency was 4 times for the total duration of rice plants

Table 2: Agronomic practices in each plot in the rice ecosystem in 2010-11.

Locations	Treated plots	Number of tillage processes during land preparation	Transplanting time (November)	Fertilizer application (g/decimal)	Weeding	Sources of waters
Sarder Para	T_1	4	Mid	Urea 250, TSP 300, MOP400	By hand	Rain and flash flood water
	T_2	4	Mid	Urea 250, TSP 300, MOP400	By hand	Rain and flash flood water
	T_3	4	Mid	Urea 250, TSP 300, MOP400	By hand	Rain and flash flood water
	T_4	4	Mid	Urea 250, TSP 500, MOP400	By hand	Rain and flash flood water
Uttar Sagardi	T_1	5	Mid	Urea 250, TSP 300, MOP300	By hand	Rain and flash flood water
	T_2	5	Mid	Urea 300, TSP 500, MOP400	By hand	Rain and flash flood water
	T_3	5	Mid	Urea 250, TSP 500, MOP300	By hand	Rain and flash flood water
	T_4	5	Mid	Urea 250, TSP 500, MOP300	By hand	Rain and flash flood water
Gabtala	T_1	5	Mid	Urea 250, TSP 300, MOP300	By hand	Rain and flash flood water
	T_2	5	Mid	Urea 300, TSP 500, MOP400	By hand	Rain and flash flood water
	T_3	5	Mid	Urea 250, TSP 500, MOP300	By hand	Rain and flash flood water
	T_4	5	Mid	Urea 250, TSP 500, MOP300	By hand	Rain and flash flood water

Data Collection

At each location we recorded the total number of hills per 10 m^2, the average number of healthy tillers at tillering and flowering, the percentage of dead heart and white head, the percentage of moisture, and grain weight in kg per 10 m^2 and yield (tonne/ha) (Figures 1–8). All agronomic and integrated farmer's pest management techniques were recorded for each plot (Tables 2 and 3). Major environmental concerns were determined through discussion with farmers and DAE personnel (Table 3).

Table 3: Record of farmers' practices in the rice ecosystem in 2010-11.

Components of integrated practices	Sarderpara	Uttarsagardi	Gabtala	Major threats/Environmental degradation
Collection of egg mass	Not applied	Not applied	Not applied	Reduction of beneficial insects, toxicity to soil, air, and water, health hazards, and outbreak of secondary pests Higher cost to benefit ratio and microbial toxicity
Sweeping	Not applied	Not applied	Not applied	
Perching	Applied at tillering stage	Applied at tillering stage	Applied at tillering stage	
Application of insecticides	Cartap	Cartap	Cartap	

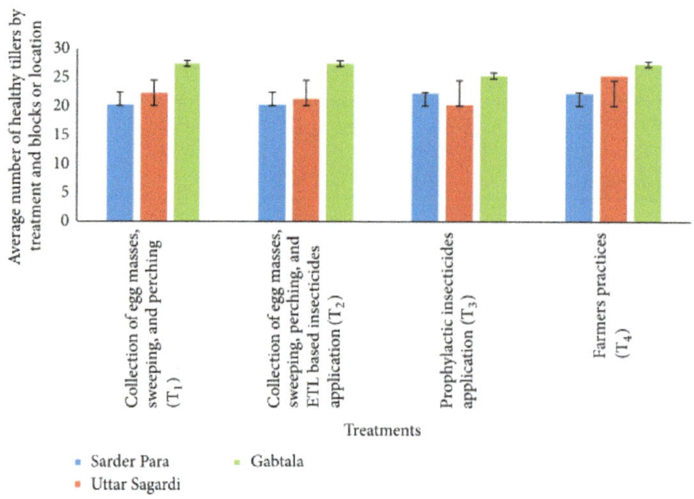

Figure 1: Outline of the different treatments.

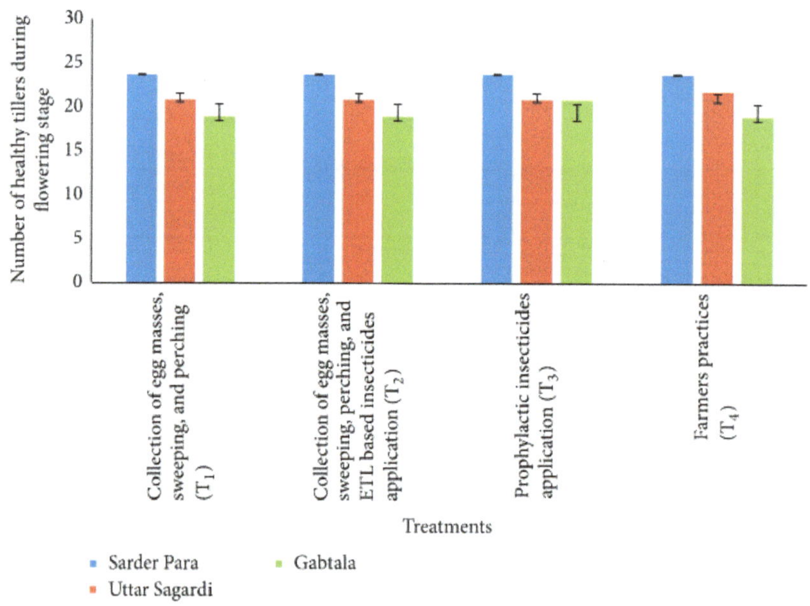

Figure 2: The number of healthy tillers by treatments.

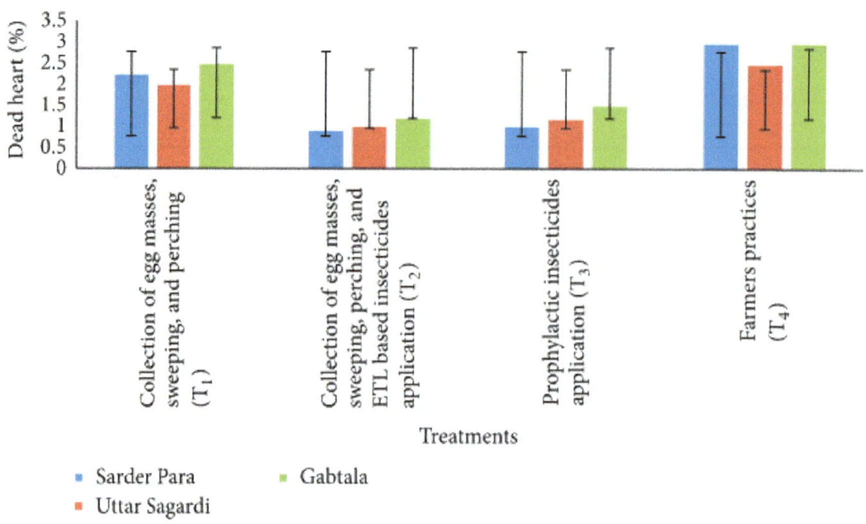

Figure 3: Percentage of dead heart by treatments.

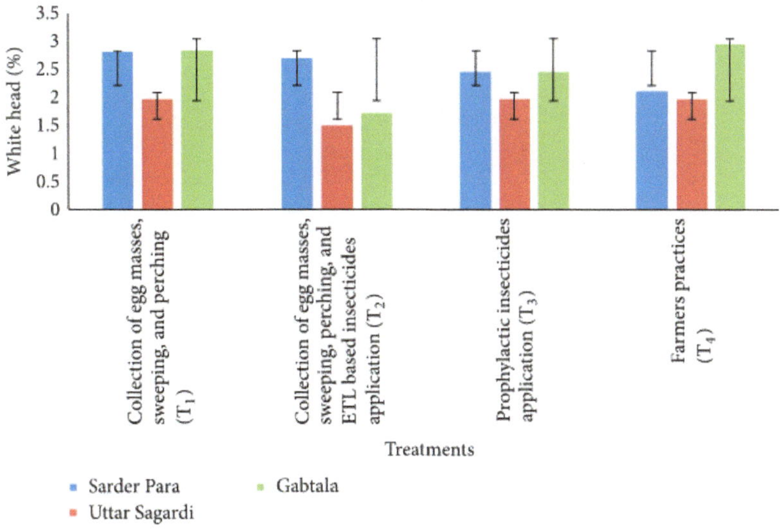

Figure 4: Percentage of white head by treatments.

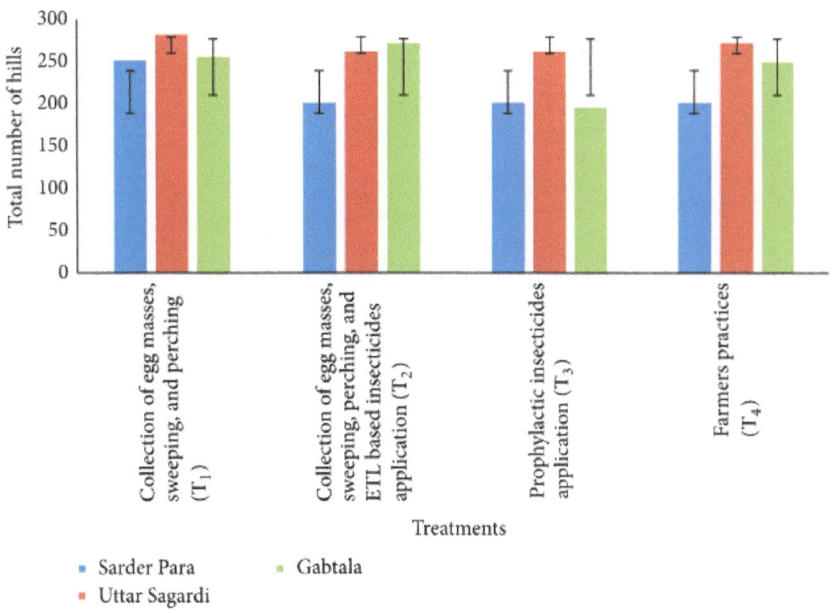

Figure 5: Total number of hills per 10 m^2.

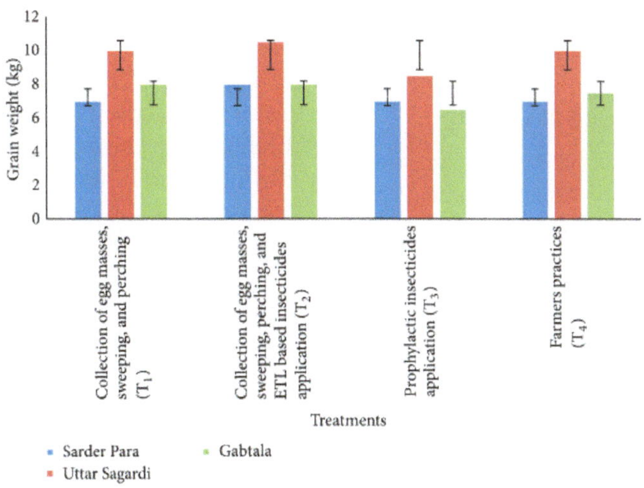

Figure 6: Average grain weight.

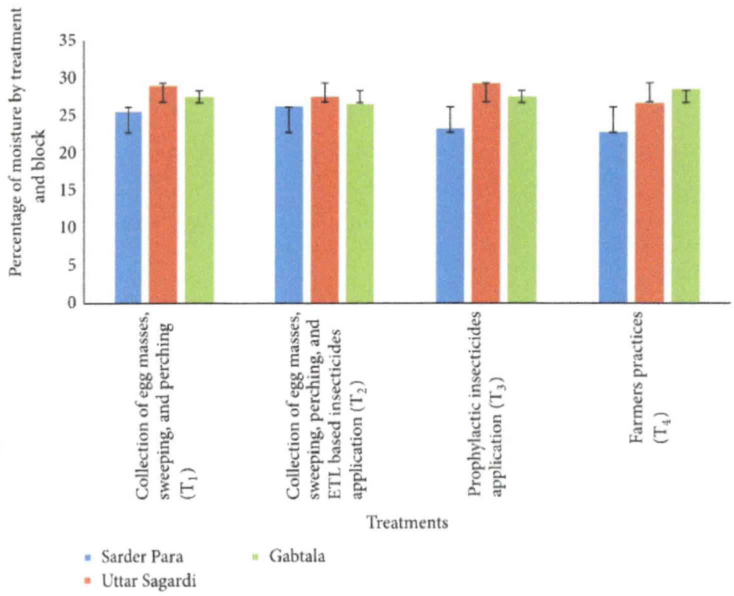

Figure 7: Moisture content.

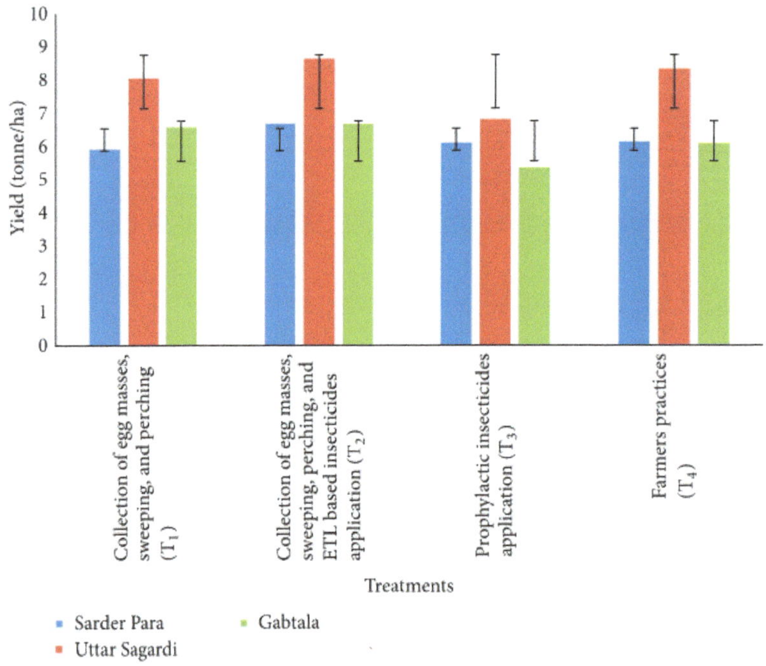

Figure 8: Yield.

Experimental Design and Statistical Analysis

The experiments were conducted during 2010-11 growing season in three locations within the Barisal region: Sarder Para, Uttar Sagardi, and Gabtala. Three fields owned by different farmers were selected from each location. The treatment plots were 12 m × 12 m and separated from other plots by 1.6 m barrier zones designed to avoid spray drift or other treatment effects. Rice seedlings raised in the nursery were transplanted to the experimental plots with 20×20 cm spacing, and recommended agronomic practices were followed. Each location was considered a replication. These three replicates, each of 4 treatments, produced a total of 12 plots for this study. Treatments were as follows: T_1 consisted of collection of egg masses, sweeping, and perching; T_2 consisted of collection of egg masses, sweeping, perching, and ETL based insecticides application; T_3 consisted of application of prophylactic insecticides; and T_4 was the farmers' "normal" pest control strategy (Table 1). The experiment was laid out as a RCBD design. Data were analyzed using Statistix 10.

RESULTS

Tiller and Hill

The number of healthy tillers was similar for all treated plots at each location (Figures 1 and 2). The total number of hills per 10 m² varied from 194 to 270 (Figure 5). This was statistically nonsignificant (Table 4).

Table 4: Impacts of treatments on yield components of rice agroecosystem in 2010-11.

Treatments	Healthy tiller-tillering	Healthy tiller-flowering	Percentage of dead heart	Percentage of white head	Total number of hills/10 m²	Grain wt. kg/10 m²	Percentage of moisture	Average yield (tonne/ha)
T_1	23.00	22.33	2.25 B	2.58	261.33	8.33 AB	27.83	6.96 AB
T_2	22.66	22.33	1.03 C	2.00	243.33	8.83 A	27.23	7.46 A
T_3	22.33	23.00	1.22 C	2.33	218.00	7.33 B	27.200	6.19 B
T_4	24.66	22.66	2.83 A	2.38	239.33	8.16 AB	26.43	6.97 AB
CV	5.93	3.04	9.00	16.69	8.60	5	5.40	5.92
SE	1.122	0.56	0.134	0.31	16.89	0.33	1.19	0.33
LSD ($p \le 0.05$)	NS	NS	0.46	NS	NS	1.15	NS	1.15

NS: nonsignificant.

Dead Heart

At Sarder Para, the average percentage of dead heart was 2.25, 0.9, 1, and 3 at, T_1, T_2, T_3, , and , T_4 respectively. For Uttar Sagardi the percentage of dead heart was 2, 1, 1.18, and 2.5 for,, T_1, T_2, T_3, , and T_4. The lowest percentage of dead heart (1.03) was found in treatment,T_2 with the highest (2.83) being recorded for T_4. For Gabtala the range of dead heart was from 1.2 to 3 percent. There was a significant difference between the percentage of dead heart and treatments ($(p \le 0.05)$). Treatments 2 and 3 were statistically identical, but T_1 treated plots were statistically different from T_4 treated plots (Figure 3 and Table 4).

White Head

At Sarder Para the percentage of white head was observed to be 2.86, 2.74, 2.5, and 2.15 for treatments,, T_1, T_2, T_3, , and T_4, respectively. In Uttar Sagardi, the percentage of white head was 2, 1.52, 2, and 2 for treatments, T_1, T_2, T_3, , and T_4. Percentage of white head observed in Gabtala was 2.88, 1.75, 2.5, and 3 for treatments T_1-T_4. The lowest percentage of white head (2.00) was found in T_2 treated plots. However this relation was statistically nonsignificant (Table 4 and Figure 4).

Grain Weight (kg/10 m²)

A 7 kg grain weight was found at T_1, T_3, and T_4 treated plots, and 8 kg was found in T_2 treated plot at Sarder Para. In Uttar Sagardi, the range of grain weight was 8.5 to 10.5 kg/10 m², but similar grain weight was found in $T_1$and T_2 treated plots. The lowest was found in the treatment of T_3 at Gabtala (Figure 6). A significantly higher average grain weight (8.83) was found for T_2 treated plots. The relationship between treatment and grain weight was significant ($p \le 0.05$). The percentage of white head in T_1 treated plots was identical with T_3 and T_4 treated plots. T_2 treated plots were statistically different from T_3 (Table 4).

Moisture Content

Moisture content in rice grains at Sarder Para was 26, 26.7, 23.7, and 23.2% in treated plots T_1, T_2, T_3, and T_4, respectively. Similarly, at Uttar Sagardi, 29.5, 28, 29.8, and 27.1% moisture contents were found in, T_1, T_2, T_3, and T_4,. The range of moisture percentage was found to be 27 to 29% at Gabtala (Figure 7). Variation in the moisture content of rice grains was not statistically significant among treatments (Table 4).

Yield (Tonne/ha)

Rice yields of 6.02, 6.81, 6.21, and 6.25 tonne/ha were found at, T_1, T_2, T_3, and T_4, treated plots in Sarder Para. In Uttar Sagardi, rice yields of 8.19, 8.79, 6.93, and 8.47 tonne/ha were recorded at, T_1, T_2, T_3, and T_4,treated plots. A similar trend was observed at Gabtala plots with 6.69, 6.79, 5.44, and 6.19 tonne/ha being recorded for, T_1, T_2, T_3, and T_4, plots, respectively. An average yield of 7.4 tonne/ha was highest in T_2 (IPM) treated plots. We found the relationship between treatment and yield to be significant ($p \le 0.05$). Rice yield in T_1 treated plots was identical with T_3 and T_4 treated plots. Treatment T_2 was statistically different from T_3 treated plots (Table 4 and Figure 8).

Major Threats Raised for a Sustainable Environment

In the last 10 years, farmers have not fully implemented all of the components of integrated pest management practices. Currently, more than 90% of farmers rely on Cartap insecticides for control of rice insects in rice agroecosystems [23]. Egg mass collection, sweeping, perching, and ETL based insecticides application are the recommended practices for the control of pest insects in the rice ecosystem. Due to the lack of integrated methods for control of major rice pest threats to the system problems include reduction of beneficial insects, outbreak of secondary pests, high cost to benefit ratio, impacts on beneficial microbes, the accumulation of toxins in soil and water bodies, and the associated health hazards [13]. Information pertaining to environmental degradation was garnered through discussions with farmers and Department of Agricultural Extension (DAE) personnel in Bangladesh (Table 3).

DISCUSSION

Integrated pest management (IPM) is a highly effective method for managing pest insect species [24]. This management system enriches ecosystem processes and can provide a healthy, sustainable food source for a range of organisms [25]. Rice agroecosystems have evolved over the last 5,000 years under human management; for this reason human beings are considered to be an integral component. An abundance of insect species are supported through a series of complex interactions that comprise the food web that at its base is maintained by rice [26]. In this agroecosystem, the food web can serve as a tool to help improve natural ecosystem functions. However, function is dependent upon the continued propagation of predator insect species. Ecosystem function in a healthy rice agroecosystem can be improved through protecting and encouraging those insects that prey on pest species [27]. In conventional pest management strategies the role of predator species is minimized or even eliminated by heavy pesticide application (Table 3). In China, 889 insect species that prey on pest insects have been identified. These predator insect species account for 64.74% of all naturally occurring enemies of pest species [28].

Increased pesticide use has led to an increase in soil, water, and air pollution. These increases have in turn had an adverse impact on human health and rice quality [29]. The incorrect use or overuse of pesticides can harm plants, aquatic organisms, livestock, and humans. Jacobsen and Hjelmsø [30] reported that pesticide use has a higher cost to benefit ratio with respect to yields and can have negative impacts on beneficial microbes. The excessive use of pesticides, in the search for higher yields, has triggered several ecological crises (Table 3). These crises include pesticide resistance, resurgence of insect pests, pesticide poisoning, environmental toxicity, destruction and elimination of natural predator species, effects on nontarget organisms, disruption in the food web, and increased prevalence of pesticide residues in food [31, 32]. These conditions could be effectively mitigated through implementation of a well-designed IPM strategy [32, 33]. Despite several IPM strategies that have been developed and promoted for rice production, adoption remains low due to the "chemical-free" strategies being less effective when compared to heavy insecticide application. Successful IPM strategies need to incorporate a combination of techniques and low to moderate insecticides application [34].

Integrated pest management and farmer practices both have a significant effect on the number of rice hills, abundance of healthy tillers, and the percentage of white head and dead heart (Table 3). In addition these practices can affect grain weight and yield. We found the incidence of dead heart and white heart both to be relatively lower in the IPM treated plots; the percentage of white head and dead heart both was observed to be lower than established critical limits [35]. Other parameters such as hills, tiller, and grain weight were variable, but yield (7.4 tonne/ha) was higher in the IPM treated plots (Table 3). Singh et al. [36] conducted similar research on the effectiveness of IPM for Rice Yellow Stem Borer (Scirpophaga incertulas) in rice-wheat cropping systems. They found similar incidence of dead heart and white head in their IPM treatments. Several other studies have shown that a well-designed IPM program can control pest insects in an ecologically friendly manner [37].

One hundred fifty-nine rice insect species are recorded in the rice agroecosystem. Among these species, 20–23 species have been found to be most damaging in Bangladesh [38]. Due to lack of knowledge pertaining to insect life cycles and pest species, farmers often apply pesticides at higher rates and at inappropriate times. A majority of farmers are not concerned with the negative impacts pesticide use may have on fish, wildlife, humans, soil, and beneficial insect species [39]. Additional education and training are needed to better inform farmers of responsible pest management techniques and appropriate pesticide application. This could be accomplished through television programming, increased contact with extension personnel, and farm credit services. An increased awareness of the correct application of agrochemicals, pest monitoring, and timely removal of weeds and diseased plants will protect the environment and decrease the financial burden on farmers [40]. By increasing the number of IPM-trained farmers, smaller quantities of pesticides will be put into the rice agroecosystem [32].

Most farmers do not understand the impacts pesticide use can have on their health. This is most often due to the lack of knowledge surrounding pesticide's mode of action. Research has been conducted on health disorders associated with the use of nonrecommended pesticides and inappropriate applications rates. One study examined negative health effects observed in rural Vietnamese communities. It was found that a majority of farmers suffer from fatigue, headache, and dizziness, most often occurring during and after pesticide application. These neurological symptoms were found to be the result of cholinesterase inhibition [41]. Insecticides are moderately hazardous due to the known health effects associated with exposure [42, 43]. Health problems associated with overuse of pesticides have risen despite increased availability and use of protective measures [43, 44].

Farmers often become dependent on pesticides without giving consideration to the environmental consequences. Unfortunately this is a common occurrence because there is less concern for environmental protection when food production takes precedence, as is often the case in the developing world. Recently there is a growing concern among scientists about the fate of pesticides in these agricultural settings [6]. The use of IPM can reduce pesticides applications up to 85% [45]. Selective pesticide use can reduce pest populations and improve ecosystem services. A study was examined on pesticide use and attitudes toward pest management strategies among rice and rice-fish farmers in the Mekong Delta

[43]. They identified patterns of pesticide use and possible integration of IPM strategies for control of major rice pests. They found that farmers who did not employ IPM methods frequently experienced negative health effects associated with pesticide use [43]. Integration of IPM practices may serve to improve farmer awareness of sustainable practices through emphasizing the importance of healthy rice ecosystems.

IPM techniques are able to increase agricultural productivity while reducing pesticide use in crop systems. Pretty et al. [46] found IPM technology to be effective at reducing pesticide use by around 71%, while increasing yields by 42%. In China, annual sweet potato production has increased by 30%, due to implementation of IPM techniques. The economic benefits of IPM have been included in annual revenue reports in the United States [47, 48]. In Africa, the introduction of an exotic wasp, Anagyrus lopezi (DeSantis), for the control of cassava mealybug, Phenacoccus manihoti Matile-Ferrero, has resulted in an average annual return of 300 dollars (US) per farmer [47, 49]. Additional reported benefits for implementation of IPM strategies range from 100 to 500 dollars (US) per hectare of productive land [50]. In the West Indies IPM has been used in banana cultivation and has led to a 65% decrease in pesticide use over the last 10 years [51]. Research was conducted on IPM and non-IPM treated farms in India for productivity and sustainability; they found that sustainable production was higher in the IPM treated farms [32]. Numerous studies have concluded that IPM techniques are extremely efficient and environmentally friendly and are the foundation of sustainable cropping systems. It is worth noting that no single technology can solve all pest problems.

Most rice farmers currently have practices that are similar to designed IPM strategies. Despite this, due to a lack of proper training on IPM practices in the rice ecosystem, farmer activities have been damaging to both the environment and human health [52]. An increased reliance on pesticides has proven to be unsustainable and not cost efficient [23, 43, 53]. Pesticides compatible with IPM strategies have been shown to be beneficial for the control of rice pests during the germination, tillering, vegetative growth, and flowering [54]. IPM strategies that employ appropriate pesticide use have consistently been shown to increase rice production [21, 55].

We conclude that to increase adoption of IPM strategies it is necessary to increase farmer awareness and knowledge concerning environmental pollution that could be mitigated with successful implementation of these techniques. It is important to improve farmers' pest management practices through examining current strategies and building upon preexisting knowledge. Additional research is needed to identify insect resistant rice varieties, analyze the patterns and use of pesticides, investigate constraints to controlling rice pests, examine the level of knowledge that farmers have concerning environmental pollution due to pesticide application, and explore alternative ways to reduce pesticide use.

CONCLUSION

Integrated pest management practices are infrequently being utilized for pest control in rice agroecosystems. The principle benefits of this technique include improved environmental conditions and healthier cropping systems. Healthy agroecosystems can provide higher rice yields than current management strategies. In our study BRRI dhan 29 rice yielded 7.4 tonne/ha in the IPM treated plots. According to BRRI (2015) the cultivar BRRI dhan 29 is capable of yields from 6 to 8 tonne/ha in Bangladesh agroclimatic zones (http://www.brri.gov.bd/). Our study indicates that an IPM strategy can have a positive effect on yield performance while lessening the environmental impacts associated with chemical intensive pest control.

COMPETING INTERESTS

The authors declare that there are no competing interests regarding the publication of this paper.

ACKNOWLEDGMENTS

The authors would like to thank the Integrated Agriculture Productivity Project (IAPP) and the Ministry of Agriculture in Bangladesh for providing funding and research facilities.

REFERENCES

1. International Rice Research Institute (IRRI), Bringing Hope, Improving Lives: Strategic Plan 2007–2015, International Rice Research Institute (IRRI), Los Banos, Laguna, Philippines, 2006.
2. Food and Agriculture Organization Stat (FAOSTAT), 2012, http://faostat.fao.org/.
3. Department of Agriculture Extension (DAE), 2010, http://www.dae.gov.bd/.
4. S. Petit, K. Haysom, R. Pywell et al., "Habitat-based models for predicting the occurrence of ground-beetles in arable landscapes: two alternative approaches," Agriculture, Ecosystems and Environment, vol. 95, no. 1, pp. 19–28, 2003.
5. B. D. Roitberg, "Why pest management needs behavioral ecology and vice versa," Entomological Research, vol. 37, no. 1, pp. 14–18, 2007.
6. F. Vinatier, F. Lescourret, P.-F. Duyck, and P. Tixier, "From IBM to IPM: using individual-based models to design the spatial arrangement of traps and crops in integrated pest management strategies,"Agriculture, Ecosystems & Environment, vol. 146, no. 1, pp. 52–59, 2012.
7. M. Rahman, "Problems and suggestions for farmers' adoption of IPM practices in rice (Oryza sativa L) Cultivation," Bangladesh Journal of Agricultural Research, vol. 37, no. 1, pp. 121–128, 2012.
8. C. Thorburn, "The rise and demise of integrated pest management in rice in Indonesia," Insects, vol. 6, no. 2, pp. 381–408, 2015.
9. J. Mangan and M. S. Mangan, "A comparison of two IPM training strategies in China: the importance of concepts of the rice ecosystem for sustainable insect pest management," Agriculture and Human Values, vol. 15, no. 3, pp. 209–221, 1998.
10. National Pesticides Information Center in USA, 2015, http://npic.orst.edu/envir/beneficial/index.html.
11. S. K. Srivastava, D. K. R. Biswas, B. K. G. Garg, N. M.

M. M. Haque, P. Ijaj, and S. N. Tiwari, Management of Stem Borers of Rice and Wheat in Rice-wheat System of Pakistan, Nepal, India and Bangladesh, vol. 17 ofRice-Wheat Consortium Paper, Rice-Wheat Consortium for the Indo-Gangetic Plains, New Delhi, India, 2004.

12. Department of Agriculture Extension (DAE), 2011, http://www.dae.gov.bd/site/search?key=Department+of+Agriculture+Extension+%28DAE%29%2C+2011.

13. A. Khaliq, M. Javed, M. Sohail, and M. Sagheer, "Environmental effects on insects and their population dynamics," Journal of Entomology and Zoology Studies, vol. 2, no. 2, pp. 1–7, 2014.

14. S. Chatterjee and P. Mondal, "Management of rice yellow stem borer, Scirpophaga incertulas Walker using some bio rational insecticides," Journal of Bio Pesticides, vol. 7, pp. 143–147, 2014.

15. M. Mohiuddin, M. M. Hossain, A. K. M. M. Rahman, and M. S. Palash, "Socio-economic study of insecticide use on vegetable cultivation at farm level in Chittagong region," Journal of the Bangladesh Agricultural University, vol. 7, no. 2, pp. 343–350, 2009.

16. M. H. Kabir and R. Rainis, "Do farmers not widely adopt environmentally friendly technologies? Lesson from Integrated Pest Management (IPM)," Modern Applied Science, vol. 9, no. 3, pp. 208–215, 2015.

17. J. Pretty and Z. P. Bharucha, "Integrated Pest Management for sustainable intensification of agriculture in Asia and Africa," Insects, vol. 6, no. 1, pp. 152–182, 2015.

18. S. E. Naranjo, P. C. Ellsworth, and G. B. Frisvold, "Economic value of biological control in integrated pest management of managed plant systems," Annual Review of Entomology, vol. 60, pp. 621–645, 2015.

19. United States Environmental Protection Agency (EPA), 2014, http://www.epa.gov/agriculture/tipm.html.

20. US Environmental Protection Agency, "Agricultural Pesticides: management improvements needed to further promote Integrated Pest Management," 2015.

21. S. Arora, I. Mukherji, A. Kumar, and R. K. Tanwar, "Pesticide residue analysis of soil, water, and grain of IPM basmati rice," Environmental Monitoring and Assessment, vol. 186, no. 12, pp. 8765–8772, 2014.

22. Bangladesh Bureau of Statistics Region Census (BBS), 2011, http://www.bbs.gov.bd/WebTestApplication/userfiles/Image/LatestReports/YB2011.pdf.

23. M. Z. Alam, "Survey and assessment of insect management technologies and environmental impact on rice ecosystem of Bangladesh," International Journal of Applied Research and Studies, vol. 2, no. 4, pp. 1–16, 2013.

24. D. Abrol, Integrated Pest Management: Current Concepts and Ecological Perspective, 2013.

25. R. J. Hillocks and J. E. Cooper, "Integrated Pest Management—can it contribute to sustainable food production in Europe with decreased reliance on conventional pesticides?" 2012, http://www.eucipm.org/docs/IPM-Can-it-Deliver-DiscussionPaperNo1b.pdf.

26. S. K. Redfern, N. Azzu, and J. S. Binamira, "Rice in Southeast Asia: facing risks and vulnerabilities to respond to climate change," 2015, http://www.fao.org/docrep/017/i3084e/i3084e18.pdf.

27. M. Allara, S. Kugbei, F. Dusunceli, and G. Gbehounou, "Coping with changes in cropping systems: plant pests and seeds," in Proceedings of the Joint FAO/OECD Workshop on Building Resilience for Adaptation to Climate Change in the Agriculture Sector, pp. 91–102, Rome, Italy, April 2012.

28. Y. G. Lou, G. R. Zhang, W. Q. Zhang, Y. Hu, and J. Zhang, "Biological control of rice insect pests in China," Biological Control, vol. 67, pp. 8–20, 2013.

29. M. A. Sattar, M. Z. Alam, and M. M. Rahaman, "The farm use of pesticides and the guidelines of pesticides research in Bangladesh," Bangladesh Journal of Environmental Science, vol. 10, pp. 313–317, 2004.

30. C. S. Jacobsen and M. H. Hjelmsø, "Agricultural soils, pesticides and microbial diversity," Current Opinion in Biotechnology, vol. 27, pp. 15–20, 2014.

31. M. W. Aktar, D. Sengupta, and A. Chowdhury, "Impact of pesticides use in agriculture: their benefits and hazards," Interdisciplinary Toxicology, vol. 2, no. 1, pp. 1–12, 2009.

32. R. Sharma, R. Peshin, U. Shankar, V. Kaul, and S. Sharma, "Impact evaluation indicators of an Integrated Pest Management program in vegetable crops in the subtropical region of Jammu and Kashmir, India,"Crop Protection, vol. 67, pp. 191–199, 2015.

33. M. J. Mohd Fuad, A. B. Junaidi, A. Habibah et al., "The impact of pesticides on paddy farmers and ecosystem," Advances in Natural and Applied Sciences, vol. 6, no. 1, pp. 65–70, 2012.

34. R. Srinivasan, "Integrating biopesticides in pest management strategies for tropical vegetable production," Journal of Bio-pesticides, vol. 5, pp. 36–45, 2012.

35. M. Bux, M. H. Khan, N. Ahmad, M. Tofique, and M. Ismail, "Field comparison of different rice (Oryza sativa L) genotypes for their resistance against rice stem borers (Pyralidae: Lepidoptera)," Pakistan Journal of Agriculture Agricultural Engineering Veterinary Sciences, vol. 29, no. 2, pp. 137–145, 2013.

36. D. Singh, A. K. Singh, and A. Kumar, "On-farm evaluation of integrated management of rice yellow stem borer (Scirpophaga incertulas Walk.) in Rice-Wheat Cropping System under Low Land Condition," Journal of AgriSearch, vol. 1, no. 1, pp. 40–44, 2014.

37. C. O. Ehi-Eromosele, O. C. Nwinyi, and O. O. Ajani, "Integrated pest management," in Weed and Pest Control—Conventional and New Challenges, S. Soloneski and M. Larramendy, Eds., chapter 5, InTech, Rijeka, Croatia, 2013.

38. S. Alam, "Checklist of rice insect pests of Bangladesh," in Literature Review of Insect Pests and Diseases of Rice in Bangladesh, pp. 79–90, Bangladesh Rice Research Institute (BRRI), Dhaka, Bangladesh, 1977.

39. P. C. Jepson, M. Guzy, K. Blaustein et al., "Measuring pesticide ecological and health risks in West African agriculture to establish an enabling environment for

sustainable intensification," Philosophical Transactions of the Royal Society of London Series B: Biological Sciences, vol. 369, no. 1639, Article ID 20130491, 2014.

40. S. Parveen, "Rice farmers› knowledge about the effects of pesticides on environmental pollution in Bangladesh," Bangladesh Research Publications Journal, vol. 3, no. 4, pp. 1214–1227, 2010.

41. S. B. Nerilo, F. A. Martins, L. B. Nerilo et al., "Pesticide use and cholinesterase inhibition in small-scale agricultural workers in southern Brazil," Brazilian Journal of Pharmaceutical Sciences, vol. 50, no. 4, pp. 783–791, 2014.

42. World Health Organization (WHO), "Preventive disease through healthy environments," 2010, http://www.who.int/ipcs/features/hazardous_pesticides.pdf.

43. H. Berga and N. T. Tamb, "Use of pesticides and attitude to pest management strategies among rice and rice-fish farmers in the Mekong Delta, Vietnam," International Journal of Pest Management, vol. 58, no. 2, pp. 153–164, 2012.

44. H. Andersson, D. Tago, and N. Treich, "Pesticides and health: a review of evidence on health effects, valuation of risks, and benefit-cost analysis," in Preference Measurement in Health, G. Blomquist and K. Bolin, Eds., Advances in Health Economics and Health Services Research, pp. 1–61, 2014, http://www2.toulouse.inra.fr/lerna/travaux/cahiers2014/14.05.406.pdf.

45. G. P. Fitt, "Critical issues in pest management for a future with sustainable biofuel cropping," Current Opinion in Environmental Sustainability, vol. 3, no. 1-2, pp. 71–74, 2011.

46. J. N. Pretty, A. D. Noble, D. Bossio et al., "Resource-conserving agriculture increases yields in developing countries," Environmental Science & Technology, vol. 40, no. 4, pp. 1114–1119, 2006. ·

47. The CGIAR, "Integrated Pest Management and Crop Health—bringing together sustainable agroecosystems and people's health," White Paper, SP-IPM Secretariat, International Institute of Tropical Agriculture (IITA), Ibadan, Nigeria, 2010.

48. J. J. Farrar, M. E. Baur, and S. Elliott, "Impacts of the regional integrated pest management competitive grants program in the Western United States," July 2015, http://westernipm.org/index.cfm/about-the-center/publications/special-reports/western-ripm-retrospective-full-pdf/.

49. International Institute of Tropical Agriculture (IITA), "Integrated Pest management," 2015, http://www.iita.org/c/document_library/get_file?uuid=60a1a863-45a5-4bcd-bb57-ef30acea07ac&groupId=25357.

50. O. Ortiz, J. Kroschel, J. Alcázar, R. Orrego, and W. Pradel, "Evaluating dissemination and impact of IPM: lessons from case studies of potato and sweet potato IPM in Peru and other Latin American countries," in Integrated Pest Management: Dissemination and Impact, R. Peshin and A. K. Dhawan, Eds., pp. 419–434, Springer, Dordrecht, Netherlands, 2009.

51. F. X. Côte, C. Abadie, R. Achard et al., "Integrated pest management approaches developed in the french west indies to reduce pesticide use in banana production systems," Acta Horticulturae, vol. 828, pp. 375–382, 2009.

52. C. Hall, B. Knight, S. Ringrose, and O. Knox, "What have been the farm-level economic impacts of the global cultivation of GM crops?" 2013, http://www.environmentalevidence.org/wp-content/uploads/2014/07/CEE11-002.pdf.

53. J. J. Rasmussen, P. Wiberg-Larsen, A. Baattrup-Pedersen, R. J. Monberg, and B. Kronvang, "Impacts of pesticides and natural stressors on leaf litter decomposition in agricultural streams," Science of the Total Environment, vol. 416, pp. 148–155, 2012.

54. R. A. da Silva, E. D. Quintela, G. M. Mascarin, J. A. F. Barrigossi, and L. M. Lião, "Compatibility of conventional agrochemicals used in rice crops with the entomopathogenic fungus Metarhizium anisopliae," Scientia Agricola, vol. 70, no. 3, pp. 152–160, 2013.

55. O. Mück, "Integrated Pest Management for Rice Production in Nigeria," 2015, http://cari-project.org/wp-content/uploads/2015/05/CARI-Rice-IPM-Nigeria-Draft-Study.pdf.

CHAPTER 7

EFFECTS OF INTEGRATED PEST MANAGEMENT ON PEST DAMAGE AND YIELD COMPONENTS IN A RICE AGRO-ECOSYSTEM

Mohammad Zahangeer Alam, Alex R. Crump, Md. Manjurul Haque, Md. Sirajul Islam, Emran Hossain, Sabiha B. Hasan, Shahela B. Hasan and Md. Sakhawat Hossain

Keywords: Rice, Dead Heart, White Head, Integrated Pest Management, pesticide, agro-ecosystem, environment

INTRODUCTION

Integrated Pest Management (IPM) is an effective, environmentally sound approach to pest management (Kabir and Rainis, 2015). It provides for the protection of beneficial insects, as well as prevention of secondary pest outbreaks, pest resurgence, and the spread of disease. IPM strategies aim to protect air, water, and soil resources while meeting specific production objectives (Mangan and Mangan, 1998; National Pesticides Information Center in USA, 2015[1]). IPM combines the use of a variety of pest-control methods in a way that facilitates biological control of pest insects in crops in order to improve economic, public-health, and environmental outcomes. Key components of effective IPM strategies are monitoring of pest populations, recognizing pest-resistant plant varieties, and modifying cultural, mechanical, chemical, and biological controls as needed to achieve production goals (Adams, 1996). Including farmers' traditional agricultural knowledge of insect behavior and life cycles is essential for developing a successful IPM plan (Petit et al., 2003; Roitberg, 2007; Vinatier et al., 2012), as is taking into account their current agricultural practices and experience in a given agro-ecosystem (Rahman, 2012; Craig, 2015).

Citation: Alam MZ, Crump AR, Haque MM, Islam MS, Hossain E, Hasan SB, Hasan SB and Hossain MS (2016) Effects of Integrated Pest Management on Pest Damage and Yield Components in a Rice Agro-Ecosystem. Front. Environ. Sci. 4:22. doi: 10.3389/fenvs.2016.00022.

The focus of IPM is to protect and encourage natural predators of pest insects (Naranjo et al., 2015). Understanding the life cycles and the interactions of pest insects and predators provides the basis for successful design and implementation of an IPM strategy. In addition to taking advantage of predator-prey interactions, IPM often includes the judicious use of pesticides. Organic food systems employ similar methods but limit the use of pesticides produced from "non-natural" or synthetic sources (U.S. Environmental Protection Agency [US-EPA], 2014[2]). Worldwide, chemical pesticides have played a vital role in providing an abundant and inexpensive food source (US-EPA, 2015[3]). Despite an increase in food production, however, the persistent overuse of chemicals has resulted in a number of adverse environmental impacts such as, outbreak of secondary pests, decreasing of beneficial insects and the accumulation of toxins in the food webs (Arora et al., 2014). Continued research and development of sustainable and effective agricultural-pest-management techniques is essential if farmers are going to successfully adopt environmentally friendly pest-management strategies (Pretty and Bharucha, 2015).

As a result of the world's unprecedented population growth, food-production systems are being pushed to maximum efficiency (International Institute for Environment and Development [IIED], 2015[4]). Among the widely cultivated food crops, cereals are recognized as an important component of food security. The Food and Agriculture Organization (FAO)[5] estimates that demand for food will more than double by the year 2050 (2009). To meet this demand, cereal-crop production in developing countries must increase by 40%

(Eze and Echezona, 2012). Rice, one of those cereal crops, is a staple food source for more than fifty percent of the world›s population (International Rice Research Institute [IRRI], 2006[6]). Rice crops either directly or indirectly sustain 3 billion people (Food and Agriculture Organization Stat. [FAOSTAT], 2012[7]). In 2010, 154 million hectares of rice was cultivated worldwide FAOSTAT, 2012. Asia comprised 137 million hectares, with 48 million hectares harvested in Southeast Asia (FAOSTAT, 2012). In Bangladesh 80 percent of agricultural land is dedicated to rice cultivation (Department of Agriculture Extension in Bangladesh [DAE], 2010[8]).

Two-hundred and sixty-six insect species have been identified in rice ecosystems (DAE, 2011[9]). Of these, 42 species are considered to be pests (Srivastava et al., 2004). Pest insects can be categorized as either minor or major pests. Different pest insects cause severe damage to rice crops at different growth stages. The degree of damage depends on the growing season and environmental conditions (Khaliq et al., 2014). One pest species, Yellow Stem Borer (YSB-*Scirpophaga incertulas;* Lepidoptera Crambidae), is considered to be a major rice pest (Chatterjee and Mondal, 2014). Rice plants damaged by this insect develop unhealthy tillers and hills, Dead Heart (central leaf drying), and White Head (grain-less panicles), all of which affect the overall yield (Satpathi et al., 2012). According to DAE (2011) YSB populations can be reduced through several IPM methods, including: light trapping (placing a light source inside a funnel-shaped trap), hand picking eggs from rice leaves (Leaves with egg masses are removed from plants by hand during tillering and booting stages), sweeping (using a funnel shaped net to capture insects), perching (installing a branch or pole which serves as a resting place for predatory birds), cultivation of insect resistant rice varieties, using bio-agents [releasing predator insects such as Long horned grasshopper *Conocephalus longipennis* (Haan)], applying Economic Threshold Level (ETL) based insecticides (The ETL is the point at which the value of the crop destroyed exceeds the cost of controlling the pest). Most rice farmers currently rely on heavy insecticide applications rather than IPM strategies for pest insect control (Mohiuddin et al., 2009).

We believe the wide-scale adoption of IPM in rice agro-ecosystems could provide a net benefit to farmers. The IPM program used in this study is from the "Farmer Field School" (FFS) model. Continuing farmer education and training play a vital role in advancing IPM practices, and this innovative IPM-education program provides training for the identification of beneficial and pest insect species, as well as the recognition of associated impacts on crop yields (Craig, 2015).

For this study our IPM strategy consisted of egg-mass collection, perching, sweeping, and ETL-based insecticide application. Non-IPM treatments consisted of egg-mass collection, perching and sweeping without insecticide application, prophylactic insecticide application only, and current pest-management practices. We studied the effects of IPM and non-IPM strategies on rice-yield components. Our working hypothesis was that fields in which we used IPM strategies would have less crop damage and higher rice yields compared with fields treated with non-IPM strategies.

MATERIALS AND METHODS

Geographical Position and Experimental Site

We chose the Barisal region of Bangladesh for our study. Because of the diversity of rice grown and the consistently high yields, it is one of the most important regions for rice cultivation in Bangladesh. For several decades farmers in this region have been plagued by a number of pest-related issues. These include the emergence of new pest-insect biotypes, pesticide resistance, the absence of predatory insects, and yield reduction (DAE, 2010). Barisal is in southern Bangladesh, and encompasses an area of approximately 3000 square kilometers (Figure 1). Our field sites are located at 22°42′00″N 90°22′00″E22.7000°N, 90.3667°E in an agricultural area that comprises approximately 324.41 km^2 (BBS, 2011[10]).

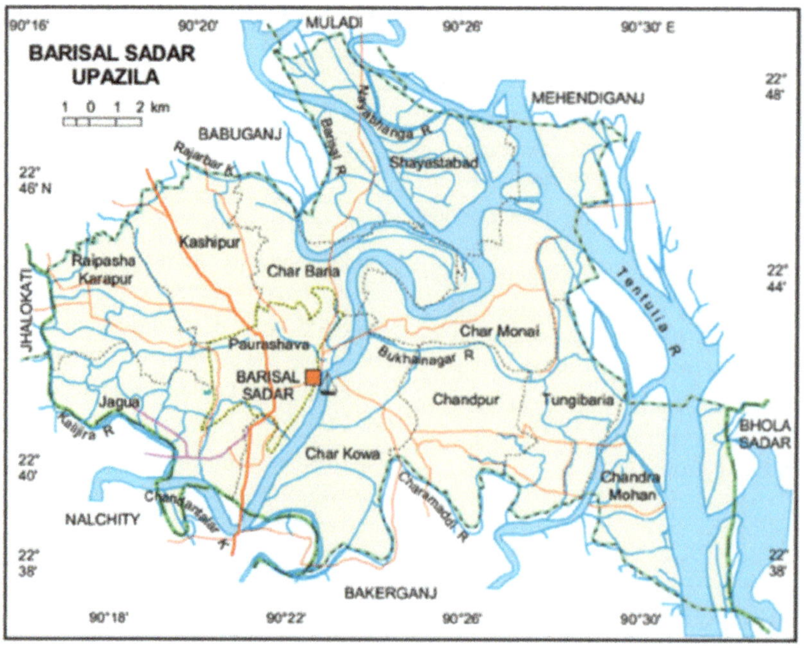

Figure 1. Study area.

In 2011 an inception meeting was organized with local farmers and the DAE at the DAE offices in Barisal Sadar, Bangladesh. The purpose of this meeting was to determine an appropriate experimental field site and target pest. It was decided that YSB (Yellow Stem Borer) would be designated as the major rice pest for this study. Site selection was based on uniformity of cropland, presence of major rice insects, cultivated rice varieties, transplanting season, and previous pest-management strategies.

Experimental Design

The experiment was conducted during November–March growing season of 2011-12 and 2012-13 at three locations in the Barisal region; (Sardarpara, Uttarsagardi, and Gaptala). At each location we had three replicate plots for each of the 4 treatments, resulting in 36 total plots. Each plot was 100 m^2 with 1.5 m buffer strips. Treatments were as follows; T_1: collection of egg masses, sweeping, and perching; T_2: collection of egg masses, sweeping, perching, and ETL-based insecticide application; T_3: application of prophylactic insecticides; and T_4: standard pest-control strategies currently employed by farmers in the Barisal region (Table 1).

Table 1. Description of all field treatments.

Treatments	Details
T_1	Collection of egg masses, sweeping and perching. Egg mass collection and sweeping were done on 7 day intervals.
T_2	Collection of egg masses, sweeping, perching and Economic Threshold Level (ETL) based insecticide application. Collection of egg masses, sweeping, and perching activities were the same as T_1. The application rate of Cartap (Suntap 50SP) insecticides was 1.4 kg/ha (0.14 g/m^2). Insecticide concentration was 0.46 mg/m^2. Insecticide was applied during tillering and booting stage of rice plants. Insecticide was applied on afternoons when the forecast called for low wind speeds.
T_3	Prophylactic insecticide application. We considered ETL at early tillering, mid-tillering, and booting stages of rice plants. During these stages we applied Cartap insecticide a total of three times with an application rate of 0.14 g/m^2. Insecticide concentration was 0.46 mg/m^2. Insecticide was applied on afternoons when the forecast called for low wind speeds.
T_4	Farmer's current practices. Pest control practices consisted of allowing farmers to apply pesticides according to their understanding of appropriate pest management practices. No consideration was given to ETL based insecticide application, egg mass collection, sweeping, insect and plant life stages, or environmental contamination. Insecticide application rate was 1.5 kg/ha (0.15 g/m^2). The concentration of Carrtap insecticide was 0.5 mg/m^2 at early, mid and late tillering and booting stages of rice plants. They did not follow weather forecasts to minimize wind drift.

Agronomic and IPM Methods

In order to maintain food security, the Bangladesh Rice Research Institute (BRRI) releases many rice cultivars to farmers. Two cultivars, BRRI Dhan 28 and BRRI Dhan 29, are popular in the Barisal region because of their yield performance. We selected BRRI Dhan 29 for our study.

Participating researchers used agronomic practices established by the Bangladesh Rice Research Institute (BRRI). These practices include: soil preparation with a motorized tiller; fertilizer application (250 g urea, 300 g Muriate of Potash (MOP) and 400 g Triple Super Phosphate (TSP)/Decimal or 40 m^2); irrigation; and hand weeding. Nursery-raised rice seedlings were transplanted in the experimental plots with 20 cm × 20 cm spacing. Our study examined common IPM strategies employed in rice agro-ecosystems including egg-mass collection, perching, sweeping, and ETL-based insecticide application. The ETL For YSB, the ETL can be defined as the presence of 2 egg masses or 1–2 moths per m^2, and 5 percent Dead Heart between planting and early tillering. Ten percent of Dead Heart, or 1 moth, or 2 egg masses per m^2 at mid tillering stage also indicates ETL. Between panicle initiation and booting stage, 2 egg masses or 1 moth per m^2 can be considered ETL (Prakash et al., 2014). Timing, frequency, application rate, and insecticide concentration for each treated field are described in Table 1.

Description of Pest Control Methods

YSB egg masses are multilayered, densely covered with hairs, and they are usually found on the leaves of rice plants. Leaves with egg masses were removed from plants by hand during tillering and booting stages during field visits in daylight hours. During weekly visits to the field sites, approximately 10 min were spent on egg-mass collection. Time dedicated to collecting egg masses was dependent on the presence and number of observed egg masses.

A sweep net is a funnel-shaped net attached to a long handle that can be swept through the foliage of the rice plants. The net is used to capture adult insects and to "scoop up" larvae and eggs. Sweeping is accomplished by quickly moving the net from side in a shallow figure eight pattern. A person practicing this method walks through the rice field while forcefully sweeping the net through the rice leaves. Sweeping continues for some distance (1–2 passes) before the net is emptied. This technique can be used to determine whether a sufficient number of pest insects are present to justify an ETL-based insecticide application. Sweeping is typically done before flowering. During weekly visits to the field site, approximately 10 min were spent sweeping each plot that required sweeping (based on the number of flying insects observed).

Perching is a treatment in which bamboo poles with branches, or perches, are placed in rice fields. These perches serve as nesting structures for predatory birds. Standard spacing for perching poles is 10 m × 10 m. We installed perching poles in each plot on a horizontal axis with an east–west orientation to lessen damage from prevailing winds. This placement also increased perch visibility for predatory birds. We installed one perch in each plot when perching was a component of the IPM treatment.

Data Collection

At each location we recorded the total number of hills (rice plants) per 10 m^2 and the average number of healthy tillers (each rice plant has approximately 20–30 tillers) during tillering and flowering stages. Measurements were taken from 20 plants that were randomly selected from diagonal transects in each plot. Percentage of Dead Heart during the vegetative stage and White Head through panicle-initiation stage were recorded from each plot. Rice yield (ton/ha) was determined post-harvest (Tables 2,7).

Table 2. Effect of location on tillers, hills, dead heart, white head, and yield of rice crops in 2011-12.

Location	Healthy tiller–tillering	Healthy tiller-flowering	Percent of Dead Heart	Percent of White Head	Total no of Hills/10 m^2	Average Yield (ton /ha)
Sarderpara	1.36 (23.16)	1.35 (22.58)	1.32 (1.83)	1.51 (2.32)	2.37 (240.50)	6.89
Uttarsagardi	1.36 (23.33)	1.36 (23.16)	1.37 (1.91)	1.59 (2.56)	2.39 (247.66)	6.95
Gabtala	1.37 (23.66)	1.36 (23.00)	1.40 (2.01)	1.49 (2.25)	2.37 (240.66)	6.89
SE	0.840 (0.450)	0.452 (0.230)	0.233 (0.608)	0.316 (0.971)	0.96 (5.00)	0.115

Values outside parenthesis are the transformed mean values. Numbers in parenthesis are the actual means.

Data Collection: Environmental Impacts

We interviewed three groups of people who had been living near each field site in 2013. Each group was comprised of 15 people at each location. Several questions were discussed with interviewees, such as the status of IPM in their rice fields and their understanding of major threats to the environment. During the course of the discussion, we noted that several people brought up their experiences with negative environmental impacts that are associated with overuse and/or inappropriate pesticide applications. Information pertaining to environmental degradation was also garnered through discussions with DAE personnel in Bangladesh (Supplementary Material).

Statistical Analysis

The experiment was laid out as a Randomized Complete Block Design (RCBD). Growing seasons were analyzed separately. Location and treatment were considered as independent variables. Dependent variables were: the number of hills, number of tillers at tillering and flowering stages, the percent of Dead Heart and White Head, and crop yields. Data was analyzed using ANOVA with a fixed-effect model to evaluate significant differences ($p \leq 0.05$) between location, treatment, and statistical interactions of treatment and location on yield components for the 2011-12 and 2012-13 growing seasons. Pairwise comparisons were made using Duncan's Multiple Range Test (DMRT) (Crop Stat 7.2, International Rice Research Institute [IRRI], Philippines; M-STAT, Michigan State University (MSU), USA). Count data was log transformed and calculated percentages were square-root transformed. Residuals were normally distributed.

RESULTS

Number of Tillers during Tillering and Flowering Stages

Field location was not significant at tillering or flowering stages for the 2011-12 or 2012-13 growing season (Tables 2, 5). In the 2011-12 growing season the number of healthy tillers at flowering was statistically similar for T_1,T_3,and T_4 treatments. During this growing season, the number of healthy tillers in T_2 was significantly different ($p \leq 0.05$; Table 3). In 2012-13 the number of healthy tillers at tillering was similar in the T_1, T_2, and T_4treated plots, with fewer tillers observed in the T_3 plots (Table 6). At flowering the number of healthy tillers in the T_2 treated plots was higher than the T_1,T_3, and T_4 treatments (Table 6). There was not a statistical interaction between treatment and field location (Tables 4, 7).

Table 3. Effect of treatment on tillers, hills, dead heart, white head and yield in 2011-12.

Treatments	Healthy tiller–tillering	Healthy tiller-flowering	Percent of Dead Heart	Percent of White Head	Total no of Hills/10 m^2	Average Yield (ton /ha)
T1	1.36 (23.22)	1.34 (22.33)b	1.53 (2.37)a	1.60 (2.60)a	2.41 (258.00)a	7.02b
T2	1.37 (24.00)	1.37 (23.66)a	1.10 (1.23)c	1.42 (2.04)c	2.40 (253.33)a	7.50a
T3	1.34 (22.33)	1.36 (23.00)ab	1.22 (1.52)b	1.56 (2.44)b	2.35 (228.66)b	6.28c
T4	1.37 (24.00)	1.35 (22.66)b	1.59 (2.56)a	1.55 (2.43)b	2.36 (231.77)b	6.83b
SE	0.970 (0.520)	0.522 (0.266)	0.269 (1.25)	0.365 (0.11)	0.11 (5.77)	0.133

Values outside parenthesis are the transformed mean values. Numbers in parenthesis are the actual means. Letters indicate significant differences between treatments on yield components.

Table 4. Keywords: Rice, Dead Heart, White Head, Integrated Pest Management, pesticide, agro-ecosystem, environment

Treatment	Healthy tiller -tillering stage			Healthy tiller- flowering Stage		
	Sarderpara	Uttarsagardi	Gabtala	Sarderpara	Uttarsagardi	Gabtala
T1	1.35 (23.00)	1.36 (23.00)	1.37 (23.66)	1.34 (22.33)	1.34 (22.33)	1.34 (22.33)
T2	1.35 (22.66)	1.39(24.66)	1.38 (24.66)	1.34 (22.33)	1.39 (24.66)	1.37 (24.00)
T3	1.34 (22.33)	1.35(22.66)	1.34 (22.00)	1.36 (23.00)	1.36 (23.00)	1.36 (23.00)
T4	1.39 (24.66)	1.36(23.00)	1.38 (24.33)	1.35 (22.66)	1.35 (22.66)	1.35 (22.66)
SE		0.168 (0.900)			0.904 (0.461)	
CV		2.1 (6.7)			1.2 (3.5)	
Treatment	**Percent of Dead Heart**			**Percent of White Head**		
	Sarderpara	Uttarsagardi	Gabtala	Sarderpara	Uttarsagardi	Gabtala
T1	1.49 (2.25)b	1.52 (2.33)ab	1.59 (2.53)a	1.60 (2.58)	1.65 (2.73)	1.57 (2.50)
T2	1.01 (1.03)cd	1.10 (1.23)cd	1.18 (1.43)c	1.40 (2.00)	1.46 (2.16)	1.39 (1.96)
T3	1.10 (1.22)cd	1.32 (1.76)c	1.25 (1.56)c	1.52 (2.33)	1.63 (2.66)	1.52 (2.33)
T4	1.68 (2.83)a	1.52 (2.33)ab	1.59 (2.53)a	1.53 (2.38)	1.64 (2.70)	1.48 (2.21)
SE		0.466 (0.121)			0.633 (0.194)	
CV		5.9 (11)			7.1 (14.1)	
Treatment	**Total no of Hills /10 m^2 area**			**Average Yield (ton/ha)**		
	Sarderpara	Uttarsagardi	Gabtala	Sarderpara	Uttarsagardi	Gabtala
T1	2.41 (261.33)	2.40 (251.33)	2.41 (261.33)	6.96	6.96	7.13
T2	2.38 (243.33)	2.40 (256.66)	2.41 (260.00)	7.46	7.56	7.49
T3	2.33 (218.00)	2.39 (250.00)	2.33 (218.00)	6.19	6.37	6.30
T4	2.37 (239.33)	2.36 (232.66)	2.34 (223.33)	6.97	6.90	6.64
SE		0.193 (10.00)			0.230	
CV		1.4 (7.1)			5.8	

Values outside parenthesis are the transformed mean values. Numbers in parenthesis are the actual means. Letters indicate significant differences for the interaction of location and treatment on yield components.

Number of Hills

The effect of field location on the average number of hills/10 m^2 was not significant for the 2011-12 or 2012-13 growing season (Tables 2, 5). For both growing seasons, the number of hills was statistically different ($p \leq 0.05$) between the T_1/T_3 and the T_2 / T_4treatments. In 2011-12 the average number of hills /10 m^2 for the T_1, T_2, T_3, and T_4 treatments was 258, 253, 228, and 231, respectively. Similarly, in 2012-13, the average number of hills was 248, 257, 226, and 224 for T_1, T_2, T_3 and T_4, respectively (Tables 3, 6). The interaction of treatment and field location was not statistically significant (Tables 4, 7).

Table 5. Effect of location on tillers, hills, dead heart, white head and yield of rice crops in 2012-13.

Location	Healthy tiller—tillering	Healthy tiller-flowering	Percent of Dead Heart	Percent of White Head	Total no of Hills/10m^2	Average Yield (ton /ha)
Sarderpara	1.37 (23.58)	1.34 (22.33)	1.47 (2.25)	1.62 (2.66)	2.37 (239.50)	6.79
Uttarsagardi	1.36 (23.08)	1.36 (23.16)	1.47 (2.20)	1.61 (2.62)	2.38 (241.58)	6.82
Gabtala	1.35 (23.00)	1.36 (23.33)	1.52 (2.34)	1.53 (2.36)	2.37 (236.41)	6.74
SE	0.89 (0.463)	0.93 (0.496)	0.311 (0.981)	0.320 (0.982)	0.78 (4.07)	0.113

Values outside parenthesis are the transformed mean values. Numbers in parenthesis are the actual means.

Table 6. Effect of treatment on tillers, hills, dead heart, white head and yield of rice crops in 2012-13.

Treatments	Healthy tiller—tillering	Healthy tiller-flowering	Percent of Dead Heart	Percent of White Head	Total no of Hills/10 m^2	Average Yield (ton /ha)
T1	1.36 (23.33)a	1.34 (22.33)b	1.61 (2.62)a	1.71 (2.94)a	2.39 (248.11)a	6.89b
T2	1.38 (24.55)a	1.39 (24.77)a	1.24 (1.55)b	1.43 (2.06)c	2.41 (257.77)a	7.32a
T3	1.33 (21.55)b	1.34 (22.22)b	1.44 (2.11)b	1.62 (2.64)a	2.35 (226.44)b	6.37cb
T4	1.36 (23.44)a	1.35 (22.44)b	1.66 (2.77)a	1.59 (2.55)ab	2.35 (224.33)b	6.56b
SE	0.103 (0.535)	0.107 (0.573)	0.359 (0.113)	0.369 (0.113)	0.901 (4.70)	0.131

Values outside parenthesis are the transformed mean values. Numbers in parenthesis are the actual means. Letters indicate significant differences between treatments on yield components.

Table 7. Interation effect of location and treatement on tillers, hills, dead heart, white head and yield of rice crops in 2012-13.

Treatment	Healthy tiller -tillering stage			Healthy tiller- flowering Stage		
	Sarderpara	Uttarsagardi	Gabtala	Sarderpara	Uttarsagardi	Gabtala
T1	1.37 (24.00)	1.34 (22.33)	1.37 (23.66)	1.34 (22.00)	1.35 (22.66)	1.34 (22.33)
T2	1.37 (24.00)	1.39 (25.00)	1.39 (24.66)	1.36 (23.33)	1.40 (25.66)	1.40 (25.33)
T3	1.34 (22.00)	1.34 (22.00)	1.31 (20.66)	1.33 (21.66)	1.33 (21.66)	1.36 (23.33)
T4	1.38 (24.33)	1.36 (23.00)	1.35 (23.00)	1.34 (22.66)	1.35 (22.66)	1.34 (22.33)
SE		0.179 (0.927)			0.186 (0.993)	
CV		2.3 (6.9)			2.4 (7.5)	
Treatment	Percent of Dead Heart			Persent of White Head		
	Sarderpara	Uttarsagardi	Gabtala	Sarderpara	Uttarsagardi	Gabtala
T1	1.72 (3.00)a	1.52 (2.33)ab	1.59 (2.53)a	1.77 (3.16)	1.70 (2.90)	1.66 (2.76)
T2	1.15 (1.33)bc	1.28 (1.66)b	1.28 (1.66)b	1.46 (2.16)	1.43 (2.06)	1.39 (1.96)
T3	1.28 (1.66)b	1.46 (2.16)b	1.57 (2.50)a	1.66 (2.76)	1.68 (2.83)	1.52 (2.33)
T4	1.72 (3.00)a	1.63 (2.66)a	1.63 (2.66)a	1.59 (2.56)	1.64 (2.70)	1.54 (2.38)
SE		0.622 (0.196)		0.64 (0.196)		
CV		7.2 (15)		7.0 (13.3)		
Treatment	Total no of Hills /10 m^2area			Average Yield (ton/ha)		
	Sarderpara	Uttarsagardi	Gabtala	Sarderpara	Uttarsagardi	Gabtala
T1	2.41 (258.33)	2.38 (244.66)	2.38 (241.33)	6.87	6.83	7.97
T2	2.42 (263.33)	2.40 (253.33)	2.40 (256.66)	7.33	7.50	7.13
T3	2.34 (224.66)	2.36 (233.33)	2.34 (221.33)	6.37	6.40	6.33
T4	2.32 (211.66)	2.37 (235.00)	2.35 (226.33)	6.60	6.56	6.53
SE		0.152 (8.14)			0.227	
CV		1.1 (5.9)			5.8	

Values outside parenthesis are the transformed mean values. Numbers in parenthesis are the actual means. Letters indicate significant differences for the interaction of location and treatment on yield components.

Dead Heart

The effect of location on Dead Heart percentages was not significant for either growing season (Tables 2, 5). The relationship between the percentage of Dead Heart and treatment was significant ($p \leq 0.05$) for both growing seasons (Tables 3, 6). During both growing seasons, the lowest percentages of Dead Heart (1.23 and 1.55) were found in the T_2treatments, and the highest were observed in the T_4 plots (2.56 and 2.77; Tables 3, 6). For both growing seasons, the percentage of Dead Heart was statistically similar in the T_1 and T_4 plots. The T_1-treated plot was statistically significant ($p \leq 0.05$) compared with the T_2 and T_3treatments in 2011-12 (Table 3). In the 2012-13 growing season, Dead Heart percentages were statistically different between T_2 and T_1 treated plots, as well as the T_3 and T_4 treated plots (Table 6). The interaction between treatment and location for Dead Heart was statistically significant ($p \leq 0.05$) for both growing seasons. The interaction between T_1 at Sarderpara was significant when compared with T_1 at Gabtala in 2011-12. Similarly, the interaction of treatment and location for T_3 at Sarderpara was significantly different when compared with T_3 at Gabtala in 2012-13 (Tables 4, 7).

White Head

The effect of location on White Head was not significant for either growing season (Tables 2, 5). Treatment was significant ($p \leq 0.05$) over both growing seasons (Tables 3, 6). For both growing seasons, the percentage of White Head (2.06 and 2.04) was significantly lower for the T_2 treatments compared with the other treatments (2.43–2.94). In 2011-12 the percentages of White Head were similar between T_3 and T_4 and was significantly higher for the T_1 treatment (Table 3). This similar result was also observed for the 2012-13 growing season (Table 6). The interaction of treatment and location for White Head was not significant (Tables 4, 7).

Yield

For both growing seasons, the effect of location on rice yields was non-significant (Tables 2, 5). The effect of treatment on yield was statistically significant ($p \leq 0.05$) for both growing seasons (Tables 3, 6). For the 2011-2012 growing season, rice yields were 7.02, 7.50, 6.28, and 6.83 ton/ha in the T_1, T_2, T_3, and T_4 treated plots, respectively. For the same plots in the 2012-2013 season, rice yields were found to be 6.89, 7.32, 6.37, and 6.56 ton/ha, respectively (Tables 3, 6). Statistically higher yields (7.3–7.5 ton/ha) were found in the T_2-treated plots compared with T_1, T_3, and T_4 treated plots (6.28–7.02 ton/ha, respectively) during both growing seasons (Tables3, 6). The interaction of treatment and location was non-significant (Tables 4, 7).

Negative Impacts

We found that farmers were overly reliant on Cartap insecticides for controlling pest insects in rice agro-ecosystems. According to interviews with community members who work and reside near the study area, as well as DAE personnel, major threats have arisen from the persistent overuse of pesticides. These threats include; lower populations of beneficial insects, outbreaks of secondary pests, high production cost-benefit ratio, noxious odors, rice-yield reduction, and an increased risk of health problems (Supplementary Material).

DISCUSSION

Rice has been cultivated by humans for 5000 years. Through those millennia farmers developed increasingly efficient systems of coping with environmental factors, including pest insects that threatened yields. When modern pesticides became available, they were marketed to farmers as a panacea for eradicating pests, saving labor, and increasing yields. Pesticides, however, came with a cost: environmental degradation and decreased crop resilience. To mitigate some of these negative impacts, IPM was developed as a pest-management strategy designed to blend traditional, knowledge-based pest-control methods and judicious use of pesticides. The goal is to improve yields, increase crop resilience, and end damage to environmental systems associated with overuse of pesticides.

When properly implemented, a pest-management strategy enriches ecosystem services and provides a sustainable food source for a range of beneficial insects (Hillocks and Cooper, 2012). Healthy rice agro-ecosystems support a diverse insect community by maintaining a complex food web (Redfern et al., 2015). The insects participating in this food web can help maintain ecosystem function as long as predator species (insects and others) continue to flourish (Allara et al., 2012). Two hundred and sixty-six insect species have been documented in rice agro-ecosystems (DAE, 2011). Predator insect species account for 64.74% of all insects, with the remainder classified as pest species (Lou et al., 2013). Since pesticides have become available, the number of predator species and populations within surviving predator species have declined. Current non-IPM pest-management strategies often involve over application of pesticides (Table 1), which has, in turn, resulted in decreased populations of beneficial insects and increased pest damage to a number of critical crop-yield components (Abrol, 2013). In this study, rice yield was typically higher in the IPM-treated plots as compared with plots treated with pesticides only (Tables 3, 6).

Rice tillers, hills, leaves, panicles, and grains are the most important yield components of rice crops. Reduction in the number of tillers, drying of central leaves (Dead Heart), and damage to the panicles (White Head) are indications of stem-borer infestation often visible in rice fields (Paul, 2007). When stem-borer damage becomes apparent, farmers often apply inappropriately large quantities of pesticides (Table 1). IPM and pesticide-only control methods both have a significant effect on rice-yield indicators. We found the incidence of Dead Heart and White Head were lower than 5% in the IPM treated plots (Tables 2, 6), which, in fact, is lower than established critical limits (5%). Egg masses were 2/m^2 or 1 moth/ m^2 from mid-tillering to booting stage.

As part of the IPM treatment, we applied ETL-based insecticides strategically, which minimized the pest population in IPM-treated plots. As a result rice yield was 7.5 ton/ha in the IPM-treated plots. These yields were significantly higher than the other treatments (Table 3). Singh et al. (2014) conducted similar research on the effectiveness of IPM for YSB in rice- and wheat-cropping systems. They found similar incidence of Dead Heart and White Head (<5%) in their IPM treatments. Other studies have shown that a well-designed IPM program can control pest insects in an ecological manner (Bux et al., 2013; Ehi-Eromosele et al., 2013). By reducing the incidence of Dead Heart and White Head in YSB-infested rice fields, IPM improves rice yields (Bux et al., 2013). It has also been demonstrated by DAE (2011) that IPM approaches minimize pest damage while maximizing yields and improving overall environmental quality (U.S-EPA, 2015).

According to Abrol (2013), IPM strategies that use appropriate ETL-based insecticide applications can be an effective method for managing pest insects. Rodrigo et al. (2013) demonstrated that appropriate pesticides used at application rates compatible with IPM strategies are beneficial for the control of rice pests during germination, vegetative stage, and flowering stage. In our study insecticides were applied at the tillering and booting stages in the IPM-treated plots (T_2) across all locations (Table 1). We found that higher than recommended rates of Cartap (0.15 g/m^2, recommended 0.14 g/m^2) were frequently applied in farmers' fields (Table 1). To mirror conventional practices, our study included treatments with three and four insecticide applications (T_3 and T_4, Table 1). We determined that IPM-treated plots (T_2) provided significantly higher yields and that conventional practices might have a negative impact on rice production. This result agrees with earlier studies, which found that pest-management systems that fail to use IPM strategies have lower average rice yields (Sattar et al., 2004; Alam, 2013).

Pest-control strategies that include appropriate pesticide use have consistently been shown to increase rice production (Arora et al., 2014; Muck, 2015). Jacobsen and Hjelmso (2014) report that heavy pesticide use often has a higher cost-to-yield ratio than IPM methods. Our IPM-treated plots had lower pesticide costs than non-IPM treatments. Net economic losses for non-IPM techniques have been documented in an annual revenue report (Consultative Group for International Agricultural Research[11] [CGIAR]), 2010; Farrar et al., 2015). Measures of productivity and sustainability are frequently higher in IPM-managed farms than non-IPM farms (Côte et al., 2009; Ortiz et al., 2009; Sharma et al., 2015). These studies conclude that IPM techniques are yield efficient, environmentally friendly, and can be the foundation of sustainable cropping systems (Hall et al., 2013). An over-reliance on pesticides has proven to be unsustainable, less cost effective, and less efficient for pest control (Berg and Tam, 2012; Rasmussen et al., 2012; Alam, 2013; Rodrigo et al., 2013).

Farmers often over-apply pesticides or spray at inappropriate times because they lack knowledge of the negative impacts associated with these practices (Jepson et al., 2014). Contributing to this is a lack of understanding about how insect life cycles correspond to the life stages of rice plants and at what stages pesticides are most effective. Additional education and training is needed to better inform farmers of responsible pest-management techniques and appropriate pesticide-application techniques (Parveen, 2010). Increasing the number of IPM-trained farmers will result in smaller quantities of pesticides being released into rice agro-ecosystems, increased rice yield, and improved human health and environmental quality (WHO, 2010[12]; Berg and Tam, 2012;Andersson et al., 2014; Nerilo et al., 2014; Sharma et al., 2015).

In the search for higher yields, the excessive use of pesticides might be the catalyst for current and future ecological crises. Potential problems include pesticide resistance, resurgence of insect pests, pesticide poisoning, environmental toxicity, elimination of predator species, negative outcomes for other non-target organisms, disruption in the food web, increased prevalence of pesticide residues in food, and reduced rice yields (Aktar et al., 2009; Sharma et al., 2015). These problems could be avoided by implementation of well-designed IPM strategies (Mohd Fuad et al., 2012; Srinivasan, 2012; Sharma et al., 2015). Despite several IPM techniques specifically designed for rice production, however, adoption remains low because of the perception that "chemical-free" strategies are less effective than heavy insecticide application.

CONCLUSIONS

IPM practices can be used for effective pest control in rice agro-ecosystems. The principle advantage of this technique is that rice yields increase as the resilience of the cropping system increases. Our study demonstrates that a well-designed IPM strategy can have a positive effect on yield components. In our study, rice variety BRRI Dhan 29 yielded 7.3–7.5 ton/ha in the IPM-treated plots. According to BRRI[13] (2015), the cultivar BRRI Dhan 29 is capable of yields from 6 to 8 ton/ha in Bangladesh agro-climatic zones. Despite the demonstrated advantages, adoption of IPM strategies might not happen quickly. It will be necessary to educate farmers about the environmental impacts associated with improper pesticide use and the consequences of those impacts on human communities and future food production. Further, in developing appropriate IPM strategies, it will be necessary to examine current pest-management practices and build on farmers› current and traditional knowledge. Additional research is needed to develop an understanding of farmers› traditional pest-management knowledge base, identify insect-resistant rice varieties, analyze the patterns of pesticide use, investigate constraints to controlling rice pests, examine farmers› awareness of environmental pollution caused by pesticide applications, and explore additional alternatives to pesticide use.

AUTHOR CONTRIBUTIONS

MA served as research team leader and was responsible for the study design, data collection, analysis and interpretation, and manuscript preparation. AC helped with manuscript preparation, edited the manuscript, and provided feedback on study design and data analysis. MH contributed to data analysis and interpretation. MI optimized the BRRI recommended agronomic practices utilized in this study. EH provided labor, hospitality, and administrative support. SaBH was responsible for data collection, tabulation, and provided an economic analysis of the yield data. ShBH helped with data collection. MH provided administrative assistance, aided in data collection, and provided information on relevant public health issues.

CONFLICT OF INTEREST STATEMENT

The authors declare that the research was conducted in the absence of any commercial or financial relationships that could be construed as a potential conflict of interest.

SUPPLEMENTARY MATERIAL

The Supplementary Material for this article can be found online at:http://journal.frontiersin.org/article/10.3389/fenvs.2016.00022

REFERENCES

1. Abrol, D. (2013). Integrated Pest Management: Current Concepts and Ecological Perspective. Available online at: http://www.elsevier.com/books/integratedpest-management/abrol/978-0-12-398529-3
2. Adams, R. G. (1996). "Introduction to integrated pest management: northeast sweet corn production and integrated pest management manual," in Cooperative Extension System, eds R. A. Adams and J. C. Clark (Storrs: University of Connecticut).
3. Aktar, M. W., Sengupta, D., and Chowdhury, A. (2009). Impact of pesticides use in agriculture: their benefits and hazards. Interdiscip. Toxicol. 2, 1–12. doi: 10.2478/v10102-009-0001-7
4. Alam, M. Z. (2013). Survey and assessment of insect management technologies and environmental impact on rice ecosystem of Bangladesh. Intl. J. Appl. Res. Stud. 2, 1–16. Available online at: http://www.ijars.ijarsgroup.com/article.php? aToken=060ad92489947d410d897474079c1477
5. Allara, M., Kugbei, S., Dusunceli, F., and Gbehounou. (2012). "Coping with changes in cropping systems: plant pests and seeds," in FAO/OECD Workshop on Building Resilience for Adaptation to Climate Change in the Agriculture Sector (23–24, April 2012) (Rome).
6. Andersson, H., Tago, F. D., and Treich, F. N. (2014). Pesticides and Health: A Review of Evidence on Health Effects, Valuation of Risks, and Benefit−Cost Analysis. Available online at: http://www2.toulouse.inra.fr/lerna/travaux/cahiers2014/14.05.406.pdf
7. Arora, S., Mukherji, I., Kumar, A., and Tanwar, R. (2014). Pesticides residue analysis of soil, water, and grain of IPM basmati rice. Environ. Monit. Asses. 186, 8765–8772. doi: 10.1007/s10661-014- 4042-9
8. Berg, H., and Tam, N. T. (2012). Use of pesticides and attitude to pest management strategies among rice and rice-fish farmers in the Mekong Delta, Vietnam. Intl. J. Pest Manag. 58, 153–164. doi: 10.1080/09670874.2012.672776
9. Bux, M., Khan, M. H., Ahmad, N., Tofique, M., and Ismail, M. (2013). Field comparison of different rice (Oryza sativa L) genotypes for their resistance against rice stem borers (Pyralidae: Lepidoptera). Pak. J. Agri. 29, 137–145. Available online at: http://www.sau.edu.pk/sau_journal/2014/ vol2014/Research%20paper%205%20Vol.pdf
10. Chatterjee, S., and Mondal, P. (2014). Management of rice yellow stem borer, Scirpophaga incertulas Walker

using some bio rational insecticides. J. Biopest 7, 143–147. Available online at: http://www.jbiopest.com/users/LW8/efiles/vol_ 7_0_143-147.pdf

11. Côte, F. X., Abadie, C., Achard, R., Cattan, P., Chabrier, C., and Tixier, P. (2009). Integrated pest management approaches developed in the French West Indies to reduce pesticide use in banana production systems. Acta Hort. 828, 375–382. doi: 10.17660/ActaHortic.2009.828.38

12. Craig, T. (2015). The rise and demise of integrated pest management in rice in Indonesia. Insects 6, 381–408. doi: 10.3390/insects6020381

13. Ehi-Eromosele, C. O., Nwinyi, O. C., and Ajani, O. O. (2013). Integrated Pest Management. Available online at: http://cdn.intechopen.com/pdfs-wm/42758. pdf doi: 10.5772/54476

14. Eze, S. C., and Echezona, B. C. (2012). Agricultural pest control programs, food security and safety. Afr. J. food Agric. Nutria. Dev. 12, 6582–6592. Available online at: http://www.bioline.org.br/pdf?nd12070

15. Farrar, J. J., Baur, M. E., and Elliott, S. (2015). Impacts of the Regional Integrated Pest Management Competitive Grants Program in the Western United States. Available online at: http://westernipm.org/index.cfm/about-the-center/ publications/special-reports/western-ripm-retrospective-full-pdf/

16. Hall, C., Knight, B., Ringrose, S., and Knox, O. (2013). What Have Been the FarmLevel Economics Impacts of the Global Cultivation of GM Crops? Available online at: http://www.environmentalevidence.org/wp-content/uploads/2014/ 07/CEE11-002.pdf

17. Hillocks, R. J., and Cooper, J. E. (2012). Integrated Pest Management—Can it Contribute to Sustainable Food Production in Europe with Decreased Reliance on Conventional Pesticides? Available online at: http://www.eucipm.org/docs/ IPM-Can-it-Deliver-DiscussionPaperNo1b.pdf

18. Jacobsen, C. S., and Hjelmso, M. H. (2014). Agricultural soils, pesticides and microbial diversity. Curr. Opin. Biotechnol. 27, 15–20. doi: 10.1016/j.copbio.2013.09.003

19. Jepson, P. C., Guzy, M., Sow, M., Sarr, M., Mineau, P., and Kegley, S. (2014). Measuring pesticide ecological and health risks in West African agriculture to establish an enabling environment for sustainable intensification. Philos. Trans. R. Soc. Land. B. Biol. Sci. 369:20130491. doi: 10.1098/rstb.2013.0491

20. Kabir, M. H., and Rainis, R. (2015). Do farmers not widely adopt environmentally friendly technologies? Lesson from Integrated Pest Management (IPM). Modern Appl. Sci. 9, 208–215. doi: 10.5539/mas.v9n3p208

21. Khaliq, A., Javed, M., Sohail, M., and Sagheer, M. (2014). Environmental effects on insects and their population dynamics. J. Entomol. Zoology Studies 2, 1–7. Available online at: http://www.entomoljournal.com/vol2Issue2/pdf/ 32old.1.pdf

22. Lou, Y. G., Zhang, G. R., Zhang, W. Q., Hu, Y., and Zhang, J. (2013). Biological control of rice insect pests in China. Bio. Control 67, 8–20. doi: 10.1016/j.biocontrol.2013.06.011

23. Mangan, M., and Mangan, M. S. (1998). A comparison of two IPM training strategies in China: the importance of concept of the rice ecosystem for sustainable insect pest management. Agril. Hum. Values 15, 209–221.

24. Mohd Fuad, M. J., Junaidi, A. B., Habibah, A., Hamzah, J., Toriman, M. E., and Azima, A. M. (2012). The Impact of pesticides on paddy farmers and ecosystem. Adv. Natl. Appl. Sci. 6, 65–70. Available online at: http://www.scopus.com/ record/display.uri?eid=2-s2.0-84857412809&origin=inward&txGid=0

25. Mohiuddin, M., Hossain, M. M., Rahman, A. K. M. M., and Palash, M. S. (2009). Socio-economic study of insecticide use on vegetable cultivation at farm level in Chittagong region. J. Bangladesh Agril. Univ. 7, 343–350. doi: 10.3329/jbau.v7i2.4745

26. Muck, O. (2015). Integrated Pest Management for Rice Production in Nigeria. Available online at: http://cari-project.org/wp-content/uploads/2015/05/ CARI-Rice-IPM-Nigeria-Draft-Study.pdf

27. Naranjo, S. E., Ellsworth, P. C., and Frisvold, G. B. (2015). Economic value of biological control in integrated pest management of managed plant systems. Annu. Rev. Entomol. 60, 621–645. doi: 10.1146/annurev-ento-010814- 021005

28. Nerilo, S. B., Martins, F. A., Nerilo, L. B., Salvadego, V. E. C., Endo, R. Y., and Junior, M. M. (2014). Pesticide use and cholinesterase inhibition in small-scale agricultural workers in southern Brazil. Brazilian J. Pharmaceutical Sci. 50, 783–791. doi: 10.1590/S1984-82502014000400014

29. Ortiz, O., Kroschel, J., Alcázar, J., Orrego, R., and Pradel, W. (2009). "Evaluating dissemination and impact of IPM: lessons from case studies of potato and sweet potato IPM in Peru and other Latin American countries," in Integrated

30. Pest Management: Dissemination and Impact, eds R. Peshin and A. K. Dhawan (Dordrecht: Springer), 419–434.

31. Parveen, S. (2010). Rice farmers' knowledge about the effects of pesticides on environmental pollution in Bangladesh. Bangladesh Res. Pub. J. 3, 1214–1227. Available online at: http://www.bdresearch.org/home/attachments/ article/nArt/Rice_Farmers_Knowledge_about_effect.pdf

32. Paul, D. A. V. N. (2007). Insect Pests and their Management. New Delhi: Indian Agricultural Research Institute IARI. 110012.

33. Petit, S., Haysom, K., Pywell, R., Warman, L., Allen, D., Booth, R., et al. (2003). Habitat-based models for predicting the occurrence of ground-beetles in arable landscapes: two alternative approaches. Agrl. Ecosyst. Environ. 95, 19–28. doi: 10.1016/S0167-8809(02)00173-1

34. Prakash, A., Bentur, J. S., Prasad, M. S., Tanwar, R. K., Sharma, O. P., Bhagat, S., et al. (2014). Integrated Pest Management for Rice. New Delhi: IARI campus: National centre for integrated pest management. 43.

35. Pretty, J., and Bharucha, Z. P. (2015). Integrated pest management for sustainable intensification of agriculture

in Asia and Africa. Insects 6, 152–182. doi: 10.3390/insects6010152

36. Rahman, M. M. (2012). Problems and suggestions for farmers' adoption of IPM practices in rice (Oryza sativa) cultivation. Bangladesh J. Agril. Res. 37, 121–128. doi: 10.3329/bjar.v37i1.11183

37. Rasmussen, J. J., Larsen, P. W., Pedersen, A. B., Monberg, R. J., and Kronvang, B. (2012). Impacts of pesticides and natural stressors on leaf litter decomposition in agricultural streams. Sci. Total Environ. 416, 148–155. doi: 10.1016/j.scitotenv.2011.11.057

38. Redfern, S. K., Azzu, N., and Binamira, J. S. (2015). Rice in Southeast Asia: Facing Risks and Vulnerabilities to Respond to Climate Change. Available online at: http://www.fao.org/docrep/017/i3084e/i3084e18.pdf

39. Rodrigo, A. D. S., Eliane, D. Q., Gabriel, M. M., José, A. F. B., and Luciano, M. L. (2013). Compatibility of conventional agrochemicals used in rice crops with the entomopathogenic fungus Metarhizium anisopliae. Sci. Agric. 70, 152–160. doi: 10.1590/S0103-90162013000300003

40. Roitberg, B. D. (2007). Why pest management needs behavioral ecology and vice versa. Entomol. Res. 37, 14–18. doi: 10.1111/j.1748-5967.2007.00045.x Satpathi, C. R., Chakraborty, K., Shikari, D., and Acharjee, P. (2012). Consequences of feeding by Yellow Stem Borer (Scirpophaga incertulas Walk.) on rice cultivar

41. Swarna mashuri (MTU 7029). World App. Sci. J. 17, 532–539. Available online at: http://idosi.org/wasj/wasj17%284%2912/19.pdf

42. Sattar, M. A., Alam, M. Z., and Rahaman, M. M. (2004). The farm use of pesticides and the guidelines of pesticides research in

43. Bangladesh. Bangladesh J. Environ. Sci. 10, 313–317. Available online at: https://www.researchgate.net/publication/272020636

44. Sharma, R., Peshin, R., Shankar, U., Kaul, V., and Sharma, S. (2015). Impact evaluation indicators of an integrated pest management program in vegetable crops in the subtropical region of Jammu and Kashmir, India. Crop Protection 67, 191. doi: 10.1016/j.cropro.2014.10.014

45. Singh, D., Singh, A. K., and Kumar, A. (2014). On-farm evaluation of integrated management of Rice Yellow Stem Borer (Scirpophaga incertulas Walk.) in RiceWheat cropping System under low land condition. J. Agri. Search 1, 40–44. Available online at: http://www.academia.edu/7038623/

46. Srinivasan, R. (2012). Integrating bio pesticides in pest management strategies for tropical vegetable production. J. Bio Pesticides 5, 36–45. Available online at: http://www.jbiopest.com/users/lw8/efiles/vol_5_0_36_45f.pdf

47. Srivastava, S. K., Biswas, D. K. R., Garg, B. K. G., Haque, N. M. M. M., Ijaj, P., and Tiwari, S. N. (2004). Management of Stem Borers of Rice and Wheat in Rice-Wheat System of Pakistan, Nepal, India and Bangladesh. Rice-Wheat Consortium paper Series 17. New Delhi: Rice-Wheat Consortium for the IndoGangetic Plains. 204.

48. Vinatier, F., Lescourret, F., Duyck, P. F., and Tixier, P. (2012). From IBM to IPM: using individual-based models to design the spatial arrangement of traps and crops in integrated pest management strategies. Agril. Ecosyst. Environ. 146, 52– 59. doi: 10.1016/j.agee.2011.10.005

49. Conflict of Interest Statement: The authors declare that the research was conducted in the absence of any commercial or financial relationships that could be construed as a potential conflict of interest.

CHAPTER 8

INTEGRATED FRUIT PRODUCTION AND PEST MANAGEMENT

Petros Damos, Lucía-Adriana Escudero Colomar and Claudio Ioriatti

Keywords: Integrated Pest Management, Technical Guidelines, Sustainable Pest Management, pest control tools

INTRODUCTION

Historically, pest control in fruit orchards was based on broad-spectrum pesticides which were associated with a diversity of problems, traditionally including environmental effects, beneficial organism extinction, and pesticide resistance. More recently, other potential negative effects caused by the use of pesticides in agriculture have drowned the attention of the public and policy makers: the threat toward the pesticide applicators and bystanders' health and the food safety endangered by pesticide residues. Fears about these issues have increased the interest in the development of alternative means for pest control that causes trivial or no impact on humans, beneficial organisms and sensitive ecosystems [1,2]. Nevertheless, most pest management programs still rely heavily upon the role of conventional insecticides which are in most cases the principal means for pest and disease control in fruit orchards [3,4,5]. In apple production, particularly, the presence of arthropod pests, several serious diseases, and high cosmetic standards for fresh fruit market (e.g., there is zero tolerance for the live codling moth) represent formidable obstacles to the adoptions of alternative means of pest control and low-spray apple production. Nevertheless, increased public fears about the role of pesticides and the potential adverse effects on human health, wildlife, soil water, and overall environmental quality have led to the development of alternative low risk control tools [6].

The solution was found in adopting and promoting a concept already formulated in the late 50s and called "Integrated Pest Management" (IPM). The turning point with regard to the definition of the concept of IPM is considered to be 1958 [7] when two separate study groups, one American and one Dutch, looking for alternative solutions to the defense of crops introduced the concept of integrated control at the 10th International Congress of Entomology held in Montreal. The document arose from concern that arises from the obvious deleterious effects caused by broad-spectrum insecticides used in agriculture [8,9]. The term "integrated control" proposed by researchers in California [10,11] was unanimously accepted by the international community. The concept worked out by Stern *et al.* in 1959 [11] and in principle is integrating the role of biological and other control measures in complementary ways. They provided the foundations for understanding the relationship between the damage and the beneficial insects in agricultural systems and how the farmer could exploit them to their advantage through the detailed description of the equilibrium position among populations of pests, the economic threshold (ET), and the economic injury level (EIL).

In Europe, noteworthy was the role played since 1956 by the International Organization of Biological Control (IOBC), and more specifically, the pioneering activity of fruit entomologists operating within the IOBC working group "integrated control in orchard" established in 1959 with the aim of promoting the development of biological control in a comprehensive concept of integrated fruit production [12].

A few years later, in 1962, van den Bosch and Stern [13] extended the conceptual framework of IPM to embrace the coordinated utilization of all available biological, cultural, and artificial practices.

Citation: Petros Damos, Lucía-Adriana Escudero Colomar and Claudio Ioriatti, Integrated Fruit Production and Pest Management, doi:10.3390/insects6030626.

In 1965 the concept was explicitly specified during a symposium sponsored by the Food and Agriculture Organization (FAO), of the United Nations, held in Rome, Italy [14,15] and with the participation of 87 delegates from 34 countries and seven international agencies [16].

Since then, national and international institutions as well as fruit cooperatives and grower associations has been encouraging farmers to move towards IPM techniques to minimize pesticide use and trim costs, supported by the continuous improvements in monitoring systems and in the basic understanding of pest biology and pest-plant interactions [6].

The development occurring during the early 1970s brought a change in the general concept and IPM become an indispensable element of a holistic system approach involving the entire orchard and defined as Integrated Fruit Production (IFP). On this virtuous route, the message of Ovronnaz, launched in 1976, was a cornerstone of the modern IFP because it introduced the need to abandon the isolated view of plant protection (IPM) and to place it in the context of the entire farm [12]. IFP and IPM offer an economical and high quality fruit production framework, giving priority to ecologically safer methods, minimizing the undesirable side effects and use of agrochemicals, and enhancing the safeguard of the environment and human health [17,18,19,20,21].

Consumer demands for certified products have a significant impact on current fruit production. Recent market studies [22] show that the European consumers are willing to pay for the reduction of pesticides and apple benefits of a premium price when produced under IPM of organic certified process. Nevertheless, there are also cautions weather IPM programs that are sustainable in the longer term and they will continue to evolve, applying an increasingly wider range of new settings [23,24,25]. Nevertheless, despite these, IFP is recognized as a model of sustainable plant protection in all European countries [19].

Until recently, the application of IPM was a voluntary approach implemented by the most advanced and environmental friendly fruit production systems. Since the approval of the EU Directive 128/2009 that established a framework for Community action to achieve the sustainable use of pesticides, the adoption of IPM becomes compulsory in Europe. In fact, member states are required to adopt National Action Plans to set up quantitative objectives, targets, measures and timetables to reduce risks and impacts of pesticide use on human health and the environment and to encourage the development and introduction of IPM and of alternative approaches or techniques in order to reduce dependency on the use of pesticides.

Within this context, IPM has changed its original goal and it is called to cover different areas of concern including worker protection, pesticide residues on fruits and overall protection of the environment and society.

The current review describes some recent advances in Integrated Fruit Production with special reference to Pest Management in Europe. The apple production systems are employed as a case study and emphasis is given on what has been changed in the apple production system since the first definitions of IPM. Moreover, stress is given in control methods used to manage apple key pest and diseases, including: the use bio-pesticides, semiochemicals and biotechnical measures, biological control, decision tools and models. Lastly, a discussion is drawn on how these tools may be transformed to proxies based on the implementation of recent guidelines, market requirements and the skeletal system of the EU directives.

CURRENT IPM TOOLS AND TECHNIQUES USED IN INTEGRATED FRUIT PRODUCTION

IPM combines knowledge of the pest and host plant with multiple tactics for long-term pest control. It is consistent with sustainable agriculture and uses total farm systems approaches to mitigate pest pressure. Sustainable agriculture provides a conceptual framework and fundamental principles to guide the future development of husbandry practices and arrangements that are economically viable, environmentally sound, and socially responsible. In this context, various authors have advocated the principle of incorporating all available control methods and measures to manage pests that are being chosen by environmental, economic and societal standards. Currently, the term IPM incorporates the wide array of pest management practices, which are adopted following certain criteria, included into a total system approach which ideally should take on all crop production objectives [20,25,26].

Pesticides

Although IPM uses all available means to keep pest populations below levels that would cause economic loss, most IPM programs still rely heavily upon the use of pesticides. Yet, recently the European commission launched a course of study which has filled in the reassessment of a list of conventional pesticides that were on the markets before 1993 [27]. The evaluation program concerned about 1000 substances and has led to removal from the market of more than two thirds of these [27,28]. Particularly, 26% of the reviewed pesticides (corresponding to approximately 250 substances), have topped the EU harmonized safety assessment, while the vast majority (67%) has been eliminated because evaluation dossiers were either not submitted, incomplete or withdrawn by the chemical industry, while nearly 70 substances failed the review and have been withdrawn from the market, because the valuation carried out did not show safe use with respect to human health and the environment. Presently, the evaluation, marketing and utilization of whole categories of pesticides (insecticides, fungicides, herbicides, *etc.*) in plant protection among the members of the European community are regulated under Council Directive 91/414/EEC. In conclusion, IPM programs encourage the use of selective pesticides as the last control option to be applied when other management tools are not available or effective.

Bio-Pesticides

The term bio-pesticide has only recently been proposed to describe those insecticides that are efficacious against the target pest but are less detrimental to natural enemies [29,30]. Traditionally, the term refers to soaps/detergents, oils and botanicals of natural origin but now includes, new systemic insecticides, insect growth regulators and products containing, micro-organisms or its components. The most important bacterial pathogen used as a biological control agent is *Bacillus thuringiensis* (Berliner), Bt., which is a gram (+)

bacterium that is pathogenic to larvae. The infection is caused by the endospore of the species having crystalline inclusions consisting of one or more insecticide protein known as d-endotoxins or Cry-proteins. Because of its high selectivity, Bt is mostly used in apple IPM programs for the management of various moth species such as leafrollers [31,32,33,34]. Bt products currently available have limited effectiveness against many orchard pests due to the pest cryptic life habits [34]. The first products against the lepidopterous larvae were based upon two subspecies of *B. thuringiensis*, var: *kurstaki* and var: *aizawai* or a compounding of the two. The second generation products are grounded on the union of the two subspecies, while the third and fourth generation products are founded on new Bt. strains based on recombinant DNA technology [32].

Virus disease play important role in the natural regulation of insect population and their potential as biocontrol agents were investigated since early 1950. The most promising virus agent for controlling apple arthropod pests is a granulovirus (family Baculoviridae), which was first isolated from dead larvae found in an orchard in Mexico [35]. Granulovirus of codling moth (CpGV) is insect-specific granulovirus that offers new means of highly selective control of the codling moth*Cydia pomonella* L in fruit orchards [36]. The commercial products that are utilized in IPM programs contain the virus in an aqueous suspension and are sprayed during the egg hatch. Thus, timing of GpCV application is critical in order to promote ingestion of the occlusion bodies into neonate larvae before they are entering into the fruits. The GpGV is highly selective and host range is actually limited to *C. pomonella* and few tortricidae [26,37]. Commercial products of CpGV are now registered and available in Europe and North America and used by orchardists worldwide [37]. In addition, the application of CpGV is compatible with the use of other compounds in IPM [37,38]. Its sensitivity to degradation by UV light is the most important limitation for its widespread use as a biological control agent and research is being carried out to solve the problem [39]. The combined use of yeast-based attractants, feeding stimulants or pear ester is proposed as novel insect control technique to enhance larval mortality when using granulovirus [40].

The insecticidal properties of nematodes have been as well tested and have therefore become increasingly popular in IPM programs [41]. Two species have been commercialized, *Steinernema* and *Heterorhabditis* (Rhabditida) which are insect specific parasitic nematodes and exert lethal action to their host due to the transmission of bacteria. *Steinernema carpocapsae* (Weiser) and *Steinernema feltiae* (Filipjev) are used for the control of *C. pomonella* and especially in controlling the overwintering larvae of the moth. However, *S. carpocapsae* expresses limited host search behavior (ambusher species), while *S. feltiae* is considered as an intermediate search strategist and expresses higher search capacity. In contrast, entomopathogenic nematodes have very little potential in providing control of the woody apple aphid *Eriosoma lanigerum* (Hausmann), probably because of the inability of their symbiotic bacteria to grow in the aphid body and to produce the lethal endotoxins [42].

The main factor limiting the effectiveness of entomopathogenic fungi is their requirement of high humidity and moderate temperature for spore germination and development. Their use as biological control agent in apple IPM is still limited to application of *Bauveria bassiana* (Basi) to control larvae of codling moth and apple clearwing moth, *Synanthedon myopaeformis*(Borkahousen) (Lepidoptera: Sesiidae) [43].

Semiochemicals, Mating Disruption and Mass Trapping

Semiochemical-mediated communication affects several types of interactions in organisms in natural habitats [44]. Likely the first and the most representative examples of semiochemicals application in IPM programs is the use of sex pheromones in detection and monitoring population emergence [45]. Captures in traps baited with synthetic pheromone lures accurately show whether a specific insect is present, and when its seasonal flight period starts. A simple and widespread strategy is to time insecticide sprays accordingly [45]. Besides, the pheromone chemical identification and the discovery of the physiological mechanisms regulating olfactory perception and behavior in insects have paved new ways to control the economically important pest species [46,47].

This know-how has been successfully used to develop direct control means whose applications outcompete conventional insecticides [48]. The primary aim is interfering with male-female sexual communication by permeating the atmosphere with synthetic pheromone and to bring down the likelihood of successful mating. Pheromones need to be formulated to prevent degradation in the field and to release them slowly into the atmosphere, providing a constant release of adequate levels of pheromone throughout the growing season while adult moths are present [46,47,48]. A wide range of pheromone delivery systems have been developed and made available for the growers (*inter alia* [49,50,51,52]). Microcapsules and hollow fibers, deployable via air-blast sprayers, were developed to fit the assumption that high density point sources were needed to uniformly permeate the atmosphere with pheromone. The main shortcoming of these formulations is the short lasting pheromone release that requires the repetition of the applications during the season. A second approach is represented by the hand-applied reservoir-type formulations, generating a plumes of high concentration of pheromone and that required to be applied at a rate of hundreds dispensers per hectare. Their season-long pheromone emission allowed us to guarantee the required amount of pheromone with a single application, but the cost of labor associated with hand application has been cited as an impediment to broader adoption of mating disruption [53].

More recently, low-density, high-release dispensing systems, emitting an aerosol spray of solvent-diluted pheromone of electronic circuit-controlled intervals, have been developed. These pheromone delivery devises are installed by hand, at a density of 2-5 units per hectare with a consistent reduction of the application labor cost compared to hand-apply hundreds of passive reservoir dispensers. Pheromone mating disruption have been shown to produce reliable results especially in area-wide programs [51,54,55] taking advantage from treatment of large contiguous blocks of orchards. In fact, potential weakness in using discrete point source of pheromone is that it may leave areas of little or no pheromone coverage, where mate finding may occur [52]. Mating disruption has been first implemented for the management of the key pest *C. pomonella*and its efficacy is widely confirmed [40,56,57,58]. However, the consequent reductions in the number of insecticide sprays needed for adequate control of the key pest have resulted sometime in outbreaks of secondary tortricid pests such as leafrollers. To tackle this problem, companies have developed multispecies disruption formulations, which are now available for growers, allowing concurrent management of co-occurring key and secondary pests [59,60,61].

Among the major advantages of pheromone mating disruption technology is that it does not exhibit adverse effects on non-target pests and it can be used to manage pest resistance. This has pushed its worldwide application; the estimated orchard area treated with codling moth mating disruption has been reported to exceed 200,000 ha of which nearly 70,000 ha are in Europe [45]. Less extended is the European orchard area treated with MD for secondary apple pests like oriental fruit moth *Grapholita molesta* (Busk), small fruit tortrix *Grapholita lobarzewskii*, (Nowicki), leafrollers *Adoxophyes orana* (Fischer von Röslerstamm), and *Pandemis heparana* (Denis and Schiffermüller), *Spilonota ocellana* (Denis & Schiffermüller), clearwing moth*Synanthedon myopaeformis* (Borkhausen), leopard moth *Zeuzera pyrina* (L.), even though either specific or multispecies disruption formulations targeting these secondary pests are commercially available.

Mass trapping (MT), using lures to catch the maximum number of fruit flies, has been devised many years ago [62] and it was occasionally practiced for many years though not always the best effect. However, the technique has been used in apple for *Ceratitis capitata* Wiedemann [63] and for pests such as *Cossus cossus* L. [64] and *S. myopaeformis* [65].

Over the last decade, the efficiency of mass trapping technique used against *C. capitata* (medfly) has been confirmed in several fruit crops [66,67,68,69,70,71] including apple [63,70]. The spread of the use of this technique has been possible thanks to the development in the last decade of the 20th century of dry food based attractants with the aim to use them to monitor male and female medfly populations in order to calculate population dynamics [71]. These attractants compete effectively with apple to attract flies, which once they have entered the trap are killed by an insecticide [72]. In areas where medfly is a key pest in apples, mass trapping is used as the main system for the pest control and it is integrated into IPM programs. More recently, a device of attract and kill using the same attractants than mass trapping but in an envelope impregnated with an insecticide in the outside, was also tested to protect apple production from medfly damage [73], although more studies are still necessary to set its efficacy in apple under different population levels and varieties. Anyway, in both systems, mass trapping and attract and kill, chemicals are utilized only when the population increases considerably, as reinforcement [67,74].

Biological Control

A wide literature documenting the effective role of natural enemies in controlling apple pests has been published and it is mainly focused in the study on conservation biological control of a wide array of arthropod pest like mites, aphids, scales, moths and leafrollers (*inter alia*, [75,76,77,78,79,80,81,82,83]). Hence, it is out of the scopes of the present work to revise these fundamental studies that constitute the conceptual cornerstone of integrated pest management. On the other hand, integrated pest management programs emphasize the more pertinent integration of biological and chemical control, to maintain pest populations below economic thresholds [11]. Combination of biological and chemical control tactics is not always compatible; insecticides can disrupt natural enemies through lethal and sub-lethal means causing pest resurgence or secondary pest outbreaks [84,85]. Compatibility between chemical and biological control depends on the chemistry and the inherent toxic properties of the pesticide in relation to the physiology of the pest and its natural enemy, on the application timing in relation to the life stage susceptibility of the natural enemy, and the actual presence of the non-target organism in the treated field or tree canopy. As combining tactics may provide better long-term sustainable pest suppression, IPM programs need to contemplate strategies able to reduce natural enemy exposure to insecticides.

Potential mitigation measures applicable at the apple farm scale include selective insecticides, low dosage, timing of application, special formulations, site-specific applications [86]. Parasitism in codling moth can be improved by better targeting of pesticide treatments [87]. Earwig populations for example are strongly affected by orchard management, suggesting that these insects could be used as a bioindicator of the intensity of the orchard management [88].

Improvement of mite biological control has been reached by replacing broad-spectrum pesticide with more selective insecticides [89,90]. Yet, the combination of a different mode of action of insecticides is a challenge in apple production which should take into account market requirements for residues, the permitted active ingredients and the need to avoid resistances. Due to these factors, the use of pyrethroids has increased the last years, mainly to control mite/insects populations close to harvest. It is well known that pyrethroids can induce outbreaks of spider mites [91] because they are harmful to phytoseids living in fruit trees [92]. This fact has challenged the natural biological control of tetranychids in apple trees in the EU countries. In the USA, New Zealand, and Canada have been reported on the effectiveness of releasing resistant strains of Phytoseiidae to control spider mites [92,93,94]. In Europe, resistant strains of *Typlodromus pyri* (Scheuten) and *Amblyseius andersoni* (Chant) to pyrethroids have been found from vineyards in the South-West of France [95,96,97]. Both species are the main predatory mites of *Panonichus ulmi* (Koch) in the European countries.

Concerning the effect of different apple management systems (organic, IPM and conventional) in beneficial mite composition, recent studies [97], related the type of management system to the Phytoseiidae composition but not to their abundance. The same study reports *A. andersoni*, *Euseius finlandicus* (Oudemans), and *T. pyri* as the three most common species, while the total phytoseiid abundance in the orchards with different pest management systems did not differ. Moreover, the relative abundance of *A. andersoni* increased with the pesticide load of the orchards whereas the relative abundance of *E. finlandicus* decreased. Nevertheless, beside the effect of different management strategies on apple mite composition [97], other factors, such as competition between species and apple variety may affect the composition of beneficial mites in apple orchards. Phytoseiidae species richness, for instance, is closely related to the presence or absence of domatia in the leaves [98]. In those cultivars where few or no phytoseids are found other strategies need to be developed to conserve the generalist predator like *Anystis baccarum* (L.) [99], a species which was reported to be the most frequent predatory mite in apples in Northern Ireland. Besides its role in controlling tetranychid pests, it has also been reported predating leafroller larvae in apple in New Zealand [100]. These new approaches to improve the biological control of mites in apples started to play a major role in the IPM design for different varieties and regions and it shows that more studies are still needed on the role of generalist predators in apple orchards.

The transition from conventional fruit production, which used broad spectrum insecticides, to integrated fruit production which was based on selective insecticides produces effects on the pests and natural enemies' populations. In New Zealand apple orchards, for example, this transition was initially accompanied by a surge in *E. lanigerum*, and this was followed by a slow colonization of its parasitoid *Aphelinus mali* (Haldemann) which gradually have reduced the aphid population to acceptable levels [101].

The efficacy of biological control in apple orchards has also been improved by the use of new spray application techniques. Low dosage of pesticide, of alternate row-middle spraying technique, left untreated refugia for *Stethorus punctum* (Casey) and was a key factor for successful biological control of mites [102].

Furthermore, recent advances in geostatistical analysis provide means to identify hot spots of high pest intensity, direct the pesticides applications and preserve natural enemies [103]. Moreover, the use of risk assessment maps, provided by the geostatistical analysis of the pest distribution, was incorporated into an apple IPM program to allow site–specific application of pesticide and minimizing direct control tactics to the codling moth. Creation of refugia and improvement of the quality of natural habitat within the farm and at the landscape scale, help to protect biodiversity and provide ecosystem services like biological control [104,105]. The use of hedgerows surrounding apple orchards as well as planting flowering plants strips improve the landscape diversity and support beneficial insects [106,107,108] as well as the abundance and diversity of predatory phytoseiid mites [107,108]. The intercropping of some aromatic plants in apple orchards can enhance the biological control of *A. orana* and other tortricids [109]. Predator complex abundance was positively correlated with increasing alternative prey availability in a manipulated habitat of an apple orchard floor [62]. Use of compost in an apple orchard ecosystem is beneficial to management of weed, fungal, and insect pests. Populations of spotted tentiform leafminer*Phyllonorycter blancardella* (Fabricius) and migrating woolly apple aphid nymphs were reduced in the compost plots [110].

Releases of natural enemies have also been used to control mites, leaf rollers, codling moth and aphids although it is labor intensive and costly and results had not always been successful [111,112].

Phenology Models, Economic Thresholds and Decision Support Systems

A typical life cycle of an insect describes the phenological forms of a species with normally distributed emergence that takes place over a growing season, in which usually more than one generation is observable. Phenology models are short abstractions of the relation between temperature and population development and are used to predict the exact time of the phenological development of pest populations. Probably the simplest case of phenology model is that which is based on a non-linear regression function that correlates cumulative moth catches to heat accumulations [113]. These models are almost useful in predicting the date of the issue of the first flight or the peaks of emergence of the successive generations and are actually empirical regression models which present the best mathematical fit and do not have any clear inherent meaning. Even so, they offer useful insights into critical time windows for which specific management activities should be packed away. Hence, they are utilized for decision making and particularly for the placement of MD pheromone dispensers, timing of pesticides and especially the application of biorational compounds that express selectivity to specific developmental stages (e.g., bio-pesticides and Insect Growth Regulators).

Moreover, phenology models can be used along with Economic Injury levels [EIL]. The EIL belongs to the most basic of the decision rules; it is a theoretical value that, if actually attained by a pest population, will result in economic damage [114,115]. However, it is noteworthy to state that many factors have limited both the design of new economic thresholds as well as the development of existing ones (for a thorough review refers to [114]. The issue may be more complex considering no tolerance for damage in fruits [116].

Nevertheless, a challenge for a wider applicability and application of models and thresholds, if available, under a routine field conditions basis is to use these mathematical descriptions across locations and real environments in an automatic manner [117]. This enables the evolution of computer software programs, as described below, to run the models and facilitate the practical application by understanding population dynamics and dissemination of pest forecasts for timely pest management decisions [118]. Thus, web-based decision support systems are becoming popular and in the future may become an absolute demand for local, regional/area-wide and international implementation of IPM systems towards sustainable Agriculture.

Improvements in Spray Technology

Despite apple production system was one of the first implementing the IPM principles, fruit protection still requires one of the highest numbers of spraying operations per year that are generally performed with air assisted spray machine. If the final goal of IPM is minimizing the use of pesticides and their potential adverse effects on human health, wildlife, soil water, and overall environmental quality, we cannot neglect the role played by the sprayer technique in optimizing the average deposit required for acceptable biological efficacy whilst minimizing spray losses and contamination of users, bystanders and the environment. The importance of the performance of spraying equipment for the safety and efficiency of pesticide use is also recognized by the enforcement of the European

Union directive on the Sustainable Use of Pesticides (EC, 2009c), that make necessary to inspect all sprayers employed for professional use in the European Union at least once by 2016 and to establish periodical mandatory inspections schemes in all EU countries [119].

Replacement of the fine or very fine hollow cone hydraulic nozzles operated at moderate to high pressures with air induction cone nozzles, which produce very coarse spray qualities, significantly decrease environmental impact by mitigating the spray drift [120,121,122]. Spray drift in the orchard strictly depends also on sprayer air setting, both in terms of velocity and direction, so a proper adjustment of air parameters is useful not only to increase treatment efficacy, but also to prevent risks of environmental contamination. In addition to adjusting the sprayer by setting it in advance to suit the particular orchard, spray equipment should be fitted with sensors so that the adjustments are done automatically in real time during spraying. Numbers of canopy sensing systems (photocells, ultrasonic sensors, spectral sensors, laser technology) have been developed in order to instruct the sprayer to avoid spraying in absence of the target or cutting down on spray volume where the crop canopy is thin or the foliage is very poor. Implementation of these innovative technologies has generally provided a considerable contraction of chemical input and reduced spray loss due to drift and run-off [123,124,125,126,127,128]. For a comprehensive review on innovative spraying technology in Europe and the future directions of the research on this topic see also [129].

Innovation in the Tree Training

Over the last 60 years, large apple trees once constituting the extensive orchards, have been replaced by dwarf trees, ideally not higher than 4 meters and planted in rows 3.5 to 4 meters apart. In these modern apple orchards, applications of chemical control are still widely based on axial fan airblast sprayers which produce a large radial spray plume resulting in high spray losses to the ground and as spray drift. A new direction in tree training is what is called "pedestrian fruit wall" that is similarly to modern vine grape plantings. This new multi axis tree training system [130,131], with more and shorter, branches than the spindle, forms close "fruit walls". The pedestrian fruit wall training system, with its narrow canopy (1 meter wide) and its reduced height (about 2.5 meters), opens entirely new possibilities in terms of mechanization of cultural practices (pruning, thinning, weeding, harvesting), thus reducing the input of growth regulators and facilitates the introduction of innovative spraying technology like different design of tunnel sprayers [132] or the Solid Set Canopy Delivery System [133]. Both these approaches are really effective in reducing spray losses, pesticide drift and number of fungicide treatments by optimizing timing applications and improving foliar coverage. Moreover this pioneering training system permits the application of the single-row multi-functional nets, named Alt'Carpò [134,135] that has been already proved to be effective in controlling codling moth, but that can be also used to replace chemical thinning when early applied during the blossom period.

INTEGRATED FRUIT PRODUCTION AND TECHNICAL GUIDELINES

Apple production is considered among the most dynamic crop production systems. Figure 1 illustrates the worldwide average regional apple production (China and US are leading countries worldwide) [136]. The total apple production in Europe reached almost more than 12 million tons (Figure 2). In most cases, such considerable production entails a very high use of pesticides. Fruit trees, actually, are the second crop in terms of pesticides quantity used by hectare and are only exceeded by vineyards [137]. This high volume of pesticides application causes frequent presence of residues in apples and has several environmental implications, such as groundwater pollution and cases of human and animal poisoning. Concerns about these consequences have increased interest in the development IFP and application guidelines which incorporate IPM and general actions that have little or no impact on humans, beneficial organisms and sensitive ecosystems [138,139].

To date, the International Organization for Biological Control has provided the leadership for developing the philosophy Integrated Production, the general guidelines, and crop specific technical guidelines [140]. Agreeing to the IOBC definition Integrated Apple Production is a farming system that produces high quality fruits by using natural resources and regulating mechanisms to replace polluting inputs and to secure sustainable farming. "Emphasis is placed on a holistic systems approach involving the entire farm as the basic unit, on the central role of agro-ecosystems, on balanced nutrient cycles. The preservation and improvement of soil fertility and of a diversified environment are essential components. Biological, technical and chemical methods are balanced carefully taking into account the protection of the environment, profitability and social requirements".

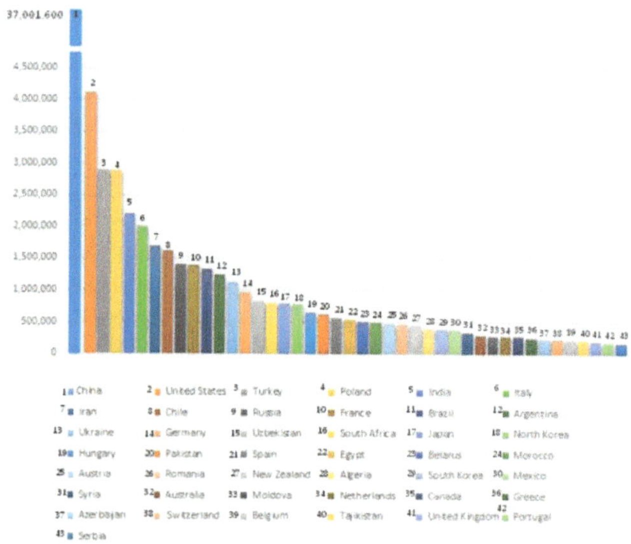

Figure 1. Worldwide leading countries in apple production. Rank is given per 'legend line' and stated as average tons for 2012 (source: FAO).

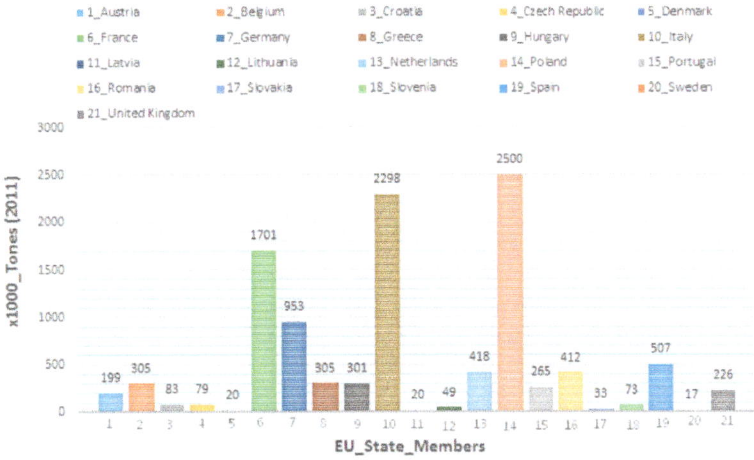

Figure 2. Apple production by country among the 27 EU state members 2011 (source: FAOSTAT, WAPA).

IOBC Guideline for IFP provided a rigorous foundation for development of sustainable utilization of pesticides legislation in the EU. Integrated Fruit Production in pome fruits has been regularly utilized to launch IFP programs in a number of European rural areas and along with pesticides properties databases can be used for environmental sound decision making [139,140]. The IOBC sets out general principles, minimum measures and guidelines for IFP in the geographic regions tracked by the IOBC/WPRS. These Guidelines are meant as a framework for the preparation of specific regional or national guidelines and measures according to IOBC standards and to facilitate their harmonization [139,140].

In general the IFP guidelines contain the following sections: definitions; professional trained and environmentally and safety conscious growers; conserving the orchard environment; site, rootstocks, cultivar and planting system for new orchards; soil management and tree nutrition; alleyways and weed-free strip; irrigation; tree training and management; fruit management; integrated plant protection; efficient and safe spray application methods; harvesting, storage and fruit quality; post-harvest chemical treatments; mode of application, controls, certification and labeling [139,140]. Nevertheless, the guidelines may be countrywide, but more often are tailored to the specific conditions of a developing area. Switzerland probably belongs among the most rigorous IFP example and about 85% of the Swiss apple farms certified as practicing integrated fruit production (IFP). Moreover, as an outcome of the latest European directives the market demand on IFP is steadily increasing and during the last years, almost 50% of the apple and the pear acreage in Western Europe are managed under an IFP program [16]. Since 1991, the cultivated apple area has increased over 40%, while the adoption of IFP in larger areas has led also the means for marketing IFP labeled fruit and which resulted up to 30% decrease in conventional pesticide application and increased usage of biorational and environmentally friendly pesticides.

However, a part of the mentioned official IFP definitions and guidelines, the regulations used by each country or organization vary and they are not always consistent with the IOBC Guidelines. The overall IPM standards, unlike organic production, are not precisely defined or legally specified in an EU or international agreement and the level of guidelines implementation varies considerably among the state members [137]. For example there are no common requirements which are imposed on growers. Instead there are many certification and market authorities which set up specific standards create IPM labels offering economic advantages to corporations and individual growers [137].

PRESENT AND FUTURE THREATS TO IPM

Pesticide Resistance

Among the possible threats that jeopardize the effective implementation of IPM in the European apple production system, the risks related to the development of pesticide resistance and the introduction of new invasive pests deserve to be discussed.

The development of resistance to pesticides by apple pests as a consequence of increasing frequencies of chemical treatments is known since long time [141]. Pesticide resistance management is an effort to slow or prevent the development of resistance; it relies on pest management and pesticide use strategies to prolong the effective life of pesticides and it is a tactical part of the IPM.

The codling moth, *C. pomonella*, is a worldwide pest in apple orchards. For many years, pesticide applications have been the dominant tools used for its control. However, increasing frequencies of chemical treatments led to the acquisition of resistance to many of the recommended pesticides belonging to various chemical groups. In this species, the first case of resistance was the resistance to arsenates reported by Hough in 1928 in the USA. Since then, new cases of resistance have been and are being reported in almost all of the main apple-growing regions worldwide. In Europe, insecticide resistance in the codling moth was firstly detected in the 1990s with the emergence of resistance to diflubenzuron in Italy and southeastern French populations [142,143,144]. Failures of control with pesticides were further observed in Switzerland and Spain [145,146]. Currently, the resistance spectrum of some of these populations has dramatically increased to include avermectins, benzoylureas, benzoylhydrazines, neonicotinoids, organophosphates, macrocyclic lactones, pyrethroids [146,147,148,149,150,151] and also biopesticide, like granulose virus [152].

Unlike the codling moth, in the European apple orchards the cases of insecticide resistance within the leafroller species are only sporadic. A reduction of susceptibility of the summer fruit tortrix moth, *A. orana*, to chlorpyrifos was first detected in England comparing a population regularly treated with the organophosphates and one collected in an orchard that never received the insecticide [153]. A larger spectrum of insecticide resistance in the same species was detected in two field populations in Switserland; this resistance concerned the benzoylureas hexaflumuron and lufenuron, the benzhydrazides tebufenozide and methoxyfenozide, as well as fenoxycarb [150,154,155] while these populations were only slightly resistant to chlorpyrifos-methyl [156]. No other cases of resistance are reported neither for A. *orana* nor for other leafroller species infesting the European apple orchards.

The implementation of integrated resistance management strategies is of the utmost importance to avoid further spread of the phenomenon. Moreover, rotational use of non-cross-resistant insecticides would benefit fitness costs associated with resistance [157]. This includes the use of semiochemicals and biocontrol agents. Evidence of the efficiency of mating disruption, both in term of pest control and reduction in selection pressure, is already reported in the European apple orchards [158]. Where conditions for effective application of mating disruption are not optimal, strategies such as treatments with CM granulosis virus have already demonstrated their efficiency to reduce the selection pressure of chemical insecticides [159].

Apart from insects, spider mites have historically demonstrated a propensity for developing resistance particularly in tree and vine crops. *Panonychus ulmi*, the European red mite (ERM), is the main Tetranychidae member that causes major damage in apples and, together with *Tetranychus urticae* (Koch), it has been reported among the top 10 resistant arthropods [159]. Resistance in ERM was reported for the first time by Newcomer and Dean in 1952 [160] to parathion and since then, many other compounds had been included in the list: Dicofol, tetradifon, propargite, cyhexatin, clofentezine, hexythiazox, pyridaben and fenpyroximate, fenbutatin oxide, tebufenpyrad and fenazaquin, parathion-methyl and bifenthrin. Nowadays, many of these compounds have been banned by EU normative, and other active ingredients are increasingly being used to control ERM populations. Resistance to chlorpyrifos and lambda-cyhalothrin has been reported in Turkey [161] and to spirodiclofen in south-western German populations [162] and in one spot in Belgium; they have also found resistance to clofentezine in Poland populations.

Other mites can also cause damage in apple, e.g., *T. urticae* and the apple rust mite *Aculus schlechtendali* (Nalepa). *Tetranychus urticae* is the most studied mite regarding its resistance to agrochemicals, mainly because it is highly polyphagous and can damage a wide variety of crops. The apple rust mite can also cause damage in apples arriving to affect the yield [163]. Up to now, only one old report on resistance to parathion has been released in Canada for the species [164].

Concerning the management of mites in apple orchards, it is clear that the few compounds that remain effective are consequence of strict resistance management strategies [165,166] and where chemical treatments have to be combined with other measures, e.g., biological control to keep mites level under economic damage threshold. Whalon *et al.*, 2008 [159] and IRAC, 2014 [167] offer a complete literature on arthropod resistance and the mechanisms involved, including the species *P. ulmi*, *T. urticae* and *A. schlechtendali* and this information has to be used to properly design the management of mites in an apple IPM program.

In the case of plant diseases resistance development may be more harsh and difficult to be managed. Because most IPM strategies for plant diseases have been adapted from insect biology models, with little considerations in biological differences, resistance management may not be as effective as in the case of arthropods. For instance, most fungal plant pathogens are haploid and with faster reproductive rates, compared with mostly diploid arthropods of economic significance, which affects inheritance of pesticide resistance traits. Beckerman *et al.* [168] argued that the actual approaches used to implement IPM, using powerful new fungicides, applied often at lower doses, have contributed to fungicide resistance problems and may still be driving that process in apple scab management.

Scab, caused by *Venturia inaequalis* (Cooke) winter, is the main disease of apple and requires numerous fungicide applications, especially where frequent rains and fairly high temperatures occur during the first growth stages. As a result, the impact of fungicide resistance in apple scab has been significant, with fungicide resistance having been identified in all classes of curative fungicides used for the control of scab. The reiterated use of fungicides to control this apple key disease threats their effectiveness and endurance over the time. Among the most recent chemical families of fungicides, activity reduction has been observed for the anilinopyrimidines (AP), the demethylation inhibitors and respiration inhibitors at the Qo site (QoI) [169,170,171,172].

To minimize potential treatment failures due to resistance events, reliable control of apple scab epidemics can be achieved by the integrated application of all types of countermeasures including the use of scab resistant varieties, biological control agents, and appropriate agronomic and hygienic practices. Contrary to arthropods pesticide resistance management, non-chemical methods of disease control are often weak or not available, so the application of fungicides of different modes of action together with the use of forecasting models, for correctly timing of treatments, is the predominant strategy in order to overcome resistance problems.

New Pest Threats

In recent years, the international trade in fruits and vegetables has been pointed out for a substantial growth in the volume. Europe together with the United States, and Japan are the largest importers of fruits and vegetables. Despite the international cooperation and regulatory systems put in place to inhibit the spread of plant pests, the international trade of fruits and plants is the main responsible for facilitating their spread.

Among the potentially invasive species endangering European apple orchards, Brown Marmorated Sting Bug (BMSB)*Halyomorpha halys* (Stål) (Heteroptera: Pentatomidae), is probably the one that deserves the highest attention [173].*Halyomorpha halys*, is an herbivorous insect species that was accidentally introduced to the United States with international commerce, *i.e.*, via bulk freight containers from either Japan, Korea, or China [174] where it has caused serious economic injury to many crops, including tree fruit [175,176]. Losses caused by *H. halys* fruit injuries in apple orchards in the mid-Atlantic states have increased dramatically in recent years and many fruit growers have implemented an aggressive management programs to control this pest resulted in up to an approximately fourfold increase in insecticide use [175,177]. Insecticide array included broad-spectrum insecticides in the postbloom period has also increased the incidence of secondary pest outbreaks [174].

Additional invasions have been detected also in Europe, namely in Switzerland, Liechtenstein, Germany, Italy, France, Hungary and Greece [176] suggesting this invasive species could emerge as a cosmopolitan pest species. *Halyomorpha halys*has been included in the EPPO Alert List for more than 3 years [177]. In 2013, it was considered that sufficient alert has been given and the pest was deleted from the Alert List. (EPPO). Even though apple injuries caused by *H. halys* has not been reported in Europe so far, damage to apples may be under-reported because it can be confused with cork spot or bitter pit, two physiological disorders caused by calcium deficiencies [174]. Based on the US experience [175,177], chemical control with broad spectrum insecticides is expected to be the essential component of the management reaction against this newly invasive species once established in the apple orchard. These aggressive management programs could have detrimental ecological consequences that could jeopardize the successfully implementation of IPM.

WHAT HAS BEEN CHANGED IN MARKET REQUIREMENTS AND RESIDUES?

The total apple production in Europe is estimated to be close to 12 million tons in 2014 [178]. While production is confirming the constant increase reported in the recent years, the consumption *per capita* is decreasing at the rate of 2% per year and it accounts now for 15 kg/person per year. The apple production system is also challenged by a loss of competitiveness due to high labor costs in relation to other regions of the world and the political goal to significantly reduce pesticides use. In addition, consumers and the markets that supply them are demanding a large reduction, ideally the elimination, of pesticide residues on fruit.

In fact, the consumer›s attention to the potential risks to health and the environment arising from the use of pesticides in agriculture is growing and despite checks by regulatory authorities attesting to the contrary, many consumers are instead concerned about the safety of fruit which are perceived as "residue carriers". Supermarket chains have interpreted this concern by imposing limits on pesticide residues on fruit well below those required by the law, introducing a new element of discrimination in the acquisition of the product [179]. These constraints itself do not have a scientific justification, but purely commercial nature, however, they have led producers to speed up the review of pest control strategies in the search of technical tools that would enable to meet the requirements of the market without jeopardizing the principles of IPM, including for instance the combination of mating disruption and biopecticides [180].

Although Integrated Production does not mean a reduction in number of pesticide residues, alternative systems to pesticide use are recommended and expected by both EU regulation and markets, to improve the quality of agricultural production and environmental safety and to meet consumers' expectations for safety apples.

As an example of how the apple production system has included the pesticide residue issue in its IFP strategies, we report here the cases of the Trentino (Italy) and Girona (Spain) regions. In Trentino apple production covers 10.500 ha and produces 500.000 tons/year representing nearly 5% of the European apple production. Guidelines for integrated fruit production are implemented since 1991 [181] with the support of a technical advisory service that is available for the growers free of charges. A technical committee oversees the updating of the guidelines and takes into consideration the results of the survey conducted annually by the growers association to detect the level of 406 pesticide residues. Fruit sampling at harvest is annually carried out in order to analyze both the number of active ingredients (a.i.) and level of residues present on the fruits. In the last three years, the average number of a.i. found in the sample collection ranged from 3.18 (2011) and 4.55 (2010). In general, given the climatic characteristics of the region, pesticide residues on fruit are mainly represented by fungicides used for the apple scab control and for the prevention of storage diseases. Table 1 shows the number of samples analyzed annually over the last seven years, the percentage of samples with residues of insecticides/acaricides, as well as the distribution of the samples in four classes according to the percentage of residual insecticides/acaricides that has been observed compared to the maximum residue limits (MRLs).

In recent years, there has been a steady reduction of insecticide/acaricide residues in so much that in 2011 two thirds of the samples were free of insecticide/acaricide residues, and if the residue was present, 99 times out of 100 it was below 30% of the MRL. This positive results are a consequence of a careful management of pests, based on field scouting and visual controls performed by the consultants, application of tolerance thresholds when available and extensive use of mating disruption for the control of the codling moth *Cydia pomonella*, the main pest in this area.

Table 1. Percentage of samples with insecticide/acaricide residues and percentage distribution of samples with respect of pesticide maximum residue levels (n: number of samples analyzed over seven years).

Year							
2005	2006	2007	2008	2009	2010	2011	
N							
Residue analysis	*560*	*725*	*584*	*603*	*607*	*617*	*575*
% of sample with residue up to 30% of MRL	96.3	95.4	96.6	95.2	97.2	99.6	99.0

% of sample with residue between 30% - 50% of MRL	3.0	2.9	1.5	2.7	2.0	0.2	0.0
% of sample with residue between 50 and 100% of MRL	0.7	1.4	1.2	1.2	0.6	0.0	0.5
% of sample with residue more than 100% of MRL	0.0	0.3	0.7	0.9	0.2	0.2	0.5

Figure 3 shows the positive influence of mating disruption on the reduction of samples with insecticide/acaricide residues; the percentage of samples with insecticide/acaricide residues decreases by increasing the MD treated area. Based on these data, it was thought useful to evaluate how feasible could be producing apple without any insecticide/acaricide residues. This goal seems to be achievable in many situations, by extending the combination of the MD with biopesticide such as GpGV [180] and/or with low residual products such as emamectin-benzoate for the control of codling moth at the end of the season [182] and by promoting the use of sulphur-based-preparations for the combined control of apple scab and rust mites (RM) *A. schlechtendali* [183].

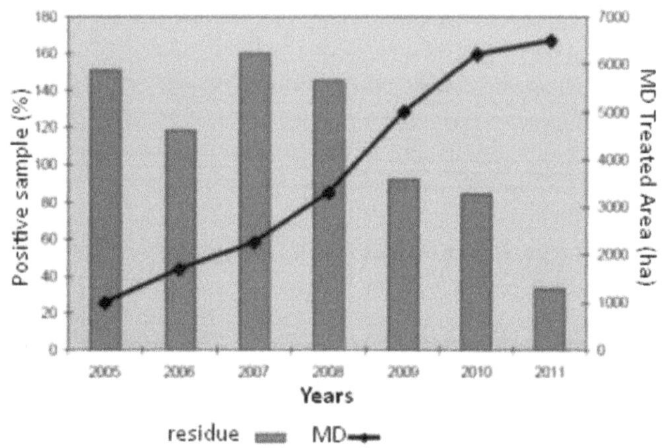

Figure 3. Percentage samples with insecticide/acaricide (gray bars) and area treated (ha) with mating disruption (black line) in Trentino.

Similarly, in the Girona region, IPM started in the early '90 and in the Catalonia region there is also an IPM committee formed by researchers, technicians and officers that update the guidelines. This committee is depended on the Government and from the beginning of IPM in the area and progressively, all the available new technologies for pests control has been incorporated at experimental orchards level until a complete program to produce apples with low residues was achieved. When results were strong and reliable regarding both pest control and residues at harvest, it was decided (2009) to implement the program in a wider area, after an agreement with the majority of the growers. In Table 2 are summarized the main arthropods and diseases and the recommended control method for each.

These control methods were chosen in function of its efficacy and also taking into account the environmental protection. The main sprays with chemicals have been concentrated when the fruits are not still present or when they are very small. In case of an outbreak close to harvest, it is recommended to use a chemical that can have an easy and quick degradation in the environment respecting the rules of resistance management. Although this program maximizes all the available tools, there still are some bottle necks; one of them is related with diseases which control are based on fungicides which in some cases can be applied as directed by the predictive models; but, in any case, these sprayings are still highly dependent on weather conditions of each year, demanding more or less sprayings. As a consequence, sometimes it is quite difficult to avoid some fungicide residues at harvest. The second bottle neck regards the new arthropods or diseases that are lately appearing. Some of them are invasive, while some others were already present in the environment and by unknown reasons started to increase their populations and damage. Nevertheless, the results of the program were highly satisfactory because it has permitted to decrease the use of insecticides up to 36.9 % and 24.5% the use of fungicides. Regarding residues level, in the 58% of the orchards in which this system was applied there were no residues and in the 42% remaining the level of residues detected was minor than the 20% of the LMR stated by the EU normative, which is very low.

Table 2. Common pests and diseases of apple in chronological order of appearance until harvest; right moment to start control and control methods for each.

Pest /Desease	**Plant, Pest or Disease Phenological Stage**	**Control Method**
Apple scab	From 07 (BBCH* scale)	Models to decide the right time to spray fungicides
San José scale	From 07 (BBCH scale)	Paraffinic oils and insecticides in pre-bloom.
Powdery mildew	From 10 (BBCH scale)	Fungicides
Fire blight	Pre-bloom Bloom	Bactericides based on copper in pre-bloom and micro-bial control during bloom
Aphids (Rosy apple aphid, Green aphid)	Pre- and/or postbloom treatments	Biological control + Insecticides
Woody Apple aphid	Pre- and/or postbloom	Biological control + Insecticides

Leafrollers	Pre- and/or postbloom treatments	Insecticides based on visual inspections (pre-bloom) and on captures in monitoring traps (post-bloom)
Codling moth	64 (BBCH scale)	Mating disruption (as a base system) plus CpGv or common insecticides based on captures in monitoring traps
Leopard moth	64 (BBCH scale)	Mating disruption (as a base system)
Mediterranean Fruit Fly	At the beginning of the flight	Mass trapping or attract and kill
European Red Mite	Pre- and/or postbloom treatments	Biological control. Miticides, only in case of not Phytoseiidae
Oriental Fruit Moth		Mating disruption (as a base system) plus insecticides based on captures in monitoring traps
Storage diseases	No fungicides	

CONCLUSIONS

From the initial development and leading implementation of the IPM in years 1950s, many other expectations were placed on this pest control approach beside the original aim of the reduction of the deleterious effects caused by broad-spectrum insecticides used in agriculture. Integrated Pest Management, is a decision-based operation, required the coordinated function of multiple tactics for optimizing the control of whole classes of pests (insects, pathogens, vertebrates and weeds) in an ecologically and economically sound manner. With the development of the concept of the holistic system approach where the orchard is playing the central role of agro-ecosystem, IPM became an indispensable element of the Integrated Fruit Production. The pest control strategies are more and more integrated with the use of natural resources where natural mechanisms are expected to replace potentially polluting inputs. Chemical treatments, accurately selected taking into account the protection of the worker and the consumer's health as well as the safeguard of the environment, are still essential to control pest and weed but they are integrated with agronomic preventive measures and biological and physical methods.

This evolution of the IFP is also referred as sustainable agriculture and involves economic, environmental and social aspects. Beside the traditional economic criteria of the market-oriented agriculture, sustainable agriculture includes the assessment of the environmental impact of agronomic practices like the carbon footprint, life cycle assessment, energy use, biodiversity. Economic and environmental sustainability are strictly interdependent, as it is the entire agricultural production process with societal context where it is held. These environmental impacts generated by the fruit production are raising consumer concerns over product standards which are now not only limited at the quality and safety standards, but they are frequently including process aspects of the production system. As a consequence, consumers and supermarket chains are requiring certification of good farm practices (e.g., Global G.A.P; TESCO NURTURE).

Europe is well placed to take advantage of sustainability issues due to its long tradition in IPM and its demonstrated ability to evolve this old concept in an innovative approach required by modern society.

ACKNOWLEDGEMENTS

We would like to thank Jesus Avilla for his suggestions to the earlier draft of the ms as well as two anonymous reviewers and the editor for their valuable comments.

AUTHOR CONTRIBUTIONS

All three authors have contributed equally to this study.

CONFLICTS OF INTEREST

The authors declare no conflict of interest.

REFERENCES

1. Stetter, J.; Folker, L. Innovation in crop protection: trends in research (review). *Angew. Chem. Int. Ed.* **2000**, *39*, 1724–1744.
2. United States Environmental Protection Agency (US-EPA). Available online: http://www.epa.gov/pesticides/factsheets/ipm.htm (accessed on 10 June 2015).
3. New York State Integrated Pest Management program. Available online: http://nysipm.cornell.edu/publications/apple_man/default.asp (accessed on 10 June 2015).
4. Hodgson, E. Pesticides: Past, present and future. In *Pesticides and the Future: Toxicological Studies of Risks and Benefits*; Hodgson, E., Roe, R.M., Motoyama, N., Eds.; North Carolina State University: Raleigh, NC, USA, 1991; pp. 3–12.
5. McHardy, W.E. Current Status of IPM in Apple orchards. *Crop Prot.* **2000**, *19*, 801–806.
6. Altieri, M.A.; Nicholls, C.I. *Agroecology and the Search for a Truly Sustainable Agriculture*; United Nations Environment Programme Environmental Training Network for Latin America and the Caribbean: Mexico City, Mexico, 2005; pp. 13–27.
7. Pickett, J.H.; Putman, W.L.; LeRoux, E.J. Progress in harmonizing biological and chemical control of orchard pests in eastern Canada. *Proc. 10th Int. Congr. Entomol.* **1958**, *3*, 169–174.
8. DeBach, P.; Bartlett, B. Effects of insecticides on biological control of insect pests of citrus. *J. Econ. Entomol.* **1951**, *44*, 372–383.

9. Smith, R.F.; Allen, W.W. Insect control and the balance of nature. *Sci. Am.* **1954**, *190*, 38–42.
10. Smith, R.F.; Hagen, K.S. Integrated control programs in the future of biological control. *Hilgardia* **1959**, *29*, 81–101.
11. Stern, V.M.; Smith, R.F.; van den Bosch, R.; Hagen, K.S. The integration of chemical and biological control of the spotted alfalfa aphid. The integrated control concept. *Hilgardia* **1959**, *29*, 81–101.
12. Boller, E.F.; van Lenteren, J.C.; Delucchi, V. *International Organization for Biological Control of Noxious Animals and Plants; History of the First 50 Years (1956–2006)*; Boller, E.F., van Lenteren, J.C., Delucchi, V., Eds.; IOBC: Zurich, Switzerland, 2006; p. 275.
13. van den Bosch, R.; Stern, V.M. The integration of chemical and biological control in arthropod pests. *Annu. Rev. Entomol.* **1962**, *7*, 367–387.
14. FAO. 1966a. Proceedings of the FAO Symposium on Integrated Pest Control Rome, Rome, Italy, 11–15 October 1965.
15. FAO. 1966b. Proceedings of FAO Symposium Integrated Pest Control Rome, Rome, Italy, 11–15 October 1965; pp. Part 1, p. 91; Part 2, p. 186; Part 3, p. 129.
16. Kogan, M.; Bajava, W.I. Integrated Pest Management: a global reality? *Annais de Sociedade Entológica do Brasil*, Available online: http://dx.doi.org/10.1590/S0301-80591999000100001 (accessed on 10 June 2015).
17. Boller, E.F.; Avilla, J.; Jörg, E.; Malavolta, C.; Wijnands, F.; Esbjerg, P. *Integrated Production: Principles and Technical Guidelines*, 3rd ed.; c/ Swiss Federal Research Station of Horticulture: Wädenswil, Switzerland, 2004.
18. Cross, J.V.; Dickler, E. Guidelines for Integrated Production of pome fruits in Europe: IOBC Technical Guideline III.*IOBC-WPRS Bull.* **1994**, *17*, 1–8.
19. Freier, B.; Boller, E.F. Integrated pest management in Europe—History, policy, achievements and implementation. In*Integrated Pest Management: Dissemination and Impact*; Peschin, R., Dhawan, A.K., Eds.; Springer: Dordrecht, The Netherland, 2009; Volume 2, pp. 435–454.
20. Maredia, K.M. *Integrated Pest management in the Global Arena*; CABI: Cambridge, MA, USA, 2003.
21. Bazoche, P.; Combris, P.; Giraud-Heraud, E.; Seabra Pinto, A.; Bunte, F.; Tsakiridou, E. Willingness to pay for pesticide reduction in the EU: nothing but organic? *Eur. Rev. Agric. Econ.* **2014**, *41*, 87–109.
22. Ehler, L.E. Integrated pest management (IPM): Definition, historical development and implementation, and the other IPM. *Pest Manag. Sci.* **2006**, *62*, 787–789.
23. Zalucki, M.P.; Adamson, D.; Furlong, M.J. The future of IPM: whither or wither? *Aust. J. Entomol.* **2009**, *48*, 85–96.
24. Weddle, P.W.; Welter, S.C.; Thomson, D. History of IPM in California pears 50 years of pesticide use and the transition to biologically intensive IPM. *Pest Manag. Sci.* **2009**, *65*, 1287–1292.
25. Damos, P. Current issues in integrated pest management of Lepidoptera pest threats in Industrial crop models. In*Lepidoptera: Ecology, Behavior and Management*; Guerritore, E., DeSare, J., Eds.; Nova Science Publications: New York, NY, USA, 2013; pp. 45–86.
26. Damos, P.; Savopoulou-Soultani, M. Microlepidoptera of Economic Significance in Fruit Production: Challenges, Constrains and Future Perspectives of Integrated Pest Management. In *Moths: Types, Ecological Significance and Control*; Cauterruccio, L., Ed.; Nova Science Publications: New York, NY, USA, 2011; pp. 75–113.
27. Pesticides-European Commission. Available online: http://ec.europa.eu/food/plant/pesticides/index_en.htm (accessed on 10 June 2015).
28. Strategy on the sustainable use of pesticides. Available online: http://ec.europa.eu/environment/index_en.htm(accessed on 10 June 2015).
29. Schuster, D.J.; Stanley, P.A. Biorational Insecticides for integrated pest management in tomatos. *Biol. Control.* **2007**, *41*, 99–109.
30. Science for Environmental Policy, EU. 2008. Available online: http://ec.europa.eu/environment/integration/research/newsalert/pdf/134na5.pdf (accessed on 10 June 2015).
31. Lacey, L.A.; Shapiro-Ilan, D.I. Microbial Control of Insect Pests in Temperate Orchard Systems: Potential for Incorporation into IPM. *Annu. Rev. Entomol.* **2008**, *53*, 121–144.
32. Andermatt, M.; Mani, E.; Wildbolz, T.H.; Lüthy, P. Susceptibility of *Cydia pomonella* to Bacillus thuringiensis under laboratory and field conditions. *Entomol. Exp. Appl.* **1988**, *49*, 291–295.
33. Ioriatti, C.; Pasqualini, E.; Delaiti, M. Effectiveness of *Bacillus thuringiensis* Berliner on three species of apple leafrollers. *Boll. Ist. Entomol. Guido Grandi› Stud Bologna* **1996**, *50*, 73–93.
34. Cross, J.V.; Solomon, M.G.; Chandler, D.; Jarrett, P.; Richardson, P.N.; Winstanley, D.; Bathon, H.; Huber, J.; Keller, B.; Langenbruch, G.A.; Zimmermann, G. Biocontrol of pest of apples and pears in northern and central Europe: 1. microbial agents and nematodes. *Biocontrol Sci. Technol.* **1999**, *9*, 125–149.
35. Tanada, J. A granulosis-virus of the codling moth, *Carpocapsa pomonella* L. (Oleuthreutidae, Lepidoptera). *J. Insect Pathol.* **1964**, *6*, 378–380.
36. Falcon, L.A.; Huber, J. Biological control of the codling moth. In *Tortricid Pests, Their Biology, Natural Enemies and Control*; van Der Geest, L.P.S., Evenhuis, H.H., Eds.; Elsevier Science Publishers: Amsterdam, The Netherland, 1991; pp. 355–369.
37. Lacey, L.A.; Thomson, D.; Vincent, C.; Arhturs, S.P. Codling moth granulovirus: A comprehensive review. *Biocontrol Sci. Technol.* **2008**, *18*, 639–663.
38. Schmidt, S.; Tomasi, C.; Pasqualini, E.; Ioriatti, C. The biological efficacy of pear ester on the activity of granulosis virus for codling moth. *J. Pest. Sci.* **2008**, *81*, 29–34.
39. Wua, Z.-W.; Fana, J.-B.; Yua, H.; Wanga, D.; Zhanga, Y.-L. Ultraviolet protection of the *Cydia pomonella* granulovirus using zinc oxide and titanium dioxide.

Biocontrol Sci. Technol. **2015**, *25*, 97–107.

40. Knight, A.; Witzgall, P. Combining mutualistic yeast and pathogenic virus—A novel method for codling moth control.*J. Chem. Ecol.* **2013**, *39*, 1019–1026.
41. Poinar, G.O., Jr. Nematode parasites. In *Tortricid Pests, their Biology, Natural Enemies and Control*; van der Geest, L.P.S., Evenhius, H.H., Eds.; Elsevier Science Publishers: Amsterdam, The Netherlands, 1991; pp. 273–281.
42. Berkvens, N.; van Vaerenbergh, J.; Maes, M.; Beliën, T.; Viaene, N. Entomopathogenic nematodes fail to parasitize the woolly apple aphid *Eriosoma lanigerum* as their symbiotic bacteria are suppressed. *J. Appl. Entomol.* **2014**, *138*, 644–655.
43. Cossentine, J.E.; Judd, G.J.R.; Bissett, J.D.; Lacey, L.A. Susceptibility of apple clearwing moth larvae, *Synanthedon myopaeformis* (Lepidoptera: Sesiidae) to *Beauveria bassiana* and *Metarhizium brunneum*. *Biocontrol Sci. Technol.* **2010**, *20*, 703–707.
44. Knight, A.L.; Howell, J.F.; McDonough, L.M.; Weiss, M. Mating disruption of codling moth (Lepidoptera: Tortricidae) with polyethylene tube dispensers: determining emission rates and the distribution of fruit injuries. *J. Agric. Entomol.***1995**, *12*, 85–100.
45. Witzgall, P.; Kirsch, P.; Cork, A. Sex Pheromones and Their Impact on Pest Management. *J Chem Ecol.* **2010**, *36*, 80–100.
46. Witzgall, P. Pheromones—Future techniques for insect control? *IOBC-WPRS Bull.* **2001**, *24*, 114–122.
47. Cardé, R.T. Using pheromones to disrupt mating of moth pests. In *Perspectives in Ecological Theory and Integrated Pest Management*; Kogan, M., Jepson, P., Eds.; Cambridge University Press: Cambridge, MA, USA, 2007; pp. 122–169.
48. Witzgall, P.; Stelinski, L.; Gut, L.; Thomson, D. Codling moth management and chemical ecology. *Annu. Rev. Entomol.***2008**, *53*, 503–522.
49. Tomaszewska, E.; Hebert, V.R.; Brunner, J.F.; Jones, V.P.; Doerr, M.; Hilton, R. Evaluation of pheromone release from commercial mating disruption dispensers. *J. Agric. Food Chem.* **2005**, *53*, 2399–2405.
50. Bohneblust, E.; Hull, L.A.; Krawczyk, G. A comparison of various mating disruption technologies for control of two internally feeding Lepidoptera in apples. *Entomol. Exp. Appl.* **2011**, *138*, 202–211.
51. Baldessari, M.; Ioriatti, C.; Angeli, G. Evaluation of Puffer® CM, a release device of pheromone to control codling moth on apple in Italy. *IOBC-WPRS Bull.* **2013**, *91*, 199–204.
52. McGhee, P.S.; Gut, L.J.; Miller, J.R. Aerosol emitters disrupt codling moth, *Cydia pomonella*, competitively. *Pest Manag Sci.* **2014**, *70*, 1859–1862.
53. Gut, L.J.; Stelinski, L.L.; Thompson, D.R.; Miller, J.R. Behavior modifying chemicals: prospects and constraints in IPM. In *Integrated Pest Management: Potential, Constraints, and Challenges*; Koul, O., Dhaliwal, G.S., Cuperus, G., Eds.; CABI Press: Wallingford, UK, 2004; pp. 73–121.
54. Waldner, W. Three years of large-scale control of codling moth by mating disruption in the South Tyrol, Italy. *IOBC-WPRS Bull.* **1997**, *20*, 35–44.
55. Knight, A. Codling moth area wide integrated pest management. In *Area Wide Pest Management: Theory and Implementation*; Koul, O., Cuperus, G.W., Elliott, N., Eds.; CABI. Press: Wallingford, UK, 2008; pp. 159–190.
56. Kovanci, O.B.; Kumral, N.A.; Larsen, T.E. High *versus* ultra-low volume spraying of a microencapsulated pheromone formulation for codling moth control in two apple cultivars. *J. Pest Manag.* **2010**, *56*, 1–7.
57. Angeli, G.; Anfora, G.; Baldessari, M.; Germinara, G.S.; Rama, F.; Cristofaro, A.D.; Ioriatti, C. Mating disruption of codling moth *Cydia pomonella* with high densities of Ecodian sex pheromone dispensers. *J. App. Ent.* **2007**, *131*, 311–318.
58. Pluciennik, Z. The control of Codling moth (*Cydia pomonella* L.) population using mating disruption method. *Jour. Hort. Res.* **2013**, *21*, 65–70.
59. Stelinski, L.L.; Mcghee, P.; Haas, M.; Il'ichev, A.L.; Gut, L.J. Sprayable Microencapsulated Sex Pheromone Formulations for Mating Disruption of Four Tortricid Species: Effects of Application Height, Rate, Frequency, and Sticker Adjuvant. *J. Econ. Entomol.* **2007**, *100*, 1360–1369.
60. Stelinski, L.L.; Gut, L.J.; Haas, M.; McGhee, P.; Epstein, D. Evaluation of aerosol devices for simultaneous disruption of sex pheromone communication in *Cydia pomonella* and *Grapholita molesta* (Lepidoptera: Tortricidae). *J. Pest Sci.* **2007**,*80*, 225–233.
61. Stelinski, L.L.; Il'ichev, A.L.; Gut, L.J. Efficacy and release rate of reservoir pheromone dispensers for simultaneous mating disruption of codling moth and oriental fruit moth (Lepidoptera: Tortricidae). *J. Econ. Entomol.* **2009**, *102*, 315–323.
62. Steiner, L.F. Fruit Fly control in Hawaii with poison-bait sprays containing protein hydrolysates. *J. Econ. Entomol.***1952**, *45*, 838–843.
63. Escudero, A.; Vilajeliu, M.; Batllori, L. Captura masiva para el control de la mosca mediterránea de la fruta (*Ceratitis capitata* Wied.) en manzano. *Phytoma España* **2005**, *171*, 26–31.
64. Faccioli, G.; Pasqualini, E.; Baronio, P. Optimal trap density in Cossus-cossus (Lepidoptera, Cossidae) Mass-Trapping.*J. Econ. Entomol.* **1993**, *86*, 850–853.
65. Bosch, D.; Sarasua, M.J.; Avilla, J. Mass trapping of *Synanthedon myopaeformis* (Borkhausen) in Lleida (Spain) with pheromone traps. *IOBC-WPRS Bull.* **2001**, *24*, 167–171.
66. Alemany, A.; Miranda, M.A.; Alonso, R.; Escorza, C.M. Efectividad del trampeo masivo de hembras de *Ceratitis capitata* (Diptera:Tephritidae) a base de atrayentes alimentarios. "Efecto-borde" y papel de los frutales abandonados como potenciadores de la plaga. *Boletín de Sanidad vegetal Plagas* **2004**, *30*, 255–264.
67. Sastre, C.; Melo, J.C.; Borreli, G. La captura de hembras: una posible salida en el control de mosca de la fruta

(*Ceratitis capitata*, Wied.) en melocotonero. *Phytoma España* **1999**, *113*, 42–47.

68. Batllori, L.; Escudero, A.; Vilajeliu, M.; Garcia, F.; Benejam, J. Area-wide mass trapping to control *Ceratitis capitata*(Wied.) on stone fruits in Girona, NE of Spain. *IOBC-WPRS Bull* **2008**, *37*, 73–82.

69. Cohen, H.; Yuval, B. Perimeter Trapping Strategy to Reduce Mediterranean Fruit Fly (Diptera: Tephritidae) Damage on Different Host Species in Israel. *J. Econ. Entomol.* **2000**, *93*, 721–725.

70. Batllori, J.L.; Vilajeliu, M; Vilardell, P.; Creixell, A.; Carbó, M.; Esteba, G.; Raset, F.; Vayreda, F.; Giné, M.; Curós, D. Área piloto de reducción de insecticidas en plantaciones comerciales de manzano. *Fruticultura Profesional* **2003**, *136*, 49–54.

71. Epsky, N.D.; Heath, R.R.; Guzman, A.; Meyer, W.L. Visual cue and chemical cue interactions in a dry trap with food-based synthetic attractant for *Ceratitis capitata* and *Anastrepha ludens* (Diptera: Tephritidae). *Environ. Entomol.* **1995**, *24*, 1387–1395.

72. Peñarrubia-María, E.; Quilici, S.; Schmitt, C.; Escudero-Colomar, L.A. Evaluation of candidate systems for mass trapping against *Ceratitis spp.* on La Réunion island. *Pest Manag. Sci.* **2014**, *70*, 448–453.

73. Escudero-Colomar, L.A.; Vilajeliu, M.; IRTA, Girona, Spain. Unpublished work. 2015.

74. Batllori, L; Vilajeliu, M.; Escudero-Colomar, L.A.; Vilardell, P.; Usall, J. Guia técnica Fruit. Net per a la producción de Poma. Available online: http://www.ruralcat.net/c/document_library/get_file?uuid=0e20413d-d0ca-43c6-81f2-31390c44efcd&groupId=10136 (accessed on 16 June 2015).

75. Adams, R.G., Jr.; Prokopy, R.J. *Aphidoletes aphidimyza* (Rondani) (Diptera: Cecidomyiidae): An effective predator of the apple aphid (Homoptera: Aphididae) in Massachusetts. *Prot. Ecol.* **1980**, *2*, 27–39.

76. Carroll, D.P.; Hoyt, S.C. Augmentation of European Earwigs (Dermaptera: Forficulidae) for Biological Control of Apple Aphid (Homoptera: Aphididae) in an Apple Orchard. *J. Econ. Entomol.* **1984**, *77*, 738–740.

77. Costa-Comelles, J.; García-Marí, F.; Laborda, R.; Marzal, C. Integrated control of the european red mite by phytoseidid predators in Lerida apple orchards. *Int. Conf. Pests Agricultura* **1987**, *3*, 17–24.

78. Brown, MW. Role of the aphid predator guild in controlling spirea aphid populations on apple in West Virginia, USA.*Biological Control* **2004**, *29*, 189–198.

79. Mathews, C.R.; Bottrell, D.G.; Brown, M.W. Habitat manipulation of the apple orchard floor to increase ground-dwelling predators and predation of *Cydia pomonella* (L.) (Lepidoptera: Tortricidae). *Biol. Control* **2004**, *30*, 265–273.

80. Avilla, J.; Bosch, D.; Escudero-Colomar, A.; Sarasua, M.J. Manzano, peral y melocotonero. In *Control Biológico de Plagas Agrícolas*; Jacas, J., Urbaneja, A., Eds.; Phytoma-España: Valencia, Spain, 2008; pp. 348–365.

81. Brown, M.W.; Mathrews, C.R. Conservation biological control of spiraea aphid, *Aphis spiraecola* (Hemiptera: Aphididae) on apple by providing natural alternative food resources. *Eur. J. Entomol.* **2008**, *105*, 537–540.

82. Boreau de Roincé, C.; Lavigne, C.; Ricard, J.-M.; Franck, P.; Bouvier, J.-C.; Gracin, A.; Sydmondson, W.O.C. Predation by generalist predators on the codling moth *versus* a closely-related emerging pest the oriental fruit moth: a molecular analysis. *Agric. For. Entomol.* **2012**, *14*, 260–269.

83. Dib, H.; Simon, S.; Sauphanor, B.; Capowiez, Y. The role of natural enemies on the population dynamics of the rosy apple aphid, *Dysaphis plantaginea* Passerini (Hemiptera: Aphididae) in organic apple orchards in south-eastern France.*Biol. Control* **2010**, *55*, 97–109.

84. Desneaux, N.; Decourtye, A.; Delpuech, J.-M. The sublethal effects of pesticides on beneficial asthropods. *Ann. Rev. Ent.* **2007**, *52*, 81–106.

85. Tricoire-Leignel, H.; Thany, S.H.; Gadenne, C.; Anton, P. Pest insect olfaction in an insecticide-contaminated environment: Info-disruption or hormesis effect. *Front. Physiol.* **2012**, *3*.

86. Roubos, C.R.; Rodriguez-Saona, C.; Isaacs, R. Mitigating the effects of insecticides on arthropod biological control at field and landscape scales. *Biol Control* **2014**, *75*, 28–38.

87. Maalouly, M.; Franck, P.; Bouvier, J.C.; Toubon, J.F.; Lavigne, C. Codling moth parasitism is affected by semi-natural habitats and agricultural practices at orchard and landscape levels. *Agric. Ecosyst. Environ.* **2013**, *169*, 33–42.

88. Malagnoux, L.; Marliac, G.; Simon, S.; Rault, M.; Capowiez, Y. Management strategies in apple orchards influence earwig community. *Chemosphere* **2015**, *124*, 156–162.

89. Wearing, C.H.; de Boer, J.A. Mortality of San Jose scale (*Diaspidiotus perniciosus* Hemiptera: Diaspididae) on an apple tree. *N. Z. Entomol.* **2014**, *37*, 107–124.

90. Agnello, A.; Reissing, W.H.; Harris, T. Management of summer populations of European red mite (Acari:Tetranychidae) on apple with horticultural oil. *J. Econ. Entomol.* **1994**, *87*, 148–161.

91. Hall, F.R. Effects of synthetic pyerethroids on major insects and mite pests of apple. *J. Econ. Entomol.* **1979**, *72*, 441–446.

92. Sentenac, G.; Bonafos, R.; Ruelle, B.; Coulon, T.; Escaffre, P.; Auger, P.; Kreiter, S. Effects non intenntionnels de certain produits phyopharmaceutiques sur *Typhlodromus pyri*, *Kampimodromus aberrans* et *Phytoseius plumifer*. *Phytoma* **2002**,*555*, 50–55.

93. Hoy, M.; Westigard, P.H.; Hoyt, S.C. Release and evaluation of a laboratory-selected, pyrethroid-resistant strain of the predaceous mite *Metaseiulus occidentalis* (Acari:Phytoseiidae) in Southern Oregon Pear orchards and a Washington apple orchard. *J. Econ. Entomol.* **1983**, *76*, 383–388.

94. Markwick, N.P.; Wearing, C.H.; Shaw, P.W. Pyrethroid insecticides for apple pest control: 1. Development of

pyrethroid-resistant predatory mites. In Proceedings of the Forty Third New Zealand Weed and Pest Control Conference, Dunedin, New Zealand, 14–16 August 1990; pp. 296–300.

95. Hardan, J.M.; Rogers, M.L.; Gaul, S.O.; Bent, E.D. Insectary rearing and initial testing in Canada of an organophosphate/pyrethroid-resistant strain of the predatos mite *Typhlodromus pyri* (Acari: Phytoseiidae) from New Zealand. *Environ Entomol.* **1997**, *26*, 1424–1436.

96. Bonafos, R.; Serrano, E.; Auger, P.; Kreiter, S. Resistance to deltamethrin, lambda-cyhalothrin and chlorpyriphos-ethyl in some populations of *Typhlodromus pyri* Scheuten and *Amblyseius andersoni* (Chant) (Acari:Phytoseiidae) from vineyards in the south-west of France. *Crop Prot.* **2007**, *26*, 169–172.

97. Szabó, A.; Pènses, B.; Sipos, P.; Hegyi, T.; Hajdú, Z.; Markó, V. Pest Management systems affect composition but not abundance of phytosiid mites (Acari: Phytoseiidae) in apple orchards. *Exp. Appl. Acarol.* **2014**, *62*, 525–537.

98. Duso, C.; Pasini, M. Distribution of the predatory mite *Amblyseius andersoni* (Acari:Phytoseiidae) on different apple cultivars. *J. Pest Sci.* **2003**, *76*, 33–40.

99. Cuthberson, A.; Qiu, B.-L.; Murchie, A.K. *Anystis baccarum*: an important generalist predatory mite to be considered in apple orchard pest management strategies. *Insects* **2014**, *5*, 615–628.

100. Baker, R.T. Predation of leafroler larvae by spiders and mites. *New Zeal. Entomol.* **1983**, *6*, 22–23.

101. Wearing, C.H.; Attfield, B.A.; Colhoun, K. Biological control of woolly apple aphid, *Eriosoma lanigerum* (Hausmann), during transition to integrated fruit production for pipfruit in Central Otago, New Zealand. *N. Z. J. Crop Hortic. Sci.***2010**, *38*, 255–273.

102. Biddinger, D.J.; Weber, D.C.; Hull, L.A. Coccinellidae as predators of mites: Stethorini in biological control. *Biol. Control* **2009**, *51*, 268–283.

103. Sciarretta, A.; Trematerra, P. Geostatistical tools for the study of insect spatial distribution: Practical implications in the integrated management of orchard and vineyard pests. *Plant Protect. Sci.* **2014**, *50*, 97–110.

104. Boller, E.F.; Hani, F.; Poehling, H.M. *Ecological Infrastructures: Ideabook on Functional Biodiversity at the Farm Level*; LBL: Lindau, Switzerland, 2004.

105. Simon, S.; Bouvier, J.C.; Debras, J.F.; Sauphanor, B. Biodiversity and pest management in orchard systems. A review.*Agron. Sustainable Dev.* **2010**, *30*, 139–152.

106. Garcia-Salazar, C.; Gut, L.J.; Whalon, M.E. Hedgerow barriers and other reduced-risk controls for managing Oriental fruit moth, *Grapholitha molesta* (Busck) (Lepidoptera: Tortricidae) in apples. *Renew. Agr. Food Syst.* **2007**, *22*, 181–188.

107. Miñarro, M.; Prida, E. Hedgerows surroundings organic apple orchards in north-west Spain: potential to conserve beneficial insects. *Agric. For. Entomol.* **2013**, *15*, 382–390.

108. Markò, V.; Jenser, G.; Mihàlyi, K.; Hegyi, T.; Balàzs, K. Flowers for better pest control? Effects of apple orchard groundcover management on mites (Acari), leafminers (Lepidoptera, Scitellidae), and fruit pests. *Biocontrol Sci. Technol.* **2012**, *22*, 39–60.

109. Song, B.; Jiao, H.; Tang, G.; Yao, Y. Combining repellent and attractive aromatic plants to enhance biological control of three tortricids species (Lepidoptera:Tortricidae) in an apple orchard. *Fla. Entomol.* **2014**, *97*, 1679–1689.

110. Brown, M.W.; Tworkoski, T. Pest management benefits of compost mulch in apple orchards. *Agric. Ecosyst. Environ.***2004**, *103*, 465–472.

111. Croft, B.A.; McRae, I.V. Persistence of *Typhlodromus pyri* and *Metaseiulus occidentalis* (Acari: Phytoseiidae) on apple after inoculative release and competition with *Zetzelia mali* (Acari: Stigmaeidae). *Environ. Entomol.* **1992**, *21*, 1168–1177.

112. Grasswitz, T.R.; Burts, E.C. Effect of native natural enemies and augmentative releases of *Chrysoperla rufilabris*Burmeister and *Aphidoletes aphidimyza* (Rondani) on the population dynamics of the green apple aphid, *Aphis pomi* De Geer. *Int. J. Pest Manag.* **1995**, *41*, 176–183.

113. Ahmad, S.; Pozzebon, A.; Duso, C. Augmentative releases of the predatory mite *Kampimodromus aberrans* in organic and conventional apple orchards. *Crop Prot.* **2013**, *52*, 47–56.

114. Damos, P.; Savopoulou-Soultani, M. Development and statistical evaluation of models in forecasting major lepidopterous peach pest complex for integrated pest management programs. *Crop Prot.* **2010**, *29*, 1190–1199.

115. Pedigo, L.P.; Hutchins, S.H.; Higley, L.G. Economic injury levels in theory and practice. *Annu. Rev. Entomol.* **1986**, *31*, 341–368.

116. Damos, P.; Savopoulou-Soultani, M. Population dynamics of *Anarsia lineatella* (Lep: Gelechiidae) in relation to crop damage and development of Economic Injury Levels. *J. Appl. Entomol.* **2009**, *134*, 105–115.

117. Damos, P.; Karabatakis, S. Real time pest modelling through the World Wide Web: Decision making from theory to praxis. Integrated protection of fruit crops. *IOBC-WPRS Bull.* **2013**, *91*, 253–258.

118. Karabatakis, S.; Damos, P. Supporting integrated pest management using open data networks and information technology through the World Wide Web. In Proceedings of ESA 60th Annual meeting, Knoxville, TN, USA, 11–14 November 2013.

119. Balsari, P.; Marucco, P. The New EU Directive Requirements and the Innovation in Pesticide Application Techniques.*J. ASTM Int.* **2011**, *8*, 1–21.

120. Polveche, V.; Crete, X.; Chapuis, G.; Douzals, J.P. Effects of nozzle types, windbreak and vegetation stages on drift performances issued from an orchard sprayer. In Proceedings of Suprofruit Workshop, Centre Ctifl Lanxade, Prigonrieux, France, 8–10 June 2011; pp. 66–67.

121. Stallinga, H.; Van de Zande, J.C.; Wenneker, M.; Michielsen, J.M.G.P. Nozzle classification for drift reduction in orchard spraying: effect of nozzle type

in full leave stage orchards. In Proceedings of 11th Workshop, Sustainable Plant Protection Techniques in Fruit Growing, Bergerac, France, 8–10 June 2011; pp. 70–71.

122. Loquet, B.; Destombes, J. Nozzles and spray application in orchards: efficiency and spray quality. In Proceedings of 11th Workshop, Sustainable Plant Protection Techniques in Fruit Growing, Bergerac, France, 8–10 June 2011; pp. 54–55.
123. Bondesan, D.; Rizzi, C.; Angeli, G.; Ioriatti, C. Evaluation of spray drift in apple orchards of Trentino: Comparison of different solutions to reduce environmental contamination. *IOBC-WPRS Bull.* **2013**, *91*, 493–499.
124. Koch, H.; Weisser, P. Sensor equipped orchard spraying – efficacy, savings and drift reduction. *Asp. Appl. Biol.* **2000**,*57*, 357–362.
125. Balsari, P.; Marucco, P.; Tamagnone, M. A crop identification system (CIS) to optimise pesticide applications in orchards. *J. Hortic. Sci. Biotech.* **2009**, *84*, 113–116.
126. Solanelles, F.; Escola, A.; Planas, S.; Rosell, J.R.; Camp, F.; Gracia, F. An Electronic Control System for Pesticide Application Proportional to the Canopy Width of Tree Crops. *Biosyst. Eng.* **2006**, *95*, 473–481.
127. Doruchowski, G.; Jaeken, P.; Hołownicki, R. Target detection as a tool of selective spray application on trees and weeds in orchards. In *Precision Agriculture and Biological Quality*; Bellingham, W.A., Meyer, G.E., DeShazer, J.A., Eds.; Publisher: Boston, MA, USA, 1998; pp. 290–301.
128. Walklate, P.J.; Cross, J.V.; Richardson, G.M.; Murray, R.A.; Baker, D.E. Comparison of Different Spray Volume Deposition Models Using LIDAR Measurements of Apple Orchards. *Biosyst. Eng.* **2002**, *82*, 253–267.
129. Cross, J.; Balsari, P.; Doruchowski, G.; Douzals, J.P.; Herbst, A.; Marucco, P.; Nuyttens, D.; Walklate, P. Orchard spray application in Europe—State of the art and research challenges. *IOBC-WPRS Bull.* **2013**, *91*, 465–475.
130. Dorigoni, A.; Lezzer, P.; Micheli, F.; Dallabetta, N.; Pasqualini, J.; Guerra, A. Parete fruttifera stretta per mele redditizie e sostenibili. *Inf. Agrar.* **2009**, *65*, 54–58.
131. Dorigoni, A.; Micheli, F. Possibilities for multi-leader trees. *Eur. Fruit Mag.* **2014**, *2*, 18–20.
132. Jamar, L.; Mostade, O.; Huyghebaert, B.; Pigeon, O. Comparative performance of recycling tunnel and conventional sprayers using standard and drift-mitigating nozzles in dwarf apple orchards. *Crop Prot.* **2010**, *29*, 561–566.
133. Agnello, A.; Landers, A. Progress in the Development of an In-Canopy Fixed Spraying System for High-Density Apple Orchards. In Proceedings of the 88th Annual Orchard Pest and Disease Management Conference, Portland, OR, USA, 8–10 January 2014.
134. Sévérac, G.; Romet, L. Des arbres bien enveloppés avec Alt'Carpo! *L'Arboriculture Frutière* **2007**, *620–621*, 24–28.
135. Sauphanor, B.; Severac, G.; Maugin, S.; Toubon, J.-F.; Capowiez, Y. Exclusion netting may alter reproduction of the codling moth (*Cydia pomonella*) and prevent associated fruit damage to apple orchards. *Entomol. Exp. Appl.* **2012**, *145*, 134–142.
136. Food and Agricultural Organization of the United Nations Statistics Department. Available online: http://faostat3.fao.org/home/E (accessed on 10 June 2015).
137. Pesticide Use Reduction Strategies in Europe. Six case studies. Available online: http://www.paneurope.info/Resources/Reports/Pesticide_Use_Reduction_Strategies_in_Europe.pdf (accessed on 10 June 2015).
138. Wauchope, R.D.; Buttler, T.M.; Hornsby, A.G.; Augustijn-Beckers, P.W.M.; Burt, J.P. SCS/ARS/CES pesticides properties database for environmental decision making. *Rev. Environ. Contam. Toxicol.* **1992**, *123*, 1–157.
139. Boller, E.F.; Avilla, J.; Gendrier, J.P.; Jörg, E.; Malavolta, C. Integrated Production in Europe. 20 years after the declaration of Ovrannaz. *IOBC-WPRS Bull.* **1998**, *21*, 41.
140. IOBC/WPRS. Guidelines for Integrated Production of Pome Fruits in Europe. *IOBC-WPRS Bull.* **2002**, *25*, 1–8.
141. Hough, W.S. Relative resistance to arsenical poisoning of two codling moth strains. *J. Econ. Entomol.* **1928**, *21*, 325–329.
142. Waldner, W. Ruckblick und Vorschau auf die Bekämpfung des Apfelwicklers. *Obstbau-Weinbau* **1993**, *12*, 355–357.
143. Riedl, H.; Zelger, R. Erste Ergebnisse der Untersuchungen zur resistenz des Apfelwickler gegen uber Diflubenzuron. *Obstbau Weinbau* **1994**, *4*, 107–109.
144. Sauphanor, B.; Benoit, M.; Bouvier, J.C.; Perron, G.; Fremond, J.C. Un cas de r´esistance du carpocapse des pommes au diflubenzuron dans le sud-est de la France. *Phytoma* **1994**, *458*, 46–49.
145. Charmillot, P.J.; Pasquier, D.; Sauphanor, B.; Bouvier, J.C.; Olivier, R. Carpocapse des pommes: premier cas de r´esistance au diflubenzuron en Suisse. *Rev Suisse Vitic Arboric Hortic* **1999**, *31*, 129–132.
146. Ioriatti, C.; Tasin, M.; Charmillot, P.J.; Reyes, M.; Sauphanor, B. Early detection of resistance to tebufenozide in field populations of *Cydia pomonella* L.: methods and mechanisms. *J. Appl. Entomol.* **2007**, *131*, 453–459.
147. Sauphanor, B.; Avilla, J.; Charmillot, P.J.; Ioriatti, C.; Michele, S.; Matias , C.; Waldner , W. Coping with insecticide resistance in fruit production: the example of codling moth in Europe. In *Book of Abstracts of the VIth European Congress of Entomology*; Brinnhofer, V., Soldan, T., Eds.; Publisher: City, County, 1998; pp. 619–620.
148. Sauphanor, B.; Brosse, V.; Bouvier, J.C.; Speich, P.; Micoud, A.; Martinet, C. Monitoring resistance to diflubenzuron and deltamethrin in French codling moth populations (*Cydia pomonella*). *Pest Manag Sci.* **2000**, *56*, 74–82.
149. Charmillot, P.J.; Pasquier, D.; Grela, C.; Genini, M.; Olivier, R.; Ioriatti, C.; Butturini, A. Résistance du

carpocapse Cydia pomonella aux insecticides. *Revue Suisse Vitic. Arboric. Hortic.* **2003**, *35*, 363–368.

150. Rodriguez, M.A.; Dolors, B.; Avilla, J. Resistance of Spanish codling moth (*Cydia pomonella*) populations to insecticides and activity of detoxifying enzymatic systems. *Entomol. Exp. Appl.* **2011**, *138*, 184–192.

151. Voudouris, C.; Sauphanor, B.; Franck, P.; Reyes, M.; Mamuris, Z.; Tsitsipis, J.A.; Vontas, J.; Margaritopoulos, J.T. Insecticide resistance status of the coodling moth *Cydia pomonella* (Lepidoptera: Tortricidae) From Greece. *Pectic Biochem. Physiol.* **2011**, *100*, 229–238.

152. Gund, N.A.; Wagner, A.; Timm, A.E.; Schulze-Bopp, S.; Jehle, J.A.; Johannesen, J.; Reineke, A. Genetic analysis of*Cydia pomonella* (Lepidoptera: Tortricidae) populations with different levels of sensitivity towards the *Cydia pomonella*granulovirus (CpGV). *Genetica* **2012**, *140*, 235–247.

153. Cross, J.V. Susceptibility of the summer fruit tortrix moth, *Adoxophyes orana* (Lepidoptera: Tortricidae), to chlorpyrifos and strategies for insecticidal control in orchards. *Ann. Appl. Biol.* **1997**, *131*, 197–212.

154. Salamin, C.; Charmillot, P.J.; Pasquier, D. Nouveau cas de resistance aux insecticides de la tordeuse de la pelure capua (*Adoxophyes orana*). *Revue Suisse Vitic. Arboric. Hortic* **2007**, *39*, 179–183.

155. Kehrli, P.; Pasquier, D.; Roux, P.A. Variabilite phenologique et sensibilite aux insecticides de capua (*Adoxophyes orana*) en Valais. *Revue Suisse Vitic. Arboric. Hortic* **2009**, *41*, 263–268.

156. Cao, G.; Han, Z. Tebufenozide resistance selected in *Plutella xylostella* and its cross-resistance and fitness cost. *Pest Manage. Sci.* **2006**, *62*, 746–751.

157. Ioriatti, C.; Charmillot, P.J.; Forno, F.; Mattedi, L.; Pasquier, D.; Rizzi, C. Control of Codling moth Cydia pomonella L. using insecticides: field efficacy in relation to the susceptibility of the insect. *IOBC-WPRS Bull.* **2005**, *28*, 259–264.

158. Charmillot, P.J.; Blanc, G.; Pasquier, D. Premier cas de résistance en Suisse de la tordeuse de la pelure capua (*Adoxophyes orana*) aux insecticides. *Revue suisse Vitic. Arboric. Hortic.* **2006**, *38*, 87–93.

159. Whalon, M.E.; Mota-Sanchez, D.; Hollingworth, R.M. Analysis of Global Pesticide Resistance. In *Arthropods*; CABI Publishing, CAB International: Wallingford, UK, 2008; pp. 5–32.

160. Newcomer, E.J.; Dean, F.P. Orchards mites resistant to Parathion in Washington. *J. Econ. Entomol.* **1952**, *45*, 1076–1078.

161. Kumral, N.A.; Susurluk, H.; Gençer, S.N.; Gűrkan, M.O. Resistance to chlorpyrifos and lambda-cyhalothrin along with detoxifying enzyme activities in field-collected female populations of European red mite. *Phytoparasitica* **2009**, *37*, 7–15.

162. Kramer, T.; Nauen, R. Monitoring of spirodiclofen susceptibility in field populations of European red mites,*Panonychus ulmi* (Koch) (Acari: Tetranychidae), and the cross-resistance pattern of a laboratory-selected strain. *Pest Manag. Sci.* **2011**, *67*, 1285–1293.

163. Duso, C.; Castagnoli, M.; Simoni, M.; Angeli, G. The impact of eriophyds on crops: recent issues on *Aculus schlechtendali*, *Calepitrimerus vitis* and *Aculops lycopersici*. *Exp. Appl. Acarol.* **2010**, *51*, 151–168.

164. Morgan, C.V.G.; Anderson, N.H. Notes on parathion-resistant strains of two phytophagous mites and a predacious mite in British Columbia. *Can. Entomol.* **1958**, *90*, 92–97.

165. Marcic, D. Acaricides in modern management of plant-feeding mites. *J. Pest Sci.* **2012**, *85*, 395–408.

166. Van Leeuwen, T.; Dermauw, W.; Grbic, M.; Tirry, L.; Feyereisen, R. Spider mite control and resistance management: Does a genome help? *Pest Manag. Sci.* **2012**, *69*, 156–159.

167. Arthropod Pesticide Resistance Database. Available online: www.pesticideresistance.com/search.php (accessed on 10 June 2015).

168. Beckerman, J.; Sundin, G.W.; Rosenberger, D.A. Do some IPM concepts contribute to the development of fungicide resistance? Lessons learned from the apple scab pathosystem in the United States. *Pest Manag. Sci.* **2015**, *71*, 331–342.

169. Dux, H.; Sierotzki, H.; Gisi, U. Sensitivity of *Venturia inaequalis* populations to anilinopyrimidyne, DMI and QoI fungicides. In Proceedings of the 14th International Reinhardsbrunn Symposium "Modern fungicides and antifungal compounds", Reinhardsbrunn, Germany, 25–29 April 2004; pp. 40–41.

170. Köller, W.; Wilcox, W.F. Evidence for predisposition of fungicide-resistant isolates of *Venturia inaequalis* to a preferential selection for resistance to other fungicides. *Phytopathology* **2001**, *91*, 776–781.

171. Köller, W.; Wilcox, W.F.; Parker, D.M. Sensitivity of *Venturia inaequalis* populations to anilinopyrimidine fungicides and their contribution to scab management in New York. *Plant Dis.* **2005**, *89*, 357–365.

172. Fiaccadori, R.; Portillo, I.; Roberti, R.; Brunelli, A. Isolation of antagonistic fungi towards *Venturia inaequalis* and preliminary applications in sanitation practice reducing ascospore inoculums. *IOBC-WPRS Bull.* **2013**, *91*, 59–66.

173. Hoebeke, E.R.; Carter, M.E. *Halyomorpha halys* (Stål) (Heteroptera: Pentatomidae): A polyphagous plant pest from Asia newly detected in North America. *Proc. Entomol. Soc. Wash.* **2003**, *105*, 225–237.

174. Nielsen, A.L.; Hamilton, G.C. Seasonal occurrence and impact of*Halyomorpha halys* (Hemiptera: Pentatomidae) in tree fruit. *J. Econ. Entomol* **2009**, *102*, 1133–1140.

175. Leskey, T.C.; Short, B.D.; Butler, B.R.; Wright, S.E. Impact of the invasive brown marmorated stink bug, *Halyomorpha halys*, in mid-Atlantic tree fruit orchards in the United States: Case studies of commercial management. *Psyche* **2012**.

176. Milonas, P.G.; Partsinevelos, G.K. First report of the brown marmorated stink bug *Halyomorpha halys* Stål (Hemiptera: Pentatomidae) in Greece. *EPPO Bull.* **2014**, *44*, 183–186.

177. Rice, K.B.; Bergh, C.J.; Bergmann, E.J.; Biddinger,

D.J.; Dieckhoff, C.; Dively, G.; Fraser, H.; Gariepy, T.; Hamilton, G.; Haye, T.; Herbert, A.; Hoelmer, K.; Hooks, C.R.; Jones, A.; Krawczyk, G.; Kuhar, T.; Martinson, H.; Mitchell, W.; Nielsen, A.L.; Pfeiffer, D.G.; Raupp, M.J.; Rodriguez-Saona, C.; Shearer, P.; Shrewsbury, P.; Venugopal, P.D.; Whalen, J.; Wiman, N.G.; Leskey, T.C.; Tooker, J.F. Biology, Ecology, and Management of Brown Marmorated Stink Bug (Hemipetra: Pentatomidae). *J. Int. Pest Manag.* **2014**, *5*, A1–A13.

178. Dalpiaz, A. Innovazione e organizzazione, le uniche risposte per uscire dalla crisi. *Frutticoltura* **2014**, *11*, 2–6.

179. Waldner, W. Le catene alimentari e la riduzione numerica di residui di agro farmaci. *Frutta e Vite* **2009**, *4*, 182–186.

180. Krawczyk, G.; Hull, L.A.; Bohnenblust, E. Utilization of mating disruption and codling moth granulosis virus (CMGV) in conventional commercial apple orchards in Pennsylvania, USA. *IOBC/WPRS Bull.* **2010**, *54*, 71–74.

181. Agnolin, C.; Ioriatti, C.; Pontalti, M.; Venturelli, M.B. IFP experiences in Trentino, Italy. *Acta Holticulturae* **2000**, *525*, 45–49.

182. Ioriatti, C.; Anfora, G.; Angeli, G.; Civolani, S.; Schmidt, S.; Pasqualini, E. Toxicity of emamectin benzoate to *Cydia pomonella* (L.) and *Cydia molesta* (Busck) (Lepidoptera: Tortricidae): laboratory and field tests. *Pest Manag. Sci.* **2009**, *65*, 306–312.

183. Baldessari, M.; Giuliani, G.; Chiesa, S.; Larcher, R.; Ioriatti, C.; Angeli, G. Pesticide residue free fruits: The aim of Trentino apple production system. *Commun. Agric. Appl. Biol. Sci.* **2013**, *78*, 133–137.

CHAPTER 9

COTTON PEST MANAGEMENT PRACTICES AND THE SELECTION OF PYRETHROID RESISTANCE IN ANOPHELES GAMBIAE POPULATION

Anges Yadouleton, Thibaud Martin, Gil Padonou, Fabrice Chandre, Alex Asidi, Luc Djogbenou, Roch Dabiré, Rock Aïkpon, Michel Boko, Isabelle Glitho, and Martin Akogbeto

BACKGROUND

Malaria remains a major public health problem in Africa. It is reported to be the most significant cause of morbidity and mortality, resulting in a critical loss of working days [1].

More than 2 billion people in the world are at risk of contracting malaria and one million deaths are recorded yearly of which 90% occur in Sub-Saharan Africa [1]. In Benin, malaria is still the most important disease leading to 67% consultations in local health centres [2]. The strategy of the National Malaria Control Programme (NMCP) is based on effective case management and vector control with Insecticide-Treated Nets (ITN) and Indoor Residual Spraying (IRS).The development of new insecticides for public health use is limited and requires enormous capital and time, making industry reluctant to embark on such ventures. Novel compounds or alternatives are to be sought in the agricultural pesticide pipeline. Several reports have recently shown evidence that the main African malaria vector, *Anopheles gambiae* s.l., has developed a high level of resistance to pyrethroid insecticides as well as to other classes of public health insecticides. While resistance is now spreading throughout Sub-Saharan Africa, reports from Benin and the West African region indicated the highest recorded frequencies of the resistance genes [3-6].

The development of pyrethroid resistance in the primary malaria vectors, *An. gambiae* s.l. and *An. funestus*[7] is a serious concern. In the last decade, the emergence of resistance in populations of *An. gambiae* to common classes of insecticides used in public health has been reported in many African countries including Kenya [8], Côte d›Ivoire [9], Benin [10-14], Niger [15], Burkina Faso [16,17], Mali[18], Nigeria [19], South Africa [20], and Cameroun [21]. In the 1960s, the role of selective treatment with organochlorines (OC) in agriculture on resistance of *An. gambiae* was observed in Mali [22]. Evidence of an association between agricultural use of insecticides and the emergence of resistance in malaria vectors has been repeatedly reported. In Côte d›Ivoire and Burkina Faso, N›Guessan *et al.* [23] reported that the level of vector resistance to pyrethroid insecticides increased during the cotton growing season. Higher frequencies of *kdr* alleles were observed in the more intensely farmed cotton production areas of Côte d›Ivoire [9]. In Bukina-Faso, a survey of *kdr* alleles in *An. gambiae* field populations showed also a higher frequency of *kdr* alleles in older cotton areas with a decreasing gradient to non treated areas [17].

Cotton crop protection represents 90% of the insecticide use in West Africa. The control strategies implemented against cotton pest especially *Helicoverpa armigera* required a regular repeated applications of insecticides during the cotton plant growing cycle. As recommended by the Institut National des Recherches Agronomiques du Benin (INRAB), six consecutive treatments are applied at two weeks interval to protect the crop against bollworms, leafworms and sucking pests. These insecticides are essentially composed of pyrethroids (PYs), organophosphates (OPs) which are also the main classes used in public health and a cyclodiene. The majority of cotton farms observed in northern Benin are located in the upland landscape

Citation: Anges Yadouleton, Thibaud Martin, Gil Padonou, Fabrice Chandre, Alex Asidi, Luc Djogbenou, Roch Dabiré, Rock Aïkpon, Michel Boko, Isabelle Glitho and Martin Akogbeto, "Cotton pest management practices and the selection of pyrethroid resistance in Anopheles gambiae population," DOI: 10.1186/1756-3305-4-60.

while the lowland covers the major mosquito breeding sites. Thus run-off has been assumed to be the mechanism by which insecticides from agricultural sites reach malaria vector breeding sites, where they exert a huge selection pressure on larval stages of *An. gambiae* s.l. The main malaria vector *An. gambiae* breeds in puddles, stagnant pools and various sites around or within the lowland. During the rainy season, insecticide residues are washed downwards into mosquito breeding sites thus affecting larval population [24]. According to Akogbeto *et al.* [25] in Benin, insecticide treatments against cotton pests are applied twice a month, for a timeframe of three consecutive months (between July and October) each year. These treatment periods coincide with the rainy season and correspond to the period of high mosquito densities. The evidence supports the hypothesis that breeding sites contamination is the result of the coincidence of agricultural pesticide application and seasonal rainfall/runoff.

Alternatively, integrated pest and vector management (IPVM) strategies based on the rational use of chemical protection, has undoubtedly reduced the negative impact of pesticides on humans, and their environment, including the breeding sites of malaria vectors.

This study aimed to assess the impact of control strategies used against cotton pests (relative amount of insecticide) on the frequency and spread of insecticide resistance in *An. gambiae* populations. The study, conducted in northern Benin, compared the BP cotton cultivation sites (absence of pesticides use) with the CCP and TICP cotton cultivation sites where insecticides are extensively used. The study focused on the investigation of agricultural practices using pesticides for the control of cotton pests and their impact on the insecticide susceptibility of *An. gambiae* populations from surrounding breeding sites.

METHODS

Cotton Pests Control Strategies

Three pest management and control strategies are officially recommended in Benin:

(1) The Calendar Control Program (CCP) is based on the conventional treatment which systematically uses the full dosage of insecticides.

(2) The Targeted Intermittent Control Program (TICP) is based on two steps of crop protection [26,27]. The first steps is a protection which follows a conventional pesticide application schedule (every 14 days from the appearance of floral organs), but only the half dose of insecticide are usually applied. The second step includes a modification of the first treatment meaning that the half-dose left over during the earlier observation made the day before treatment would suggest that the pest populations exceed the economic thresholds of damage. The program was established five years ago.

(3) In the biological control program (BP), no chemical is used for plant protection. That program started over the past five years.

The area of the farms applying CCP and TICP was about 4 hectares and usually farmers ploughed sometime individually or work in groups of farmers' organizations. However, in BP sites of 1 hectare, farmers worked under the supervision of technicians from the Beninese Organization for Organic Farming Promotion (OBEPAP) who assisted in the implementation and the survey of good agricultural practices on organic cotton. These areas were characterised by a continual production of cotton crop.

Study sites

The study was conducted in the cotton areas around 8 cities in Benin (Figure 1). The choice of these areas took into account the various strategies of pest control. Semi permanent breeding sites were found in cotton fields where farmers used:

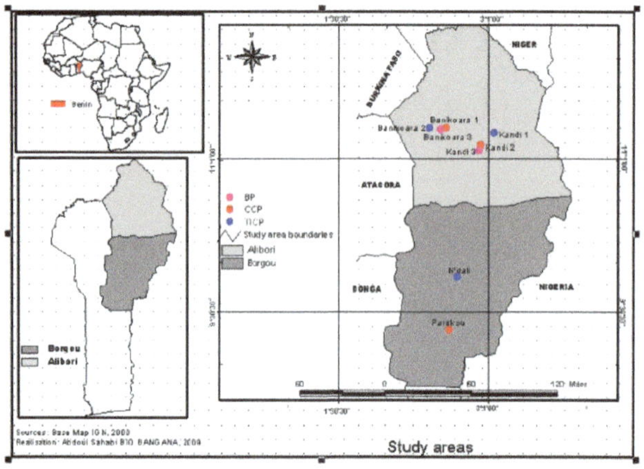

Figure 1: Map of Benin showing the study sites.

- the CCP around Parakou (2°62 E, 9°33 N), Kandi (2°95 E, 11°16 N) and Banikoara (2°59 E, 11°31 N). This pest management program started thirty years ago and was the main strategy against pest control used by more than 95% of the cotton farmers.
- the TICP around N'dali (2°70 E, 9°84 N), Kandi (3°08 E, 11°27 N) and Banikoara (2°41 E, 11°31 N) started five years ago and applied by 4% of cotton farmers.
- the BP around Kandi (2°92 E, 11°09 N) and Banikoara (2°52 E, 11°29 N) started five years ago and practiced by 1% of cotton farmers.

The annual mean rainfall recorded was about 1,300 mm yearly and characterized by a Sudanian climate with one rainy season (middle of May to October) and one dry season (November-May).

KAP Study on the use of Insecticides in Cotton Farms

To generate adequate information on the use of insecticide on cotton fields, Knowledge Attitude-Practice (KAP) surveys were organized in the study sites. In each site, leaders of farmer's organizations were interviewed using semi-structured questionnaires that focused on the treatment strategies, and the use of insecticides in the farms. Further, qualitative data was collected through direct observations, in-depth interviews and focus group discussions.

Mosquito Collections

Mosquito larvae were collected during the rainy season as well as before and during the period of insecticide treatments. The treatment periods started from July to October. During pest control, insecticide residues contaminate mosquito breeding sites whereby they diffuse into the water applying a selection pressure on mosquito larvae. Larvae were collected in the breeding sites of each site and transported to the laboratory of the Centre de Recherche Entomologique de Cotonou, Benin (CREC) for resistance testing. The adults were tested after emergence. A laboratory susceptible strain of *An. gambiae* Kisumu was used as a reference strain to compare the susceptibility levels of the field populations.

Insecticide Susceptibility Tests

Mosquitoes collected were assayed using WHO discriminating dosages of four insecticides: permethrin (0.75%), DDT (4%), deltamethrin (0.05%) and bendiocarb (0.1%). Four batches of 25 unfed females, aged 2-5 days, were exposed to the diagnostic doses of insecticide treated papers for 1 hour. The twenty five females of *An. gambiae* were introduced into each tube and monitored at different time intervals (10, 15, 20, 30, 45, 60 minutes) the number «knocked-down» recorded. After one hour exposure, mosquitoes were transferred into holding tubes and provided with cotton wools wet with a 10% honey solution. Mortalities were recorded after 24 hours and the susceptibility status of the population was graded according to the WHO protocol [28]. Dead and surviving mosquitoes from this bioassay were separately kept in Carnoy solution at -20°C for further molecular analysis.

Molecular Characterization

All *An. gambiae* s.l. were identified to species using PCR [29] and as M and S forms by PCR-RFLP [30]. To detect the presence of *Kdr* mutation in the samples collected from each study site, polymerase chain reaction diagnostic test for detection of *kdr* «Leu-phe» genes was carried out on *An. gambiae* mosquitoes from each study site as described by Martinez-Torres et *al.* [31].The PCR-RFLP diagnostic test was used to detect the presence of G119S gene (*Ace.1* gene) as described by Weill et *al.* [32].

Screening Of Pesticide Residues in Soil from Agricultural Settings

This investigation was performed using indirect bioassays focused on factors which can affect the normal growth of mosquito larvae in cotton breeding sites. The hatching rates of *An. gambiae* eggs and the larval survival during rearing period were assessed in artificial breeding sites made of soil samples collected from different cotton areas. The pyrethroid-susceptible Kisumu strain and the resistant VKPER strain were used to test the presence of insecticide residues in the soil samples by means of mortality rates after exposure. The testing method was based on an artificial breeding site made of a mixture of soil from cotton sites (BP, TICP and CCP programs) and CREC soil used as a control. 100 g of each soil was weight and mixed in 1,000 ml of water. 200 eggs of the susceptible *Kisumu* strain were placed in each artificial breeding site and compared with 200 eggs of VKPER.

Larvae in all artificial breeding sites were fed with similar quantity and type of food (well ground cat biscuits mixed with yeast powder). Daily observation was done in order to record the number and the instars of larvae. This experiment was replicated three times per month.

Data Interpretation

The resistant status of mosquito samples was determined according to the WHO criteria [33]:

- Mortality rates is > 97%: the population was considered fully susceptible
- Mortality rates ranged between 80 > × < 97%: resistance suspected in the population
- Mortality rates < 80%, the population was considered resistant to the tested insecticides.

The knockdown times for 50% and 95% of tested mosquitoes (*KdT50* and *KdT95*) were estimated using a log-time probit model [34].

The resistance allele frequency at the *kdr* and *Ace*-1 locus was calculated using Genepop software (version 3.3) as described by Raymond and Rousset [35].

A Fisher's exact test was performed to compare the resistance allele frequency at the *kdr* and *Ace*-1 among the mosquitoes from the different strategies.

An analysis of variance (ANOVA) was performed to compare the percentage of hatching eggs in the different treatments in order to know the impact of insecticide on the normal growth of mosquito larvae in cotton breeding sites.

RESULTS

Knowledge-Attitude -Practice (KAP) Investigations

Results from our KAP investigations from June to September 2008 in the cotton growing areas showed a common use by farmers of mineral fertilizer NPK at about 200 kg.ha-1 and urea at about 50 kg.ha-1 in both the CCP and the TICP sites. By contrast, in BP areas, all farmers in this group used organic and natural fertilizers which included animal excreta. In the CCP sites, about 6 pesticide treatments were applied by farmers 45 days after seeding and at two week intervals from flowering. Endosulfan or Tihan® (mixture of spirotetramat + flubendiamide), were sprayed in the first two treatments followed by the mixtures of cyfluthrin + chlorpyrifos ethyl for the 3rd and the 4th treatment and then cypermethrin + dimethoate applied for the last two treatments. In TICP sites, the same treatments at intervals of two weeks were made either at half dose or at a complete full dosage when the threshold of infestation was reached (5 of *H. armigera* larvae observed on 50 plants). In areas practicing Biological Program, farmers apply a mixture of neem or papaya leaves with added chilli and local soap three times before the harvest.

Resistance to Insecticides

A total of 1,313 females of *An. gambiae* collected from different sites around Parakou, N›dali, Kandi, and Banikoara were exposed to papers impregnated with discriminating doses of permethrin (0.75%), deltamethrin (0.05%), DDT (4%) and bendiocarb (0.1%).

The knockdown times (*KdT50, KdT95*) of *An. gambiae* populations from CCP and TICP sites were significantly longer than that of the susceptible strain Kisumu ($p < 0.05$). However, the *KdT50* for *An. gambiae* from BP site around Kandi was not significantly different from Kisumu (Table 1). Data recorded before and during the period of treatments showed a higher resistance to DDT and permethrin in populations from the CCP and TICP sites compared with those from the BP sites (Figure 2 and Figure 3). All populations of *An. gambiae* mosquitoes were resistant to DDT with an average of 33%, 42% and 65% of mortality respectively for CCP, TICP and BP sites. The mortality difference associated with the different pesticide application strategies was highly significant between BP and CCP programs ($P < 0.05$) but not significant between BP and TICP program ($P = 0.56$). However, *An. gambiae* populations from BP, CCP and TICP sites were fully susceptible to deltamethrin and bendiocarb (100% of mortality). Permethrin resistance was found in *An. gambiae* populations from CCP and TICP sites with an average mortality of 50% and 58% respectively. However, *An. gambiae* collected from BP sites were more susceptible to permethrin with 94% mortality.

Table 1: Knockdown times (KdT_{50} and KdT_{95}) and mortality of *Anopheles gambiae* s.l. populations from 3 cotton sites after exposure to DDT 4% and permethrin 0.75% and their resistance status

Sites/ Strains	Program	Insecticides	N	kdT50 [CI95] (min)	kdT95 [CI95] (min)	% Mortality [Conf lim 95]	Resistance status
		DDT	40	65.1 [57.5-73.4]	152.1 [118.4-228.1]	38 [29.31-46.7]	R
Parakou	CCP	Permethrin	75	35.3 [32.1-38.4]	112.1 [88.2-151.9]	54 [47.49-60.5]	R
		DDT	60	38.1 [29.4-36.5]	65.1 [57.5-86.5]	45 [37.73-52.3]	R
N'dali	TICP	Permethrin	60	19.3 [15.6-22.4]	67.1 [53.2-87.4]	60 [52.8-67.16]	R
		DDT	70	63.1 [60.2-72.3]	135.1 [5.2-184.5]	32 [25.7-38.31]	R
Kandi1	CCP	Permethrin	80	19.3 [15.6-22.4]	87.1 [63.5-138.4]	50 [43.67-56.32]	R
		DDT	50	35.1 [27.1-35.2]	62.5 [54.2-79.1]	43 [35.07-50.92]	R
Kandi 2	TICP	Permethrin	88	15.1 [13.6-20.1]	56.5 [43.2-77.2]	58 [52.00-64]	R
		DDT	60	30.3 [25.4-38.9]	72.6 [50.2-90.5]	66 [59.08-72.92]	R
Kandi 3	BP	Permethrin	80	11.0 [8.7 - 17.6]	23.5 [18.4-35.8]	94 [91.01-96.98]	S
		DDT	65	56.5 [51.7-63.2]	186.9 [146.0-265.2]	35 [28.3041.69]	R
Banikoara 1	CCP	Permethrin	90	24.6 [20.3-29.0]	105.2 [92-130.8]	51[45.04-56.96]	R
		DDT	70	32.4 [25.8-30.2]	60.1 [50.1-78.6]	41 [34.34-47.65]	R
Banikoara 2	TICP	Permethrin	70	18.5 [16.1-22.4]	58.1 [47.4-78.4]	59 [52.35-65.65]	R
		DDT	75	29.1 [24.4-37.6]	70.1 [48.2-88.2]	64 [57.73-70.27]	R
Banikoara 3	BP	Permethrin	80	13.2 [9.4 - 18.2]	25.5 [20.2-37.1]	95 [92.24-97.75]	S
		DDT	100	25.7 [24.3-27.0]	40.7 [38.0-44.5]	98 [96.41-99.58]	S
An. gambiaes.l	*	Permethrin	100	10.9 [9.7-12.0]	18.1[16.0-21.6]	99 [97.87100.13]	S

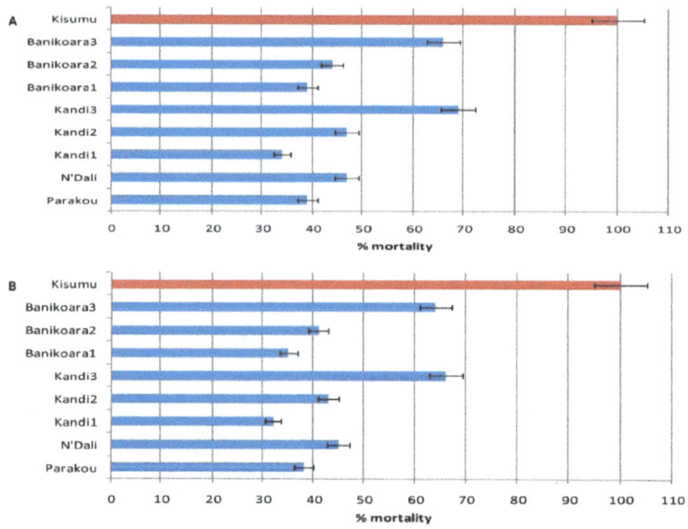

Figure 2: Mortality rates and resistance status of *Anopheles gambiae s.l.* collected before (A) and after (B) cotton treatments to DDT 4% using WHO bioassay tests.

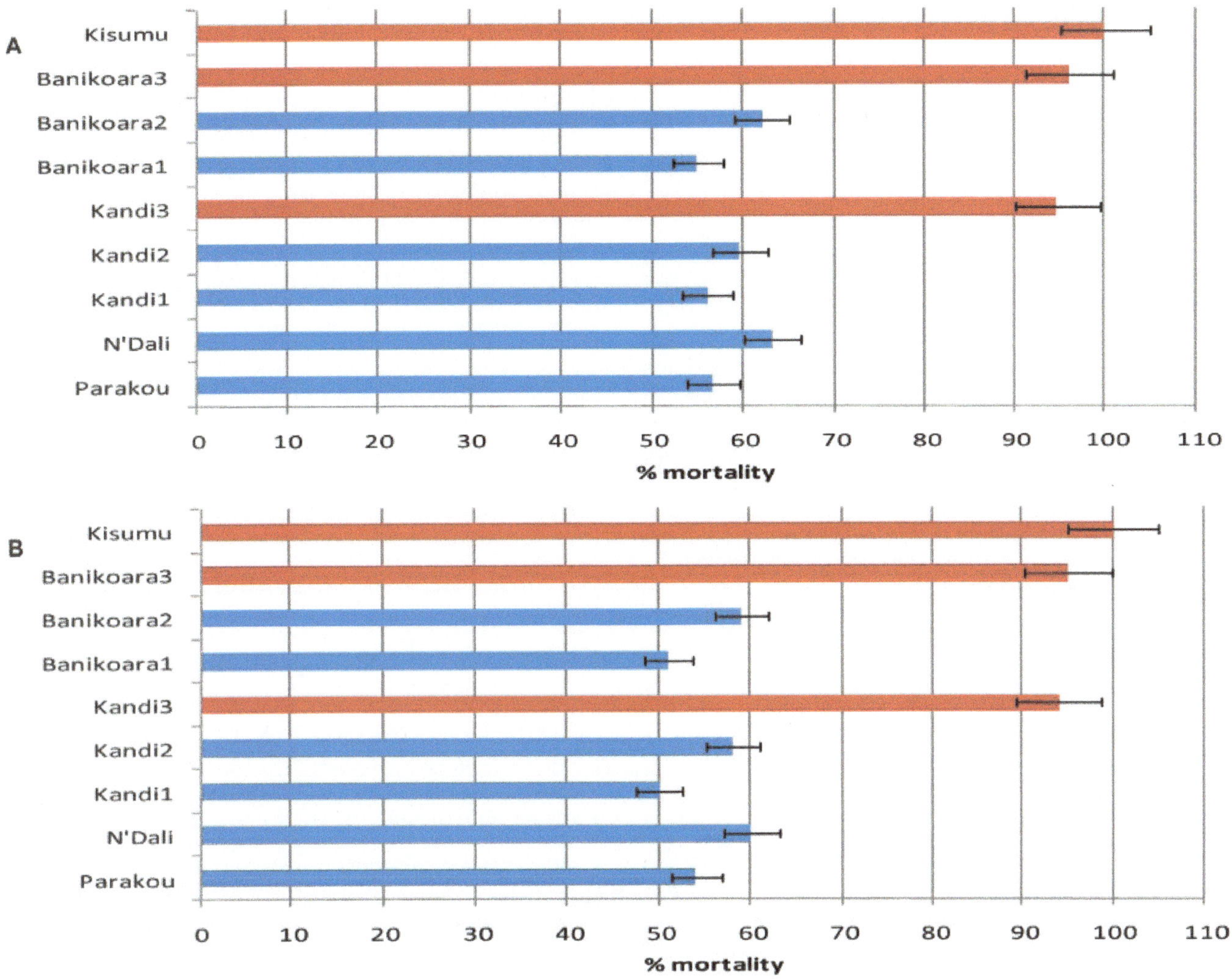

Figure 3: Mortality rates and resistance status of *Anopheles gambiae s.l.*collected before(A) and after (B) cotton treatments to permethrin 0.75% using WHO bioassay tests.

Species Identification

A total of 850 *An. gambiae* adults were analysed for species and molecular forms. Most of the mosquitoes collected from all the sites were *An. gambiae* s.s. (90%), which was found in sympatric with a low proportion of *An. arabiensis* (0 to 5%) except in Parakou and N›dali where they were more present (22 to 30% respectively) (Table 2). In *An. gambiae s.s*, the M and S forms were always found in sympatric but the S form was mostly predominant (96%).

Table 2: Species and molecular forms identification within *Anopheles gambiae*complex and the frequency of *Kdr* and *Ace-1R* mutations in *Anopheles gambiae* s.s. in Benin

	Species[a]		Mol. Form		*Kdr* mutation				*Ace.1* mutation			
Locality	**%Aa**	**%Ag**	**%M**	**%S**	**SS**	**RS**	**RR**	**F(R)**	**SS**	**RS**	**RR**	**F(R)**
Parakou[1](92)	22	78	5	95	12	35	45	0.68	30	02	0	0.06
N'dali[2](92)	30	70	3	97	22	40	30	0.54	25	0	0	0.00
Kandi[1](92)	5	95	0	100	7	45	40	0.67	30	02	0	0,06
Kandi[2](92)	4	96	0	100	28	34	30	0.51	30	01	0	0.03
Kandi[3](90)	0	100	0	100	45	25	20	0.35	30	0	0	0.00
Banikoara[1](102)	3	97	0	100	12	36	54	0.78	30	02	0	0.06
Banikoara[2](92)	02	98	0	100	14	48	30	0.59	30	0	0	0.00
Banikoara[3](96)	02	98	2	98	48	36	12	0.32	25	0	0	0.00

Resistance Mutations

The *kdr* genotype was scored for 1,400 individuals (100 mosquitoes consistently failed to amplify). The *kdr*gene occurred in S forms (Table 2). The highest frequency of *Kdr* mutation was recorded for the populations from three CCP sites (67-78%) and the lowest (35 and 32%) were found in the populations from BP sites around Kandi and Banikora respectively.

The *Ace-1R* gene was found at very low frequency ranging (from 0.00 to 0.06) in heterozygote *An. gambiae*s.s from the three CCP sites (Table 2). Among *An. gambiae* s.s, there was no mosquito of the M molecular form carrying the *ace-1R* gene.

The resistance allele frequency at the *kdr* was significantly higher in areas where farmers used insecticide for pest control (CCP and TICP) than in those no insecticide is not request (BP) ($p < 0.05$.). However, there is no difference between the resistance allele frequency at the *kdr* from mosquitoes in CCP and TICP strategies ($p > 0.05$).

Pesticide Residues

Results of soil samples for pesticide residues analysis showed that artificial breeding sites made with soil from CREC (control) and soil from Biological program (BP) sites were similar with no effect on hatching of*An. gambiae* Kisumu and VKPER strains (Figure 4). Tests with the susceptible *An. gambiae* Kisumu strain gave percentages of hatching equivalent to 80% in control soil (no contact with pesticides) and 75% with soil from BP sites. However, with the pyrethroid resistant strain VKPER the percentages of hatching were 83% and 77% with the control soil and soil from BP sites respectively. The hatching percentages of both strains decreased significantly when soil samples from CCP and TICP sites were used. With the TICP soil, *VKPER*hatching was 45% against 25% for the Kisumu strain, whiles the CCP soil gave 34% hatching for VKPER and 11% for Kisumu. In both cases the results showed that the hatching rates were significantly higher ($P < 0.05$) with VKPER than Kisumu when the soil samples tested were from TICP and CCP sites.

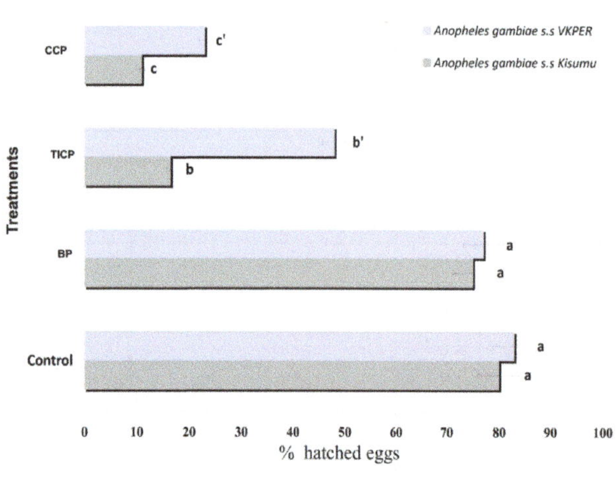

Figure 4: Percentage of eggs hatching of *An. gambiae* in water collected from breeding sites inside CCP program (in Parakou), TICP program (in N›Dali), BP program (in Banikoara) compared with distilled water (as control).

Similar results were obtained with the emergence of adults of *VKPER* and Kisumu strains from eggs placed in artificial breeding sites consisting of water and soil samples from CCP and TICP relative to the control (Figure 4). There was no significant difference between the emergence of adults of *VKPER* and *Kisumu*strains breeding in artificial sites made with the soil samples from BP compared with the control (Figure 5).

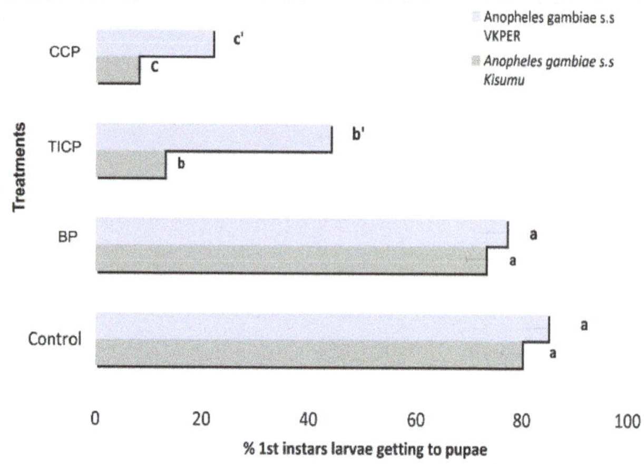

Figure 5: Percentage of emerging adults of *An gambiae* reared in water collected from breeding sites inside CCP program (in Parakou), TICP program (in N›Dali), BP program (in Banikoara) compared with distilled water (as control). - A: before treatments - B: after ...

However, a significant difference ($P < 0.05$) was observed between the emergence of *VKPER* on artificial sites made of soils from TICP (43%) and CCP (35%) sites and Kisumu which gave 20% emergence on TICP and 13% on CCP.

DISCUSSION

The information generated during interviews with cotton farmers and the observations made in cotton fields has confirmed a common use of fertilizers and insecticides in cotton fields. Cotton cultivation requires intensive use of pesticides including insecticides belonging to the two main classes recommended for vector control in public health: organophosphates and pyrethroids. In West Africa, pyrethroid-treated bed nets remain one of the effective tools for malaria vector control and it provides personal protection to individuals who sleep under them. When used by the whole community, bed nets protect collectively against infective mosquito bites by a mass killing effect of the vectors [36].

In Benin, pyrethroids have been extensively introduced in agriculture since 1980s [25]. This factor is probably one of the causes of the selection of strong resistance in *An. gambiae* to permethrin and DDT, particularly in cotton growing areas. Based on recent results, several authors [8-12; 37] have reported that past and current agricultural use of DDT then pyrethroids for crop protection have led to the selection of resistant mosquitoes through insecticide residues accumulated in breeding sites around cotton growing areas. This hypothesis was recently confirmed by Akogbeto *et al.* [25] showing indirectly the presence of pesticide residues in soil and water from vegetable farms and other agricultural activities in Benin that delay or reduce the emergence rates of mosquito larvae.

The use of insecticides in households for public health purposes and massive quantities of pesticides in agricultural settings has been highlighted as a key factor contributing to the emergence of vector resistance. A recent report by Yadouleton et *al*.[37] showed that agricultural practices in urban areas seem to have contributed to the emergence of insecticide resistance in *Anopheles* populations. Our study in vegetable farming systems in Benin demonstrated that improper use of insecticides to control vegetable pests in urban areas directly exerted a huge selection pressure on mosquito larval populations. The high mortality observed with mosquitoes reared on soils from CCP, TICP sites can be explained by the presence of DDT residues in the soil from those sites and the extensive use of pyrethroids by farmers [38]. Our results showed that the high level of the *kdr-west (*Leu-Phe) gene seemed to be the main resistance mechanism and responsible for the decrease of mortality rates to DDT and permethrin and is more of an ongoing process in *An. gambiae*populations from CCP and TICP sites. The *kdr* gene in the main malaria vector *An. gambiae* was found at high frequency in samples from the sites using insecticide (CCP and TICP) than those with no use of insecticide (BP program). The low frequency of *Kdr* gene in BP localities compared with those from CCP and TICP could be due to the fact that in the past these farmers in BP sites used insecticide to control cotton pests. According to reports by Akogbeto et *al* [25], Djogbénou et *al* [14], and Yadouleton et *al.* [38], the presence of *Kdr* genes in mosquito can be due to external factors that affect mosquitoes as larvae or adults. In 2000, a study in Burkina Faso by Diabate *et al.* [17] reported higher levels of *kdr* alleles frequency in *An. gambiae* collected from cotton-growing areas constantly subjected to insecticide treatments, as compared to the low frequency of *kdr* recorded in rural areas where farmers are restricted to low or no use of pesticides. Despite the use of insecticide in both CCP and TICP sites, the difference in adult mortality rates between CCP and TICP program can be explained by the fact that CCP program uses more insecticide than TICP program.

This study provides clear evidence of the association between the use of insecticides in agriculture and the widespread emergence of insecticide resistance in *Anopheles* species.

Indeed, in Benin, insecticide treatments against pests in cotton plantations are carried out twice a month, for an average period time of three months (between July and October) per year. That treatment period during the rainy season correspond with the period of high mosquito densities because *Anopheles* populations have numerous breeding sites particularly in cultivated areas. As reported by Akogbeto *et al* [26], some populations of *An. gambiae* may lay their eggs in breeding sites containing insecticide residues. These eggs undergo a selection pressure from agricultural pesticides, which leads to the emergence of resistant strains. There is clear evidence on the implication of agricultural use of insecticides in the selection of resistance in the major malaria vectors. Our results agree with the work of Akogbeto *et al* [25] and confirm once again the impact of the extensive use of insecticides in cotton crop protection on the emergence of insecticide resistance in *An. gambiae* populations. Moreover, in CCP and TICP program, some farmers used insecticides belonging to the organophosphate classes. *Ace-1R* gene is the main resistance mechanism of *An. gambiae* s.l. to organophosphates and carbamates also. The present study, has shown that the *ace-1R* gene is present at low frequency (ranging from 0.01 to 0.09), but only in CCP and TICP program. However, previous field surveys on *An. gambiae s.l.* populations of South-Western

Burkina Faso by Djogbenou *et al* [14] in cotton fields showed that *ace-1R* gene was the main resistance mechanism in *An. gambiae* s.l.

However the National Malaria Control Program (NMCP) in Benin has started scaling up Long Lasting Insecticidal Nets (LLINs) and carbamate for Indoor Residual Spraying (IRS) countrywide for malaria control. The challenge to find effective strategies to manage insecticide resistance in *Anopheles gambiae*remains a high priority and an urgent need particularly in Benin where pyrethroid resistance has been reported with a clear evidence in experimental huts of reduced efficacy of ITNs and IRS [39]. One of the strategies will be to remove pyrethroids from agricultural pest control and leave these classes of insecticides for public health purposes and promote other classes of insecticides such as Spinosad which does not show cross resistance to pyrethroids (i.e. the *kdr* gene).

CONCLUSIONS

With the spread of *Kdr* allele frequency from CCP and TICP programmes, to reduce the emergence of insecticide resistance in *An. gambiae* population, African governments would be better advised to promote the BP cotton or genetically modified cotton such as *Bt* Cotton (*Bacillus thurengiensis*) which require lower pesticide than the cotton with CCP and TICP programme and would permit to suppress the massive use of pyrethroid insecticides.

COMPETING INTERESTS

The authors declare that they have no competing interests.

AUTHORS' CONTRIBUTIONS

AY carried out field experiments, collected, analysed, interpreted data and wrote the manuscript. TM, GP reviewed the manuscript and contributed to the design of the study and substantially helped in drafting the manuscript, and revised the manuscript. FC, LD, RD, IG and MB contributed to the design of the study. AA helped in drafting and reviewing the manuscript and RA helped with the activities. TM, MA conceived and designed the study, supervised fields and laboratory procedures, and reads the manuscripts. All authors read and approved the final manuscript.

ACKNOWLEDGEMENTS

This work was financially supported by the French government (Corus 6015 project). I am grateful to CREC's staff particularly Sebastien Koudenoukpe and Géraldo Houndéton for technical assistance during laboratory bioassays and field collections.

REFERENCES

1. WHO. World Malaria Report 2009. World Health Organzation, Geneva; 2009.
2. Ministère de la santé Annuaire des statistiques sanitaires 2006. Cotonou; 2009.

3. Reimer L, Fondjo E, Patchoke S, Diallo B, Lee YNg, Ndjemai HM, Atangana J, Traore SF, Lanzaro G, Cornel AJ. Relationship between *kdr* mutation and resistance to pyrethroid and DDT insecticides in natural populations of *Anopheles gambiae*. J Med Entomol. 2008;45:260–266. doi: 10.1603/0022-2585(2008)45[260:RBKMAR]2.0.CO;2.
4. Ramphul U, Boase T, Bass C, Okedi LM, Donnelly MJ, Muller P. Insecticide resistance and its association with target-site mutations in natural populations of *Anopheles gambiae* from eastern Uganda. Trans R Soc Trop Med Hyg. 2009;103:1121–1126. doi: 10.1016/j.trstmh.2009.02.014.
5. Chandre F, Darrier F, Manga L, Akogbeto M, Faye O, Mouchet J, Guillet P. Status of pyrethroid resistance in *Anopheles gambiae* sensu lato. Bull World Health Organ. 1999;77:230–234.
6. Santolamazza F, Calzetta M, Etang J, Barrese E, Dia I, Caccone A, Donnelly MJ, Petrarca V, Simard F, Pinto J, Torre A. Distribution of knock-down resistance mutations in *Anopheles gambiae* molecular forms in west and west-central Africa. Malar J. 2008;7:74. doi: 10.1186/1475-2875-7-74.
7. Djènontin A, Sahabi B, Moiroux N, Henry MC, Bousari O, Chabi J, Ossè R, Koudénoukpo S, Corbel V, Akogbéto M, Chandre F. Culicidae diversity, malaria transmission and insecticide resistance alleles in malaria vectors in Ouidah Kpomasse-Tori district from Benin (West Africa): A pre-intervention study. Parasit Vectors. 2010;3:83.
8. Vulule JM, Beach RF, Atieli FK, Mcallister JC, Brogdon WG, Roberts JM, Mwangi RW, Hawley WA. Elevated oxidase and esterase levels associated with permethrin tolerance in *Anopheles gambiae*from Kenyan villages using permethrin impregnated nets. Med Vet Entomol. 1999;13:239–244. doi: 10.1046/j.1365-2915.1999.00177.x.
9. Elissa N, Mouchet J, Rivière F, Meunier JY, Yao K. Resistance of *Anopheles gambiae* s.s. to pyrethroids in Côte d›Ivoire. Ann Soc Belge Méd Trop. 1993;73:291–294.
10. Akogbéto M, Yakoubou S. Résistance des vecteurs du paludisme vis-à-vis des pyréthrinoïdes utilisés pour l›imprégnation des moustiquaires au Bénin, Afrique de l›Ouest. Bull Soc Pathol Exot.1999;92:123–130.
11. Yadouleton AW, Asidi A, Rousseau FD, Braïma J, Agossou CD, Akogbeto MC. Development of vegetable farming: a cause of the emergence of insecticide resistance in populations of *Anopheles gambiae* in urban areas of Benin. Malar J. 2009;8:103. doi: 10.1186/1475-2875-8-103.
12. Corbel V, N›Guessan R, Brengues C, Chandre F, Djogbenou L, Martin T, Akogbeto M, Hougard JM, Rowland M. Multiple insecticide resistance mechanisms in Anopheles gambiae and Culex quinquefasciatus from Benin, West Africa. Acta Trop. 2007;101:207–216. doi: 10.1016/j.actatropica.2007.01.005.
13. N›Guessan R, Corbel V, Akogbeto M, Rowland M. Reduced efficacy of insecticidetreated nets and indoor residual spraying for malaria control in pyrethroid resistance area, Benin. Emerg infect diseases.2007;13:199–206.
14. Djogbénou L, Dabire R, Diabate A, Kengne P, Akogbeto M, Hougard JM, Chandre F. Identification and Geographic Distribution of the *Ace-1R* Mutation in the Malaria Vector *Anopheles gambiae* in South-Western Burkina Faso, West Africa. Am J trop med hyg. 2008;78:298–302.
15. Czeher C, Labbo R, Arzika I, Duchemin JB. Evidence of increasing *Leu-Phe* knockdown resistance mutation in *Anopheles gambiae* from Niger following a nationwide long-lasting insecticide-treated nets implementation. Malar J. 2008;7:189. doi: 10.1186/1475-2875-7-189.
16. Diabate A, Baldet T, Chandre F, Guiguemde RT, Brengues C, Guillet P, Hemingway J, Hougard JM. First report of the *kdr* mutation in *Anopheles gambiae* M form from Burkina Faso, West Africa.Parassitologia. 2002;44:157–158.
17. Diabate A, Baldet T, Chandre F, Akogbéto M, Guiguemde RT, Darriet F, Brengues C, Guillet P, Hemingway J, Graham JS, Hougard JM. The role of agricultural use of insecticides in resistance to pyrethroids in *Anopheles gambiae* s.l in Burkina Faso. Am J Trop Med Hyg. 2002;67:617–622.
18. Fanello C, Petrarca V, Torre D, Santolamazza A, Dolo F, Coulibaly G, Alloueche MA, Curtis CG, Toure YT, Coluzzi M. The pyrethroid knock-down resistance gene in the *Anopheles gambiae* complex in Mali and further indication of incipient speciation within *An. gambiae s.s.* Insect Mol Biol.2003;12:241–245. doi: 10.1046/j.1365-2583.2003.00407.x.
19. Awolola TS, Brooke BD, Koekemoer LL, Coetzee M. Resistance of the malaria vector *Anopheles gambiae* s.s. to pyrethroid insecticides, in south-western Nigeria. Ann Trop Med Parasitol.2002;96:849–852. doi: 10.1179/000349802125002581.
20. Hargreaves K, Koerkemoer LL, Brooke B, Hunt RH, Mthembu J, Coetzee M. *Anopheles funestus*resistant to pyrethroid insecticides in South Africa. Med Vet Entomol. 2000;14:181–189. doi: 10.1046/j.1365-2915.2000.00234.x.
21. Etang J, Manga L, Chandre F, Guillet P, Fondjo E, Mimpfoundi R, Toto JC, Fontenille D. Insecticide susceptibility status of *Anopheles gambiae* s.l. (*Diptera: Culicidae*) in the Republic of Cameroon. J Med Entomol. 2003;40:491–497. doi: 10.1603/0022-2585-40.4.491.
22. WABI. The cross-border integration of the cotton sub-sector. West African Borders and Integration Network. Coton et Fibres Tropicales. 2003;112:54–71.
23. N'Guessan F, Darriet P, Guillet P, Carnevalle M, Lamizana T, Corbel V, Chandre F. Resistance to carbosulfan in *Anopheles gambiae* base on reduced of acethylcholinesterase. Med vet Entomol.2003;17:19–25.
24. Akogbéto M, Djouaka R, Gazard DA. Screening of pesticide residues in soil and water samples from agricultural settings. Malar J. 2006;5:22.
25. Akogbéto M, Djouaka R, Noukpo H. Use of agricultural

insecticides in Benin. Bull Soc Pathol Exot.2005;98:400–405.

26. Deguine JP, Ekukole G, Amiote P. La lutte étagée ciblée: un nouveau programme de protection insecticide en culture cotonnière au Cameroun. Coton et Fibres Tropicales. 1993;48:99–119.
27. Prudent P, Loko S, Deybe D, Vaissayre M. Factors limiting the adoption of IPM practices by cotton farmers in Benin: a participatory approach. Cambridge University Press. 2006;10:10–17.
28. WHO. World Malaria Report 2008. World Health Organzation, Geneva; 2008.
29. Scott JA, Brogdon WG, Collins FH. Identification of single specimens of the *Anopheles gambiae*complex by the polymerase chain reaction. Am J Trop Med Hyg. 1993;49:520–529.
30. Favia G, Torre D, Bagayoko M, Lanfrancotti M, Sagnon NF, Toure Y, Coluzzi M. Molecular identification of sympatric chromosomal forms of *Anopheles gambiae* and further evidence of their reproductive isolation. Insect Mol Biol. 1997;6:377–383. doi: 10.1046/j.1365-2583.1997.00189.x.
31. Martinez TD, Chandre F, Williamson MS, Darriet F, Berge JB, Devonshire AL, Guillet P, Pasteur N, Pauron D. Molecular characterization of pyrethroid knockdown resistance (*kdr*) in the major malaria vector *Anopheles gambiae s.s.* Insect Mol Biol. 1998;7:179–184. doi: 10.1046/j.1365-2583.1998.72062.x.
32. Weill M, Malcolm C, Chandre, Mogensen K, Berthomieu A, Marquine M, Raymond M. The unique mutation in *ace-1* giving high insecticide resistance is easily detectable in mosquito vectors. Insect Mol.2004;9:1–7. doi: 10.1111/j.1365-2583.2004.00452.x.
33. WHO. World Malaria Report 1998. World Health Organzation, Geneva; 1998.
34. Finney DJ. Probit analysis. Cambridge University Press Cambridge; 1971.
35. Raymond M, Rousset F. Genepop (version 1.2). A population genetics software exact tests and ecumenicism. Heredity J. 1995;86:248–249.
36. Henry MC, Assi SB, Rogier C, Dossou-Yovo J, Chandre F, Guillet P, Carnevale P. Protective Efficacy of Lambda-Cyhalothrin Treated Nets in *Anopheles gambiae* Pyrethroid Resistance areas of Cote d›ivoire. Am J Trop Med Hyg. 2005;73:859–864.
37. Hargreaves K, Koekemoer LL, Brooke BD, Hunt RH, Mthembu J, Coetzee M. *Anopheles funestus*resistant to pyrethroid insecticides in South Africa. Med Vet Entomol. 2000;14:1–9. doi: 10.1046/j.1365-2915.2000.00234.x.
38. Yadouleton AW, Padonou G, Asidi A, Moiroux N, Sahabi B, Corbel V, N'guessan R, Gbenou D, Yacoubou I, Kinde G, Akogbeto MC. Insecticide resistance status in *Anopheles gambiae* in southern Benin. Malar J. 2010;9:83. doi: 10.1186/1475-2875-9-83.
39. Akogbéto MC, Padonou G, Gbénou D, Irish S, Yadouleton AW. Bendiocarb, a potential alternative against pyrethroid resistant *Anopheles gambiae* in Benin,West Africa. Malar J. 2010;9:204.

CHAPTER 10

ADOPTION OF INTEGRATED PEST MANAGEMENT

S. A. Dhenge and D.M.Mankar

Keywords: Suggestion, Integrated Pest Management Practices and paddy grower

INTRODUCTION

Paddy (Oryza sativa L.) is one of the important cereal crops of the world and it forms the staple food for more than 50 per cent of population, and also it is known as "king of cereals". The United Nations General Assembly, in a resolution declared the year 2004 as the "International Year of Rice". Which has tremendous significance in food security. It is very eloquently upheld the need to heighten awareness about the role of rice in alleviating poverty and malnutrition (Barath and Pandey, 2005). In Asia, India has the largest area under the rice accounting 28.5 per cent of the global rice area. Rice is the staple food over half of the world population. Rice provides 21% energy and 15% protein for human globaly. Calories from rice are particularly important in Asia, especially among the poor, where it accounts 50-80% daily caloric intake. Asia accounts for over 90% of the world's production of rice, with China, India and Indonesia are higher producer. The United States produces 1.5% the world's rice crop with Arkansas, California and Louisiana producing 80% of the U.S. rice crop. One of the most important reasons for low level of yields is the incidence of pest and disease. Losses in rice yield the tune 10-25 percent occurs due to the attack of insect pest and disease (Singh 2001). With a view to keep the infestation of pest and disease within normal limits the Integrated Pest Management Practices are recommended for adoption by the rice growers. According to FAO (1972) Integrated Pest Management is a pest management system that in the context of the associated environment of the pest practice, utilizes all suitable techniques and method as compatible a manner as possible and maintain the pest population at level below those acouring economic injure.

Citation: S. A. Dhenge, D. M. Mankar, "Adoption of Integrated Pest Management," DOI: 10.15373/2249555X.

Material and Methods

The present study was carried out in Bhandara district of Vidarbha region in Maharashtra State because it considers as a progressive agricultural belt of Vidarbha region and the University head quarter is also located in the study area. Lakhani ,Sakoli and Lakhandur Tahsil of Bhandara district were selected purposively for study because having maximum area under cultivation of paddy Crops. Villages were selected on the basis of maximum area under cultivation of paddy crop. Four villages from Lakhani, Four villages from Sakoli and Four villages from Lakhandur Panchayat Samiti were selected randomly. Ten respondents were selected from each village by random sampling method, making a sample size of 120 in total.

Techniques of sampling

1. Random sampling method

Statistical method use

1. Mean
2. standard deviation
3. coefficient correlation
4. multiple coefficient correlation

RESULTS AND DISCUSSION

It is seen from Table 1 that in respect of Input supply, (71.67%) farmers suggested guidance for availability plant protection appliances and (91.67%) technical guidance for preparation of organic manure. As regard to technical suggestion (62.50%) of farmer suggested technical guidance for seed treatment with thirum,(75%) farmers suggested to Demonstration regarding Installation of Pheromone traps,(56.67%) ,(60.83%) and (95.83%) farmers suggested to information should be regarding pest and disease resistant varieties, Demonstration regarding preparer brine solution and to aware the people through booklets to identify of friendly insect respectively. In case of financial suggestion (55.83%) of farmers suggested availing the crop loan facility from bank or financial institution for routine agricultural operation

Table 1: Distributions of the respondents according to their Suggestion expressed about integrated pest management practices by paddy growers

Sr. No.	Suggestion	Respondents (n=120)	
		Frequency	Percentage
1	Guidance for availability plant protection appliances	86	71.67
2	Technical guidance for preparation of organic manure	110	91.67
3	Technical guidance for seed treatment with thirum	75	62.50

Sr. No.	Suggestion	Respondents (n=120)	
		Frequency	Percentage
4	Demonstration regarding Installation of Pheromone traps	90	75.00
5	Information should be regarding pest and disease resistant varieties	68	56.67
6	Demonstration regarding preparer brine solution	73	60.83
7	To aware the people through booklets to identify of friendly insect	115	95.83
8	To avail the crop loan facility from bank or financial institution for routine agricultural operation	67	55.83

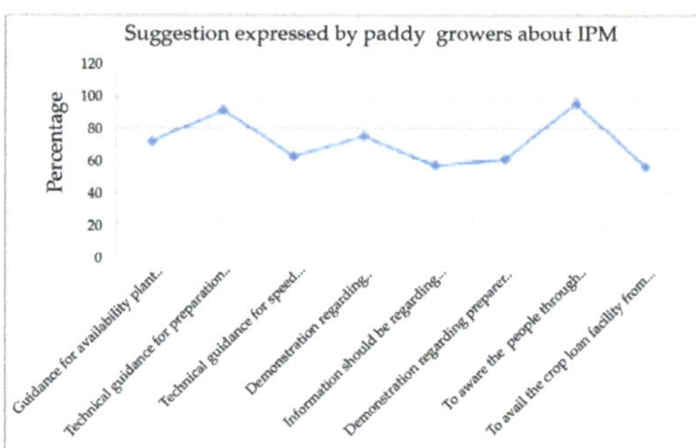

Figure 1: Distributions of the respondents according to their Suggestion expressed about integrated pest management practices by paddy growers.

CONCLUSIONS

IPM is a holistic approach of pest control, based on sound ecological factors. It was obvious from the results that the farmers suggested various suggestions that encouraged them to adopt IPM practices for rice cultivation. It may be concluded that Majority of the respondents suggested that the Input supply, guidance for availability of plant protection appliances and technical guidance for preparation of organic manure. Hence, there is need to put more importance and emphasis on adoption of integrated pest management practices, extension personnel suggested technical guidance for seed treatment with thirum,, demonstration regarding installation of pheromone traps, information should be regarding pest and disease resistant varieties, Demonstration regarding preparer brine solution and to aware the people through booklets to identify the friendly insect respectively and availing the crop loan facility from bank or financial institution for routine agricultural operation. Were suggested by paddy growers about adoption of integrated pest management practices arrange the instructional programme for the farmers by the implementing agencies.

REFERENCES

1. Rahman, M. M. (2012). Problem and suggestion for farmers adoption of IPM practices in rice (Oryza Sativa L.) cultivation. Bangladesh J.Agril. Res. 37(1):121- 128.
2. Tripathi, M. K. and A. P. Kanungo, (2008). Demonstration of integrated pest management technology in rice at farmers' fields. J. Plant Protection Environ. 5 (1):30- 33.
3. Arun M. Balamakatti (1993). A study on paddy cultivation pattern of siddi farmers and their socio economic chrematistics
4. Yellapur. Karnataka. M. Sc. (Agri.) Thesis, Uni. Agri. Sci., Dharwad, Karnataka (India). | Bheemappa, A. (2001). A comparative analysis of knowledge and technological gap in adoption of paddy and cotton cultivation practices between migrant and nonmigrant farmers of TBP command area. Ph. D. Thesis (Unpub.), Univ. Agric. Sci.
5. Dharwad, Karnataka (India). | Joshi, B. V. (1992). A study on paddy cultivation pattern of Tibetian rehabilitants and their socio economic characteristic, Mungod, Karnataka. MSc Thesis (Unpub.), Univ. Agric.Sci., Dharwad, Karnataka (India). |

CHAPTER 11

THE INFLUENCE OF SYNTHETIC PYRETHROIDS ON MEMORY PROCESSES, MOVEMENT ACTIVITY AND CO-ORDINATION IN MICE

Barbara Nieradko-Iwanicka

INTRODUCTION

In plants of Chrysanthemum species (Tancetum cinerariefolium, or Chrysanthemum cinerariefolium, or Pyrethrum cinerariefolium) there are natural compounds called pyrethrins of insecticidal properties. Since the beginning of the XIXth century the flowers of Chrysanthemum were used to fight nets and lice. In 1828 massive production of Chrysanthemum began in order to make an insecticidal formulas. Before World War II the main centers of Chrysanthemum production were Dalmatia and Japan. After World War II production of Chrysanthemum was launched in Africa (Kenia, Tanzania, Rwanda). Until 1920s kerosene extract of Chrysanthemum flowers was commonly used to fight nets, lice and mosquitoes. Initially, it was believed to be absolutely safe for users and it was recommended to take 10-20 mg of the formula orally in adults or 5-10 mg in children to get rid of intestinal parasites. Adverse events of such use, especially skin reactions to pyrethrins were described afterwards (Casida J.E. 1980). The chemical structure of pyrethrins was discovered by Herman Staudinger and Leopold Ruzicka between 1910 and 1916 (Casida J.E. 1980). Thre six natural pyrethrins (pyrethrin I, pyrethrin II, cinerin I, cinerin II, jasmolin I and jasmolin II) are esters of two cyclopropanecarboxylic acids and three cyclopentenolone alcohols (piretrol, jasmolol or cynerol) (Różański, 1998). The main drawback pf pyrethrins was their instability in light and air and lack of residual activity after single use, which made them unsuitable for use in agriculture for crop protection from pest insects (Soderlund 2002). In 1949 allethrin –the first syntheic pyrethroid was produced (Bradberry et al., 2005). It was followed by resmethrin, tetramethrin, bifenthrin and tefluthrin. Permethrin was the first pyrethroid of enough photostability to be used in agriculture. In current use there are mainly permethrin, deltamethrin, cypermethrin cyfluthrin, cyhalothrin and fenpropathrin (Soderlund 2002). These compounds exhibit strong insecticidal activity and photostability. Since the 1970s pyrethroids have been used in public health to prevent vector born diseases like malaria (Soderlund 2002). Malaria affects people in over 110 tropical and subtropical countries. About one million people die every year because of it. Clinical symptoms of malaria develop in almost 500 million people per year (Kajfasz 2011). It is estimated, that about 15 thousand Europeans import malaria from tropical countries every year (Knap & Myjak 2009). Current climate change is expected to have a substantial effect on the burden of mosquitoes such as anopheles species, which transmit malaria and other mosquito-borne diseases like West Nile virus infection (Shuman 2010). With global temperatures increase by 2 to 3°C, the population at risk for malaria is expected to increase by 3-5%, i. e. millions of additional people would probably become infected with malaria each year (Shuman 2010). Therefore people use bednets soaked with pyrethroids or spray them indoors or outdoors to protect themselves from insect vectors carrying diseases. This means prolonged (subchronic or chronic) exposure of humans (considered to be the non-target organisms) to pyrethroid insecticides. Side effects of such uses are not fully understood and need further studies. Pyrethroids are also commonly used in veterinary medicine, for agricultural and horticultural purposes. It was estimated, that pyrethroid use has increased to reach about 23% of the world insecticide market (Soderlund 2002). Even though organophosphates still are in wide use as insecticides (Casida & Quistad, 1998), there is evidence

Citation: Barbara Nieradko-Iwanicka, The Influence of Synthetic Pyrethroids on Memory Processes, Movement Activity and Co-Ordination in Mice, DOI: 10.5772/28777.

that they contaminate groundwaters (Badach et al., 2007; Drożdżyński 2008) , are able to accumulate in fatty tissue of living organisms (Molina et al., 2005) and can irreversibly damage the hippocampal structure in the central nervous system of mammals (Mitra et al., 2008). As hippocampus plays a key role in memory processes, it might impair memory in humans exposed to organophosphates. Therefore there is a growing interest in pyrethoids and so is their use. However, the knowledge of pyrethroids' effects on memory processes and movement activity in non-target organisms is incomplete. Pyrethoids act as acute neurotoxins (Soderlund 2002). They alter functioning of nerves in target animals by modifying the kinetic characteristics of voltage sensitive sodium channels (Soderlund & Bloomquist, 1989; Bloomquist 1993). Before 1970s little was known about mammalian toxicity of pyrethroids.The first study documenting modest oral toxicity of pyrethrins and pyrethoids in rats was published in 1972 (Verschoyle & Barnes, 1972). Acute mammalian toxicity of pyrethroids and structuretoxicity relationships were elucidated after numerous experiments with intracerebral and intravenous administration of these neurotoxins to mice and rats (Verschoylea & Aldridge, 1980; Lawrence & Casida,1982). The milestone in pyrethroid improvement was the discovery of the fact that the presence of an α-cyano substituent in S configuration in the 3- phenoxybenzyl alcohol moiety greatly enhanced neurotoxicity in mammals as well as in insects (Elliot et al. 1974; Soderlund 2002). There were found pyrethroid structure-toxicity relationships in mammals, which were congruent with those found in insects (Soderlund 2002). The studies from 1970s provided the first descriptions of signs of pyrethroid intoxication in mammals after oral and intravenous dosing (Verschoyle & Barnes 1972). The authors described hypersensitivity and aggression followed by tremor, coma, an eventually death after exposure of the experiment rats to bioallethrin and resmethrin orally or intravenously (Verschoyle & Barnes 1972). The first pyrethroid containing the α-cyano-3-phenoxybenzyl moiety was deltamethrin (Elliot et al. 1974). Oral or intravenous exposure of rats to deltamethrin produced salivation, jerking leg movements and choreoatetosis (Elliot et al. 1974). In 1980 pyrethroids were divided into two groups: those producing T–syndrome (tremor) and those producing CS-syndrome (choreoatetosis and salivation) (Verschoyle & Aldridge 1980). It was found then, that majority of pyrethroids containing the α-cyano-3-phenoxybenzyl moiety produced the CSsyndrome, and those without the α-cyano-3-phenoxybenzyl moiety produced the Tsyndrome.

In 1982 a new nomenclature for pyrethroids (Type I/ II) was proposed basing on the syndromes of intoxication in mammals (Lawrence & Casida 1982). This nomenclature is used parallel to the T/CS nomenclature in many publications. Type I pyrethroids are considered to produce the T-syndrome, and Type II the CS-syndrome. However, there were some pyrethroids that were tested neither in the study about Type I/ II nor T/CS pyrethroids, for example bifentrin and cyhalothrin. Also, in the classifying studies pyrethroids were administered orally, intravenously or intracerebrally (Verschoyle & Aldridge 1980; Lawrence & Casida 1982). However the most likely routes of pyrethroid intake by humans is orally, by inhalation and transdermally. The transdermal absorption of pyrethroids is most likely in greenhouse workers and farmers. It is also possible in holidaymakers or campers who use ' mosquito repellent' containing pyrethroids all evening or day and night. Pyrethroid containing formulas available on the market usually have enclosed leaflet warning the users about skin reactions to the chemicals and suggest symptomatic treatment if they develop. There is little or no information about their main mechanism of toxic action- about neurotoxicity. The aim of this chapter is to analyze the influence of synthetic pyrethroids injected intraperitioneally to female Albino Swiss mice at the doe of 0.1LD_{50} on memory processes, movement activity and co-ordination. The pyrethroids chosen were lambda-cyhalothrin, deltamethrin, cypermethrin, fenpropathrin and bifenthrin.

MATERIAL AND METHODS

Animals

Non-gravid female albino Swiss mice weighing 18-24g approximately 6 weeks of age purchased from a licensed dealer (T. Górzkowski, Warsaw, Poland) were used in the study. All animals were given a 7-day acclimation period and maintained on a 12 hr light/dark cycle. Food and tap water were provided ad libitum. Temperature was maintained at 21 ± 2°C.

Groups of Animals

There were two groups of animals in each experiment: I injected with 0.9% NaCl i.p. and II injected with a pyrethroid (lambda-cyhalothrin, deltamethrin, cypermethrin, fenpropathrin or bifenthrin) at the dose of 0.1LD_{50} i.p. The injections were made 15 min. before beginning each experiment. In passive avoidance task the injections were given once only -15 min. before training.

Opinion of the Local Ethics Committee

The Local Ethics Committee for Animal Experiments in Lublin approved the experiment (Opinion No. 30/2000, dated 24.11.2000).

Materials

Bifenthrin (powder 99,6±0,2%) in glasses 0,1g each was purchased from the manufacturer – Institute of Industrial Organic Chemistry, Annopol 6, 03-236 Warsaw, Poland. Cypermethhrin (powder 99,7%) in glasses 0,25g each was purchased from the manufacturer – Institute of Industrial Organic Chemistry, Annopol 6, 03-236 Warsaw, Poland. Deltamethrin (powder 99,7%) in glasses 0,25g each was purchased from the manufacturer – Institute of Industrial Organic Chemistry, Annopol 6, 03-236 Warsaw, Poland.

Fenpropathrin (powder 99,4 ±0,3%) in glasses 0,25g each was purchased from the manufacturer – Institute of Industrial Organic Chemistry, Annopol 6, 03-236 Warsaw, Poland. Lambda-cyhalothrin (Karate 025EC containing 25g of lambda-cyhalothrin per 1 litre) was purchased from the manufacturer Syngentia Limited, United Kingdom. In order to suspend the pyrethroids in bidistilled water Tween 60 (poloxyethylene sorbitan monostearate 100% in glasses 250g each was purchased from the manufacturer – Laboratorium Reagenzien, Germany) was used. Analytic weighing scale manufactured at Radwag, Poland was used. Water was bidistilled at the Hygiene

Department of Medical University of Lublin. 0.9% NaCl for control animals was prepared at the Hygiene Department of Medical University of Lublin.

Dosing

LD_{50} of each pyrethroid was calculated with Lichtfield and Wilcoxon's method (Lichtfield & Wilcoxon 1949).

LD_{50} for bifenthrin in mice was calculated to be 16.1 mg /kg of bw [13.1-19.7].

LD_{50} for cypermethrin in mice was calculated to be 169.9 mg /kg of bw [151.9-190.1]. LD50 for deltamethrin in mice was calculated to be 83 mg /kg of bw [79.2-87].

LD_{50} for fenpropathrin in mice was calculated to be 23.8 mg / kg of bw [21.2-26.7].

LD_{50} for lambda-cyhalothrin in mice was calculated to be 6.9 mg /kg of bw [5.5-8.5]. At the beginning of each experiment 0.1 of LD_{50} of the tested pyrethroid was injected i.p. to each animal from group II. Controls were mice from group I, that received respective volume of 0.9% Na Cl i.p.

Behavioral Tests

Passive Avoidance

A step-through passive avoidance (PA) task was used in the study. The task relies on the innate preference of rodents for dark, enclosed spaces and it is regarded as a measure of long-term memory retention. Avoidance training consisted of a single trial in which each animal was placed in an illuminated box (15 x 12 x 15cm) adjacent to a darkened one (the same size) with an electric grid floor. A 4 x 5 cm doorway was located at floor level in the center of the wall separating the boxes. Thirty second after placing the animal in the centre of the illuminated box a passage joining the two boxes was opened. After entering the dark box the animal was punished with an electric foot shock (2 mA for 2s). Twenty four hours after the training trial memory retention test was conducted in which the same animals were placed in the illuminated box and the latency to enter the darkened box was recorded. The test ended when the mouse entered the darkened box or when 180s has elapsed. Mice that did not enter in the time allotted received latency 180s. Administration of the tested pesticide before training may impair or improve learning by affecting memory acquisition and/or recalling.

Y-maze task

Spontaneous alternation was assessed in a Y-maze, which is used as a measure of working spatial memory. The total number of arm entries in Y-maze can be also considered a measure of exploratory locomotor activity. The Y- maze consists of three 10 x 10 x 10 cm compartments without floor joined together with 4-cm long corridors at 120° in such a way that each corridor opens to one compartment only. The maze is placed on a clean sheet of paper on a table-top. In order to prevent odor cues, the maze was cleaned between the trials of different mice and a clean sheet of paper was used for each animal. Mice naturally tend to explore the maze by systematically entering each arm. The ability to alternate requires that the animals knew which arm they have already visited. In the task, each mice was placed at the end of one arm and was allowed to move through the maze for 8 min. The percentage of alternation, defined as consecutive entries into all three arms without repetitions in overlapping triplet sets, to all possible alternations x 100 % was counted. For example, if the arms were marked as X, Y and Z and the animal entered the arms in a following order XYZXZYZXYXYZXZ, the actual alternation would be 7, the total number of arm entries would be 14 and the percent alternation would be 58.33%.

Movement Activity

Horizontal spontaneous locomotor activity was assessed with an automated device consisting of a circular box (32 cm in diameter) with two photocells mounted horizontally 2 cm above the floor at the angle of 90°. The photo-beam was activated when the mouse interrupted the beam. In the task the animals were not habituated for the apparatus, therefore they were placed individually in the actometers for 1 hour (two subsequent 30-min periods: 0.-30. min, 31.-60. min). The number of impulses was recorded after 30 and 60 min. The first period was considered as a rate of exploratory locomotor activity. The second period was considered as a rate of spontaneous locomotor activity.

Movement Co-ordination

Movement co-ordination was examined on a rod rotating at the rate of 10 cycles per min. The animals were placed on the rod (1 cm diameter) 50 cm above the ground for 120 sec. The trial ended when the mouse fell off the rod or 120 s had elapsed, whichever occurred first.

Statistical Analyses

A Kruskall-Wallis non- parametric ANOVA test was used to analyze the data from passive avoidance task. PA results were expressed as median values with the 25th and 75th percentiles. The results of spontaneous alternation in Y-maze task, movement co-ordination and spontaneous motor activity were shown as means ± SEM, and evaluated by one-way analysis of variance ANOVA followed by Student- Newman-Keuls test. The p value < 0.05 was considered statistically significant.

RESULTS

Effect of Bifenthrin on Memory Retention in Passive Avoidance Task

No statistically significant difference was observed in memory retention between group I and II. Median values of latency (with the 25th and 75th percentiles) were: 180 sec. (180, 180) in group I (control) and 180 sec. (180, 180) in group II (0.1 LD_{50} of bifenthrin i.p.) . Post test were not calculated; $p > 0.05$. There were 10 mice in each group.

Influence of Bifenthrin on Working Spatial Memory in Y - maze Task

There was a statistically significant difference between the examined groups in working spatial memory. Results obtained were (% of logical alternation behaviour expressed as mean ± SEM) : group I (control) 62.2 ± 1.601; group II (0.1 LD_{50} of bifenthrin i.p.) 52.59 ± 2.932; p= 0.0116 considered significant. There were 10 mice in each group (Fig.1.).

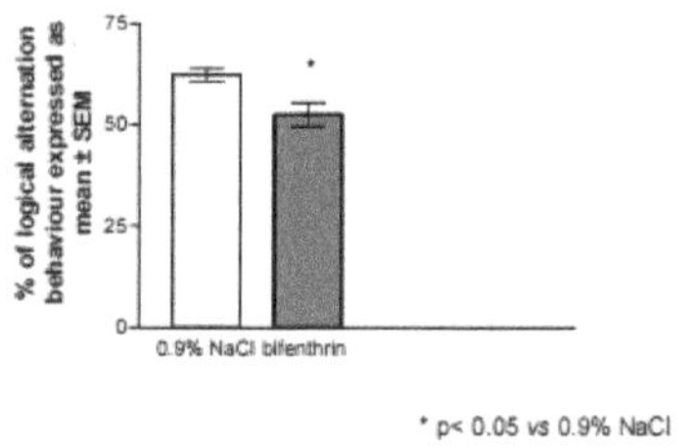

Fig. 1. The influence of bifenthrin (0.1 LD_{50} i.p.) on working spatial memory in Y- maze task. Columns represent the means ± SEM. Number of mice in each group was 10. * p < 0.05 vs 0.9% NaCl (ANOVA).

Influence of Bifenthrin on Movement Activity in the Actometer

The results of movement activity assessed within the 0-30 min. of observation were significantly different in the groups. Mean values ± SEM were: group I (control) 521.4 ± 33.903; group II (0.1 LD_{50} of bifenthrin i.p.) 664.9 ± 37.721; p= 0.0116 considered significant. There were 10 mice in each group. The movement activity assessed within the 31-60 min. of observation did not significantly differ. Mean values ± SEM were: group I (control) 342.9 ± 39.672; group II (0.1 LD_{50} of bifenthrin i.p.) 346.6 ± 33.532.p> 0.05 considered not significant. There were 10 mice in each group (Fig.2.).

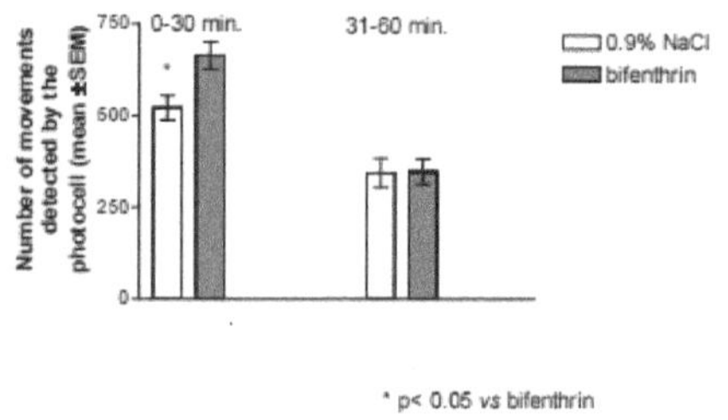

Fig. 2. The influence of bifenthrin (0.1 LD_{50} i.p.) on movement activity in the actometer. Columns represent the means ± SEM. Number of mice in each group was 10. * p< 0.05 vs bifenthrin (ANOVA).

Influence of Bifenthrin on Movement Coordination

There was observed a statistically significant difference between the examined groups in movement coordination. The mean times of fully coordinated gait on the rotating rod in sec. (± SEM) were: group I (control) 119 ± 1.0; group II (0.1 LD_{50} of bifenthrin i.p.) 78.2 ± 17.285. p= 0.03 considered significant. There were 10 mice in each group (Fig.3).

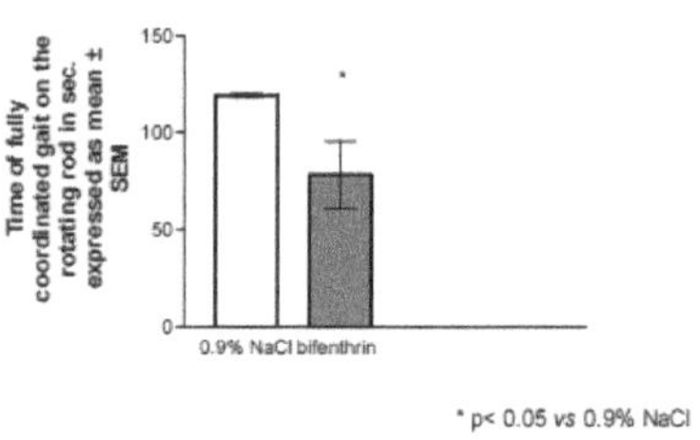

Fig. 3. The influence of bifenthrin (0.1 LD_{50} i.p.) on movement coordination. Columns represent the means ± SEM. Number of mice in each group was 10. * p< 0.05 vs 0.9% NaCl (ANOVA).

Effect of cypermethrin on memory retention in passive avoidance task

No statistically significant difference was observed in memory retention between group I and II. Median values of latency (with the 25th and 75th percentiles) were: 180 sec. (180, 180) in group I (control) and 172.5 sec. (40, 180) in group II (0.1 LD_{50} of cypermethrin i.p.). Post test were not calculated; p> 0.05. There were 10 mice in each group.

Influence of cypermethrin on working spatial memory in Y- maze task

There was no statistically significant difference between the examined groups in working spatial memory. Results obtained were (% of logical alternation behaviour expressed as mean ± SEM): group I (control) 64.47 ± 2.361; group II (0.1 LD_{50} of cypermethrin i.p.) 59.41 ± 2.853 ; p > 0.5 considered not significant. There were 10 mice in each group.

Influence of cypermethrin on movement activity in the actometer

The movement activity assessed within the 0-30 min. of observation were not significantly different in the groups. Mean values ± SEM were: group I (control) 552.44 ± 66.165 ; group II (0.1 LD_{50} of cypermethrin i.p.) 368.125 ± 87.653; p >0.05 considered not significant. There were 10 mice in each group. The movement activity assessed within the 31-60 min. of observation did not significantly differ. Mean values ± SEM were: group I (control) 394.66 ± 40.48; group II (0.1 LD_{50} of cypermethrin i.p.) 260 ± 107.11; p > 0.05 considered not significant. There were 10 mice in each group.

Influence of cypermethrin on movement coordination

There was observed no statistically significant difference between the examined groups in movement coordination. The mean times of fully coordinated gait on the rotating rod in sec. (± SEM) were: group I (control) 120 ± 0.0 ; group II (0.1 LD_{50} of cypermethrin i.p.) 120 ± 0.0. There were 10 mice in each group.

Effect of deltamethrin on memory retention in passive avoidance task

No statistically significant difference was observed in memory retention between group I and II. Median values of latency (with the 25th and 75th percentiles) were: 180 sec. (180, 180) in group I (control) and 180 sec. (65, 180) in group II (0.1 LD_{50} of deltamethrin i.p.). Post test were not calculated; p> 0.05. There were 10 mice in each group.

Influence of deltamethrin on working spatial memory in Y- maze task

There was no statistically significant difference between the examined groups in working spatial memory. Results obtained were (% of logical alternation behaviour expressed as mean ± SEM): group I (control) 61.84 ± 1.492; group II (0.1 LD_{50} of deltamethrin i.p.) 56.22 ± 3.274 ; $p > 0.5$ considered not significant. There were 10 mice in each group.

Influence of deltamethrin on movement activity in the actometer

The movement activity results analyzed within the 0-30 min. of observation were not significantly different in the groups. Mean values ± SEM were: group I (control) 728.66 ± 288.62; group II (0.1 LD_{50} of deltamethrin i.p.) 417 ± 50.964; $p > 0.05$ considered not significant. The movement activity assessed within the 31-60 min. of observation did not significantly differ between the examined groups . Mean values ± SEM were: group I (control) 608.66 ± 371.16; group II (0.1 LD of deltamethrin i.p.) 215.66 ± 41.571; $p > 0.05$ considered not significant

Influence of deltamethrin on movement coordination

There was observed no statistically significant difference between the examined groups in movement coordination. The mean times of fully coordinated gait on the rotating rod in sec. (± SEM) were: group I (control) 112.2 ± 7.8 ; group II (0.1 LD50 of deltamethrin i.p.) 82 ± 15.721 ; $p > 0.05$ considered not significant. There were 10 mice in each group.

Effect of deltamethrin on memory retention in passive avoidance task

No statistically significant difference was observed in memory retention between group I and II. Median values of latency (with the 25th and 75th percentiles) were: 180 sec. (180, 180) in group I (control) and 180 sec. (65, 180) in group II (0.1 LD_{50} of deltamethrin i.p.). Post test were not calculated; $p > 0.05$. There were 10 mice in each group.

Influence of deltamethrin on working spatial memory in Y- maze task

There was no statistically significant difference between the examined groups in working spatial memory. Results obtained were (% of logical alternation behaviour expressed as mean ± SEM): group I (control) 61.84 ± 1.492; group II (0.1 LD_{50} of deltamethrin i.p.) 56.22 ± 3.274 ; $p > 0.5$ considered not significant. There were 10 mice in each group.

Influence of deltamethrin on movement activity in the actometer

The movement activity results analyzed within the 0-30 min. of observation were not significantly different in the groups. Mean values ± SEM were: group I (control) 728.66 ± 288.62; group II (0.1 LD_{50} of deltamethrin i.p.) 417 ± 50.964; $p > 0.05$ considered not significant. The movement activity assessed within the 31-60 min. of observation did not significantly differ between the examined groups . Mean values ± SEM were: group I (control) 608.66 ± 371.16; group II (0.1 LD_{50} of deltamethrin i.p.) 215.66 ± 41.571; $p > 0.05$ considered not significant.

Influence of deltamethrin on movement coordination

There was observed no statistically significant difference between the examined groups in movement coordination. The mean times of fully coordinated gait on the rotating rod in sec. (± SEM) were: group I (control) 112.2 ± 7.8 ; group II (0.1 LD_{50} of deltamethrin i.p.) 82 ± 15.721 ; $p > 0.05$ considered not significant. There were 10 mice in each group.

Effect of fenpropathrin on memory retention in passive avoidance task

No statistically significant difference was observed in memory retention between group I and II. Median values of latency (with the 25th and 75th percentiles) were: 180 sec. (180, 180) in group I (control) and 30 sec. 135, 180) in group II (0.1 LD_{50} of fenpropathrin i.p.). Post test were not calculated. There were 10 mice in each group

Influence of fenpropathrin on working spatial memory in Y- maze task

There was no statistically significant difference between the examined groups in working spatial memory. Results obtained were (% of logical alternation behaviour expressed as mean ± SEM): group I (control) 62.38 ± 1.709; group II (0.1 LD_{50} of fenpropathrin i.p.) 61.88 ± 3.379 ; $p > 0.5$ considered not significant. There were 10 mice in each group.

Influence of fenpropathrin on movement activity in the actometer

The results of movement activity measurement analyzed within the 0-30 min. of observation were not significantly different in the examined groups. Mean values ± SEM were: group I (control) 524.44 ± 64.61 ; group II (0.1 LD_{50} of fenpropathrin i.p.) 456 ± 45.128; $p > 0.05$ considered not significant. The movement activity assessed within the 31-60 min. of observation did not significantly differ between the examined groups . Mean values ± SEM were: group I (control) 360 ± 51.465 ; group II (0.1 LD_{50} of fenpropathrin i.p.) 316.77 ± 34.44 ; $p > 0.05$ considered not significant. There were 10 mice in each group.

Influence of fenpropathrin on movement coordination

There was observed no statistically significant difference between the examined groups in movement coordination. The mean times of fully coordinated gait on the rotating rod in sec. (± SEM) were: group I (control) 112.3 ± 7.7 ; group II (0.1 LD_{50} of fenpropathrin i.p.) 105.8 ± 9.881 ; $p > 0.05$ considered not significant. There were 10 mice in each group.

Effect of lambda-cyhalothrin on memory retention in passive avoidance task

No statistically significant difference was observed in memory retention between group I and II. Median values of latency (with the 25th and 75th percentiles) were: 180 sec. (180, 180) in group I (control) and 180 sec. (15, 180) in group II (0.1 LD_{50} of lambda-cyhalothrin i.p.). $p > 0.05$ considered not significant.

Post test were not calculated. There were 10 mice in each group.

Influence of lambda-cyhalothrin on working spatial memory in Y- maze task

There was no statistically significant difference between the examined groups in working spatial memory. Results obtained were (% of logical alternation behaviour expressed as mean ± SEM): group I (control) 62.57 ± 2.875; group II (0.1 LD_{50} of lambdacyhalothrin i.p.) 61.83 ± 1.865; $p > 0.5$ considered not significant. There were 10 mice in each group.

Influence of lambda-cyhalothrin on movement activity in the actometer

The movement activity results analyzed within the 0-30 min. of observation were not significantly different in the examined groups. Mean values ± SEM were: group I (control) 546.8 ± 28.171 ; group II (0.1 LD_{50} of lambda-cyhalothrin i.p.) 491.9 ± 28.917; $p > 0.05$ considered not significant. The movement activity assessed within the 31-60 min. of observation did not significantly differ between the examined groups. Mean values ± SEM were: group I (control) 357.8 ± 31.48 ; group II (0.1 LD_{50} of lambda-cyhalothrin i.p.) 291.4 ± 22.935 ; $p > 0.05$ considered not significant. There were 10 mice in each group.

Influence of lambda-cyhalothrin on movement coordination

There was observed no statistically significant difference between the examined groups in movement coordination. The mean times of fully coordinated gait on the rotating rod in sec. (± SEM) were: group I (control) 119 ± 1.0; group II (0.1 LD_{50} of lambdacyhalothrin i.p.) 107.6 ± 9.48 ; $p > 0.05$ considered not significant. There were 10 mice in each group.

DISCUSSION

In this work bifenthrin administered i.p. at the dose of 0.1 LD_{50} was the only pesticide tested, that produced significant changes in mice bahaviour. Bifenthrin administered once 15 min. before testing in the Y-maze has significantly impaired the spatial working memory in comparison with the control group (Fig.1.). Bifenthrin was also found to impair movement coordination (Fig.3.). Bifenthrin has significantly increased movement activity within 0-30 min. of examination in the actometer (Fig.2.). In the previous acute neurotoxicity studies bifenthrin administered orally to male and female rats at the dose of 75mg/kg bw was found to produce whole body tremors, twitching, staggered gait, uncoordinated movement, ataxia, splayed hindlimbs, abnormal posture, clonic convulsions and abdominogenital staining (Watt 1998). Bifenthrin is an non-cyano pyrethroid, which was not included in the original studies estabilishing the T/CS classification. At present bifenthrin is classified as Type I (producing T-syndrome) (Breckendridge et al. 2009). All the above data show its toxicity and suggest possibe side effects of use in non-target organisms, like humans. Cypermethrin administered i.p. at low dose in this work to female mice did dot significantly affect memory and movement. In previous studies cypermethrin was identified as producing the CS syndrome (Verschoyle & Aldridge 1980; Lawrence & Casida 1982). In experiments with male and female rats given 100 or 200 mg of cypermethrin /kg bw in corn oil salivation, oral discharge, abdominogenital staining, ataxia, staggered gait, decreased locomotor activity and mortality were observed (Freeman 1993). Deltamethrin did not affect movement non memory in this set of experiments. Deltamethrin is classified as producing the CS intoxication syndrome (Barnes & Verschoyle 1974).The acute neurotoxicity studies of deltamethrin were conducted on male and female rats (Nemec 1998). Deltamethrin was administered orally in corn oil at the dose of 50mg/kg bw. Deltamethrin administered this way produced salivation, a flattened posture, limb extension, clonic and tonic convulsions, tremor, biting the cage, eyelid ptosis, decreased reaction to removal and handling, lacrimation, decreased arousal, wave-like movements of the abdomen dragging hindlimbs, decreased response to stimuli, increased auditory startle response, reduced forelimb strength, decreased fore- and hindlimb extensor strength, hypothermia and mortality (Nemec 1998). Lambda-cyhalothrin did not produce any statistically significant effect on memory processes nor movement activity in mice tested in this experiment. Cyhalothrin is a member of the α-cyano-3-phenoxybenzyl subfamily of pyrethroids (Soderlund 2002). However, it was not included in the T/CS classification as the effect of acute intoxication with lambdacyhalothrin were not specific to any of theses types. Lambda-cyhalothrin administered orally at the dose of 35mg/kg bw in corn oil to rats caused decreased activity, ataxia, reduced stability, tiptoe gait, decreased landing foot splay and decreased tail flick response (Barmmer 1999). There is data, that pyrethroids (cypermethrin, deltamethrin) administered orally in corn oil to rats produce dose-dependent decrease in motor activity (Crofton & Reiter 1984, 1988 a,b). There was evidence that i.p. administration of deltamethrin to rats caused a dose-dependent reduction in frequency of a prevuoisly learned behaviour (Bloom et al. 1983; Stein et al. 1987) which is congruent with data obtained in this work. Deltamethrin administered orally to rats for 15 days reduced learning and memory measured in a Y- maze task (Husain et al. 1996). Oral administration of cypermethrin to rats caused a similar reduction in learned bahaviour (Peele & Crofton 1987). Fenpropathrin does not fit into the traditional classification of pyrethroids (T/CS). In our former study (Nieradko-Iwanicka & Borzęcki 2010) fenpropathrin together with transient incomplete brain ischemia were found to reduce latency in the passive avoidance task compared to controls.

CONCLUSION

Pyrethroids, especially bifenthrin, should be used with caution as insecticidal formulas containing it may impair memory and movement in non target animals.

REFERENCES

1. Badach H. et al. (2007) Pesticide content in drinking water samples collected from orchard areas in central Poland. Ann Agric Environ Med, 14, pp. 109-114.
2. Barmmer A. (1999) Lambda-cyhalothrin: acute neurotoxicity study in rats. Zeneca Agricultural Products Laboratory Study No. CTL/P/6151: AR 6699.
3. Barnes J.M., Verschoyle R.D. (1974). Toxicity of new

pyrethroid insecticide. Nature 248, pp. 711.

4. Bloom A.S., Staatz C.G., Dieringer T. (1983) Pyrethroid Effects on operant responding and condiotioning) Neurobehav Toxicol Teratol 5, pp. 321-324.

5. Bloomquist J.R. (1993) Neuroreceptor mechanisms in pyrethroid mode of action and resistance In: Reviews in Pesticide Toxicology Roe, M., KuhrR.J. (Eds.), , Toxicology Communications, Raleigh, NC, pp. 181-226.

6. Bradberry S.M., et al. (2005) Poisonong due to pyrethroids. Toxicol Rev, 24 (2)pp. 93-106.

7. Breckendridge C.B., Holden L., Struggerss N., Weiner M., Sheets L., Sargent D., Soderlund D.M., Choi J.-S., Symington S., Clark J.M., Burr S., Ray D. (2009) Evidence for separate mechanism of toxicity for the Type I and the Type II pyrethroid insecticides. Neurotocicology, 30, pp.17-31.

8. Casida J.E et al. (1980) Metabolic chemistry of pyrethroid insecticides. Pestic Sci, 11, pp. 257- 269.

9. Casida J.E., Quistad G.B. (1998) Golden age of insecticide research: past, present, or future? Annu Rev Entomol 43, pp.1-16.

10. Crofton K.M., Reiter L.W. (1984) Effects of two pyrethroid insecticides on motor activity and acoustic startle response in the rat. Toxicol Appl Phasrmacol 75, pp. 318-328.

11. Crofton K.M., Reiter L.W. (1988a) The effects of Type I and II pyrethroids on motor activity and acoustic startle response in the rat. Fund Appl Toxicol, 10, pp. 624-634.

12. Crofton K.M., Reiter L.W. (1988b) Pyrthroid insecticides and gamma-aminobutyric acid receptor complex. Mptor activity and acoustic startle response in the rat. J Pharmacol Exp Ther 243, pp. 946-954.

13. Drożdzyński D. (2008) Studies on residues of pesticides used in rape plants protection in surface waters of intensively exploited arable lands in Wielkopolska Province of Poland. Ann Agric Environ Med, 15, pp. 231-235.

14. Elliot M., Farnham A.W., Janes N.F., Needham P.H., Pulman D.A., (1974) Synthetic insecticide with a new order of activity. Nature 248, pp.710-711.

15. Freeman C. (1993) Cypermethrin technical: acute neurotoxicity screen in rats. FMC Corporation study No. A92-3542.

16. Husain R., Adhami V.M., Seth P.K. (1996) Behavioral, neurochemical, and neuromorphological effects of deltamethrin in adult rats. J Toxicol Environ Health 48, pp. 515-526.

17. Kajfasz P. (2011) (Pol.) Profilaktyka chorób tropikalnych. Medycyna po Dyplomie 6 (18)3, pp. 34-41.

18. Knap J.P & Myjak P. (2009) (Pol.) Malaria w Polsce i na świecie- wczoraj i dziś. α-medica Press Wydanie II (uzupełnione). Bielsko-Biała , pp.36.

19. Lawrence L.J., Casida J.E. (1982) Pyrethroid tocxicology: Mouse intracerebral structuretoxicity relationships. Pestic Biochem Physiol, 18, pp.9-14.

20. Lichtfield J.T., Wilcoxon F.A. (1949) Simplified method of evaluating dose-effect experiments. J Pharmacol Exp Ther, 96. pp.99-113.

21. Mitra N.K. et al.(2008) Evaluation of neurotoxicity of repeated dermal application of chloprioryfos on hippocampus of adult mice. Ann Agric Environ Med, 15,pp. 211- 216.

22. Molina C. et al.(2005) HCH and DDT residues in human fat in the population of Murcia (Spain). Ann Agric Environ Med, 12 , pp. 133-136.

23. Nemec M (1998) An acute neurotoxicity study of deltamethrin in rats AgroEvo USA Company Study No. 53014.

24. Nieradko-Iwanicka B., Borzęcki A. (2010) Infl;uence of fenpropathrin on memory and movement in mice after transient incomplete cerebral ischemia. J Toxicol Environ

25. Health A 73, pp. 1166-1172.

26. Peele D.B., crofton K.M. (1987) Pyrethroid effects on schedule- controlled behavior: time and dosage relationships. Neurotoxicol Teratol 9, pp. 387-394.

27. Różański L. (1998) (Pol.) Przemiany pestycydów w organizmach żywych i środowisku. AGRAENVIRO LAB. Poznań.

28. Shuman E.K. (2010) Global climate change and infectious diseases. N Engl J Med, 362; 12, pp. 1061-1063.

29. Soderlund D.M., Bloomquist J.R. (1989) Neurotoxic actions of pyrethroid insecticides. Annu Rev Entiomol, 34, pp.77-96.

30. Soderlund D.M. et al. (2002) Mechanisms of pyrethroid neurotoxicity: implications for cumulative risk assessment. Toxicology, 171, pp. 3-59.

31. Stein E.A., Washburn M., Walczak C., Bloom A.S. (1987) Effects of pyrethroid insecticides on operant responding maintained by food. Neurotoxicol Teratol 9, pp. 27-31.

32. Verschoyle R.D., Aldridgwe W.N. (1980) Structure-activity relationships of some pyrethroids in rats. Arch Toxicol 45, pp. 325-329.

33. Verschoyle R.D, Barnes J.M. (1972) Toxicity of natural and synthetic pyrethrins to rats . Pestic Biochem Physiol, 2, pp. 308-311.

34. Watt B. (1998) FMC 548000 technical: acute neurotoxicity screen in rats. FMC Corporation Study No. A97-4642.

CHAPTER 12

SIMPLE ADAPTIVE CONTROL FOR POSITIVE LINEAR SYSTEMS WITH APPLICATIONS TO PEST MANAGEMENT

Chris Guiver, Christina Edholm Yu Jin, Markus Mueller, Jim Powell, Richard Rebarber, Brigitte Tenhumberg, Stuart Townley,

Keywords: Adaptive control, high-gain control, pest management, positive state linear system

INTRODUCTION

The 21st century is likely to see increasing pressure on world food production because of growing population and per capita consumption, greater competition for land use, and climate change [1]. One of the challenges in meeting the increasing demand for affordable food is to reduce potential losses in crop yield to animal pests and weeds, while minimizing the environmental impact of agriculture. The prevalent method for controlling pest populations is the use of chemical pesticides. Worldwide, in both 2006 and 2007, it is estimated that farmers applied about 2.5 billion kg of pesticides [2] and spent nearly $80 billion in doing so. Despite widespread pesticide application, yield loss is still significant. Indeed, in the early 1990s it was estimated that 37% of all potential food crops grown in the U.S. was destroyed by pests [3]. More recently, during 2001–2003 total global potential loss varied from about 50% in wheat to more than 80% in cotton production [4]. From the farmers' perspective pesticide application is still economically beneficial (generally each dollar invested in pesticide control returns about $4 in protected crops [5]). With such huge social and economic costs associated with crop losses to pests, it is clear that significant research and development in pest management must continue. The academic literature contains a myriad of theoretical approaches to pest control and the related issue of management of unwanted species. Examples include detailed analysis of predator-prey–type differential [6] and difference [7] equation models, linear programming [8], linear quadratic optimal control [9], and stochastic dynamic programming applied to Markov decision processes [10, 11, 12]. To our knowledge, the prevailing approaches are mostly based on deterministic or stochastic optimal controltype tools where a control, denoting a management or intervention strategy, is chosen to achieve some desired dynamic behavior as well as to minimize a specified cost. Although minimizing costs is of course appealing, there are disadvantages. To name but three, the computation of optimal controls is typically intractable analytically or expensive computationally (suffering, for example, from the "curse of dimensionality," coined in [13]; see more recently [14]), and so optimal controls can be impractical to implement. Second, and often overlooked, in applied contexts it is not always clear that "off-the-shelf" optimal control approaches will respect positivity of the system states. Third, and most pressingly, models of pest dynamics are inevitably highly uncertain as they seek to describe often novel species in novel environments. Uncertainty is a broad term in ecology and ecological modelling, although codices of the term may be found in [15] or [16]. In our study uncertainty encompasses choice of model structure (for example, type of model, number of stage-classes, or any modelled density-dependence), parametric uncertainty (for instance, how to accurately fit vital rates for a chosen model), and unknown disturbances of the dynamics (such as sampling error). Control engineers observed in the 1970s that optimal control techniques need not be robust to the above sources of uncertainty and overoptimization leads to fragility [17]. Therefore, owing to the sources of uncertainty expected to be present in models for pest species, approaches to pest management must be robust and must qualify and quantify the uncertainty they can handle. When seeking strategies to stabilize highly uncertain systems, and to avoid fragility, control engineering has focused on two

Citation: Chris Guiver, Christina Edholm, Yu Jin, Markus Mueller, Jim Powell, Richard Rebarber, Brigitte Tenhumberg, and Stuart Townley, "Simple Adaptive Control for Positive Linear Systems with Applications to Pest Management," DOI. 10.1137/140996926.

approaches: H∞-control theory, based on [18], and adaptive control theory [19, 20, 21, 22]. We argue that there is considerable scope and value in exploring adaptive control approaches in pest management. The pest dynamics and management strategy may be modelled as a control system with input, state, and output denoting the pest control, the pest population, and some measurement of that population, respectively. The aim is to design a management strategy, or course of management action, to eradicate a pest population (or at least reduce its abundance) which corresponds to a control objective of determining an input that stabilizes the model, that is, drives the state variables to zero. Importantly, both the state variables and the model parameters are uncertain, if not entirely unknown, and thus an input is desired that achieves stabilization in the presence of such a paucity of information. Within the framework of adaptive control, much attention has been devoted to so-called simple (also known as nonidentifier based) adaptive controllers originating in the 1980s in the work of Willems and Byrnes [23, 24], Mareels [25], M°artensson [26, 27], Morse [28], Nussbaum [29], and others. We refer the reader to the survey papers [30, 31] for more background and development of simple adaptive controllers. The power of these adaptive controllers is twofold: (a) they achieve global asymptotic stability of the state (here denoting the pest population) around zero, with minimal knowledge of the to-be-controlled system, often just a measured variable, and (b) they are straightforward to compute (and thus, in theory, implement). The nomenclature "simple" refers to the nonidentifier property of the controller, meaning that it does not seek to update the underlying dynamical model over time (for instance, by inferring or estimating parameters). That said, in an applied context the management strategy determined by a simple adaptive controller changes over time in response to how the measured variable changes. In that sense, there are parallels between adaptive control, an example of feedback control more generally, and (at least the technical, modelling aspects of) the academic discipline of adaptive management [32, 33, 34], as known in the resource and ecological management literature. The connections between robust feedback control and adaptive management have been explored by other authors as well: Heinimann [35] proposes principles from control theory as a concept for scholars and practitioners in adaptive ecosystem management. The application of simple adaptive control to pest management is to the best of our knowledge novel and, owing to its potential societal and economic value, worthy of consideration. The present mathematical investigation is additionally motivated by the observation that existing simple adaptive controllers are typically not designed for models where the state variables are constrained to take nonnegative values, as is evidently the case not only in applied pest management contexts, but also in models for harvesting, scavenging, culling, or predation. One exception that we are aware of is [36, Chap. 15], where adaptive control for positive dynamical systems is considered, and we compare those results with ours. Dynamical systems that leave a positive cone invariant—such as the nonnegative orthant in n-dimensional Euclidean space—are called positive dynamical systems. Correspondingly, the state variable of positive dynamical systems take only nonnegative values, typically denoting abundances, densities, or concentrations. Positive dynamical systems arise as models in a diverse range of fields from biology, chemistry, ecology, and economics to genetics, medicine, and engineering [36, p. xv], and are understandably well studied, with textbooks [37, 38, 39]. Control of positive dynamical systems leads to positive input control systems [40], where the input variables are assumed to be positive as well. In a pest management context, the input variable is naturally allowed to take negative values, provided that a nonnegative number or distribution of pests remains, leading to the concept of so-called positive state systems [41]. Such a framework allows the modelling of control actions that are essential for pest management but, importantly, fall outside the existing positive input systems theory. The purpose of the present paper is twofold. On the one hand, for those interested in applications, but possibly not familiar with adaptive control theory, we seek to describe how ideas from adaptive control may be used in pest management. We present a suite of tools that explore the utility of adaptive control techniques in pest management to eradicate or reduce pest abundance. Ultimately, we seek to use adaptive control methods to reduce pesticide usage, which in turn reduces the cost of food production and decreases consequential but undesirable costs of pesticide usage such as the humanitarian and economic cost of human pesticide poisoning and other pesticide-related illnesses, the destruction of beneficial insects (predators, pollinators), pesticide resistance evolution, and contamination of ground and surface water [42]. On the other hand, for those familiar with classical non identifier-based adaptive control, we describe how the typical structural assumptions on the set of (unknown) systems, such as minimum phase, relative degree, minimality, and knowledge of the high gain response, can essentially be replaced by component wise non negativity and an irreducibility or primitivity assumption. Therefore, we present and develop two ideas from adaptive control in the con text of pest management, developed for discrete-time, positive state linear systems. First, we derive a so-called high-gain result (section 2). We present several extensions, including a continuous-time analogue and an input-to-state stability-type estimate. Second, we present a so-called switching or universal adaptive control scheme (section 3). The results are applied to a model for Diaprepes root weevil, an invasive species threatening U.S. citrus production (section 4). Section 5 contains some concluding remarks. Proofs of certain results appear in the appendix. Notation. We collect mathematical notation and terminology used in what follows. The symbols N and R denote the sets of positive integers and real numbers, respectively. The symbol N_0 denotes the set of nonnegative integers. For n, m ∈ N, Rn and $R^{n\times m}$ denote usual n-dimensional Euclidean space and the set of n×m matrices with real entries, respectively. The superscript ^{T} denotes both matrix and vector transposition. With the conventions that $R^{n\times 1}$ = Rn and R^1 = R, for M,N ∈ $R^{n\times m}$ with entries m_{ij} and n_{ij} , respectively, we write

$M \ll N$ (or $N \gg M$) if $m_{ij} < n_{ij}$ (or $n_{ij} > m_{ij}$) $\forall i \in \{1,\dots,n\}, j \in \{1,\dots,m\}$.

$M < N$ (or $N > M$) if $M \leq N$ and $M \neq N$ (or $N \geq M$ and $M \neq N$).

$M \leq N$ (or $N \geq M$) if $m_{ij} \leq n_{ij}$ (or $n_{ij} \geq m_{ij}$) $\forall i \in \{1,\dots,n\}, j \in \{1,\dots,m\}$.

We let $R^{n\times m}{}_+$ denote the set of nonnegative matrices, that is, M ∈ $R^{n\times m}{}_+$ if M ≥ 0. We call M positive or strictly positive if M > 0 or M >> 0, respectively, noting that there are different conventions present in the academic literature for the term positive matrix.

We recall that a square matrix M ∈ $R^{n\times n}$ with entries m_{ij} is said to be reducible if there exist nonempty, disjoint subsets J_1, J_2 ⊆ {1,...,n} such that $J_1 \cup J_2$ = {1,...,n} and m_{ij} = 0 for all (i, j)

$\in J_1 \times J_2$. If M is not reducible, then it is said to be irreducible. If M is additionally nonnegative, then M is irreducible if and only if for each i, j ∈ {1, 2,...,n} there exists k ∈ N such that the (i, j)th entry of Mk is positive. A nonnegative matrix M is primitive if there exists k ∈ N such that $M^k >> 0$ and the index of primitivity is the smallest such k. The matrix M is called Metzler (also known as quasi-positive or essentially nonnegative) if every off-diagonal entry is nonnegative (see, for example, [39,]). We let r(M) and α(M) denote the spectral radius and spectral abscissa of M, respectively, which we recall are given by

$$r(M) := \max\{|\lambda| \, : \, \lambda \in \sigma(M)\} \quad \text{and} \quad \alpha(M) := \max\{\mathrm{Re}\,\lambda \, : \, \lambda \in \sigma(M)\},$$

where σ(M) denotes the set of eigenvalues of M, its spectrum. For $v \in \mathbb{R}^n$, $\|v\|$ and $\|M\|$ denote the one-norm of v and the operator induced one-norm of M, respectively. We let K denote the set of so-called comparison functions

$$\mathcal{K} := \{f : \mathbb{R}_+ \to \mathbb{R}_+ \, : \, f \text{ is continuous, strictly increasing and } f(0) = 0\};$$

see, for example, [43, p. 172].

Adaptive control schemes for pest management: In the spirit of high-gain. We assume a discrete-time, age- or stage-structured model for the pest population, which, when the state space is finite-dimensional, is called a matrix Population Projection Model (PPM) in the ecology literature. We refer the reader to the monograph [44] for more background on matrix PPMs. Specifically, the structured population x, effect of the pest control v, and measurement y are modelled as

$$\left.\begin{array}{l} x(t+1) = Ax(t) + v(t), \quad x(0) = x^0, \\ y(t) = Cx(t), \end{array}\right\} \quad t \in \mathbb{N}_0, \tag{2.1}$$

where n, p ∈ N, $A \in R^{n\times n}_+$ is a population projection1 matrix, and $C \in R^{p\times n}_+$ records the stages (or combinations of stages) measured. In pest control applications, r(A) > 1 so that uncontrolled, the pest population is increasing exponentially, at least asymptotically. The variable y(t) in (2.1) denotes some measured portion of the population and is the variable that is available to inform management strategies—in applications we do not expect to have exact knowledge of the state variables x(t) at all times owing to, for example, the difficulty or unreliability of measuring certain stage-classes.

For a biologically realistic and meaningful model, the control effect v satisfies the following three properties:

(i) v is targeted in that it affects only specific life-stages;

(ii) for those stages targeted, v is proportional to the pest population present;

(iii) v is a (possibly nonlinear but) increasing function of some applied control effort u.

Properties (i)–(iii) motivate a v of the form

$$v(t) = -B\Phi(u(t))Fx(t), \quad t \in \mathbb{N}_0, \tag{2.2}$$

where $B \in R^{n\times m}_+$, $F \in R^{m\times n}+$ for some m ∈ N and u(t) ∈ R+ denotes the applied control effort. The B in (2.2) is determined by which age- or stage-classes are affected by the pest control, and F is determined by the underlying biology. The inclusion of the function $\Phi : R+ \to R^{m\times m}_+$ in (2.2) models the efficacy on various life stages of the control effort at time-step t, i.e., u(t), which, for example, could denote the mass of active pesticide ingredient used. For what follows we record an assumption of functions Φ arising in (2.2):

(A1) $\Phi : \mathbb{R}_+ \to \mathbb{R}_+^{m\times m}$ has nonzero components belonging to K and mapping $\mathbb{R}_+ \to [0,1]$.

We note that if Φ satisfies (A1), then the limit $\Phi\infty := \lim_{\omega\to\infty} \Phi(\omega)$ exists. Example 2.1. If a pest management strategy (2.2) targets only the first stageclass, usually denoting juveniles in an animal model, seeds or seedlings in a plant model, or egg or larval stages in an insect model, then m = 1 and B = $[1\ 0 \ldots 0]^T$ in (2.2). The row vector F in (2.2) has entries taken from the corresponding entries in the first row of A. Possible $\Phi: R_+ \to R_+$ are sigmoid-type functions, taking values between zero and one, such as

$$\Phi(u) = 1 - e^{-k_1 u}, \quad \Phi(u) = \tanh(k_2 u), \quad \text{or} \quad \Phi(u) = \frac{k_3 u}{1 + k_3 u} \tag{2.3}$$

for constants k_1, k_2, $k_3 > 0$. We comment further on possible choices of Φ in applications in section 4.

The combination of (2.1) and (2.2) yields

$$\left.\begin{array}{l} x(t+1) = (A - B\Phi(u(t))F)x(t), \quad x(0) = x^0, \\ y(t) = Cx(t), \end{array}\right\} \quad t \in \mathbb{N}_0, \tag{2.4}$$

where the sequence of control efforts u is still to be determined. The premise of the present contribution is that A, B, F, and Φ are uncertain or unknown, and so it is generally not possible to determine a constant choice ˆu > 0 in (2.2) or (2.4) such that

$$r(A - B\Phi(\hat{u})F) < 1, \tag{2.5}$$

which, in pest management applications, would lead to eradication of the pest. However, we seek to exploit the idea that the choice of pest management (2.2) is able to eradicate the pest, at least for a sufficiently large, but crucially unknown, control effort. To capture this property mathematically, we record a second assumption which pertains to the triple (A, B, F) $\in R^{n\times n}_+ \times R^{n\times m}_+ \times R^{m\times n}_+$ that appears in (2.4) and a function $\Phi : R_+ \to R^{m\times m}_+$ satisfying assumption (A1):

(A2) For all ω ∈ [0, ∞), the matrix A − BΦ(ω)F is nonnegative and irreducible, and $r(A - B\Phi_\infty F) < 1$, where $\Phi_\infty := \lim_{\omega\to\infty} \Phi(\omega)$.

Remark 2.2. Assumption (A2) is in the spirit of "high-gain" in the control engineering literature as it posits that when u(t) (the "gain") is sufficiently large, the "closed loop" matrix A − BΦ(u(t))F is stable. Assuming that A − BΦ(ω)F is irreducible for every positive, finite ω is not restrictive for reasonable ecological models [45]; in other words, the directed graph (see, for example, [39, Definition 2.4, p. 29]) associated with the population projection matrix A − BΦ(ω)F is strongly connected for every ω. Recalling that the directed graph of a population projection matrix models the life-cycle of a stage-structured population (see [44, p. 57]), loosely speaking, a consequence of the irreducibility assumption in (A2) is that effects of the pest management strategy are (eventually) felt by every stage-class (although they perhaps target only a single stage-class). To exploit the biologically reasonable assumptions (A1) and

(A2) in pest management we propose an adaptive control scheme:

$$\left.\begin{aligned} x(t+1) &= (A - B\Phi(u(t))F)x(t), & x(0) &= x^0, \\ y(t) &= Cx(t), & & \\ u(t+1) &= u(t) + \psi(\|y(t)\|), & u(0) &= u^0, \end{aligned}\right\} \quad t \in \mathbb{N}_0, \tag{2.6}$$

where $u^0 \geq 0$ and $\psi \in K$ are design parameters. The term u^0 is the initial control effort, and the function ψ provides a recipe for converting current measured pest abundance $\|y(t)\|$ into an increase of control effort at the next time-step and dictates "how fast" the control effort u increases. In practice, ψ will depend on the units of the (currently dimensionless) control effort u, such as mass or volume. The upshot is that the control scheme (2.6) continues to increase the control effort at time-step t, i.e., u(t), so long as the measured variable y(t) is nonzero. Our first result demonstrates that, on the basis of only (A1) and (A2) holding and a minor constraint on ψ, the adaptive control scheme (2.6) is globally asymptotically stable and that rate of convergence of the state to zero is exponential. In particular, the sequence of nondecreasing control efforts u converges.

Theorem 2.3. Given the adaptive control scheme (2.6), assume that Φ satisfies (A1), $(A, B, F) \in R^{n\times n}_+ \times R^{n\times m}_+ \times R^{m\times n}_+$ and Φ satisfy (A2), $\psi \in K$, and $C \in R^{p\times n}_+$, $C \gneq 0$. For positive x^0 and nonnegative u^0, let (x, u) denote the solution of (2.6). Then, there exist $M > 0$ and $\gamma \in (0, 1)$ such that

$$\|x(t)\| \leq M\gamma^t\|x^0\|, \quad t \in \mathbb{N}_0. \tag{2.7}$$

If additionally ψ has the property that

$$\forall\, \delta > 0,\ \forall\, \lambda \in (0,1) : \sum_{j\in\mathbb{N}_0} \psi(\delta\lambda^j) < \infty, \tag{2.8}$$

then u converges, and its limit $u_\infty \geq 0$ (which depends on x^0, u^0, and ψ) is stabilizing, that is,

$$r(A - B\Phi(u_\infty)F) < 1. \tag{2.9}$$

Proof. For any nonnegative sequence u, the state x is bounded by

$$x(t+1) \leq Ax(t), \quad x(0) = x^0, \quad t \in \mathbb{N}_0,$$

and thus (2.7) trivially holds if $r(A) < 1$. We therefore assume that $r(A) \geq 1$. Assumption (A2) now implies that $\Phi \gneq 0$, and thus by (A1), Φ is componentwise strictly increasing in its nonzero components, so that

$$0 \leq u_1 < u_2 \quad \Rightarrow \quad \Phi(u_1) < \Phi(u_2). \tag{2.10}$$

In the arguments that follow we shall make use of monotonicity of the spectral radius:

$$A_1, A_2 \in \mathbb{R}^{n\times n}, \quad 0 \leq A_2 < A_1 \text{ and } A_2 \text{ irreducible} \quad \Rightarrow \quad 0 \leq r(A_2) < r(A_1); \tag{2.11}$$

see, for example, [39, Corollary 1.5, p. 27]. Let $u_* \geq 0$ be such that $A_* := A - B\Phi(u_*)F$ has $r(A_*) = 1$. The existence of such a $u*$ is ensured by the intermediate value theorem applied to the continuous, strictly decreasing function

$$[0, \infty) \ni \omega \mapsto \kappa(\omega) := r(A - B\Phi(\omega)F) - 1.$$

That κ is continuous follows from the continuity of the spectral radius and the continuity of the nonzero components of Φ. The monotonicity of κ follows from (2.10) and (2.11). Clearly, $\kappa(0) = r(A) - 1 \geq 0$, and, by assumption (A2), $\lim_{\omega\to\infty} \kappa(\omega) < 0$. We consider two exhaustive cases: either (a) $u(t) \leq u_*$ for all $t \in N_0$, or (b) there exists some $\tau \in N$ such that $u(\tau) > u_*$. Seeking a contradiction, we assume that (a) holds. By its construction in (2.6), u is a nondecreasing sequence that is assumed to be bounded from above and hence convergent to some $z \leq u_*$. Consequently, the sequence $(\psi(\|y(j)\|))_{j\in\mathbb{N}_0}$ is summable, that is, belongs to l^1 as

$$\sum_{j=0}^{N} \psi(\|y(j)\|) = \sum_{j=0}^{N} [u(j+1) - u(j)] = u(N+1) - u(0) \to z - u^0 < \infty \quad \text{as } N \to \infty,$$

necessarily implying that

$$\psi(\|y(j)\|) \to 0 \quad \text{as } j \to \infty \tag{2.12}$$

Since $\psi \in K$, the convergence in (2.12) yields that

$$y(j) \to 0 \quad \text{as } j \to \infty. \tag{2.13}$$

The assumption that $u(t) \leq u_*$ for every $t \in N_0$ and (2.10) together imply that

$$-B\Phi(u(t))F \geq -B\Phi(u_*)F, \quad t \in \mathbb{N}_0,$$

and thus

$$x(t+1) = (A - B\Phi(u(t))F)x(t) \geq (A - B\Phi(u_*)F)x(t) = A_*x(t), \quad t \in \mathbb{N}_0. \tag{2.14}$$

From the convergence in (2.13) and difference inequality (2.14) we infer the following:

$$\begin{aligned} x(t) \geq A_*^t x^0,\ t \in \mathbb{N}_0. \quad &\Rightarrow \quad y(t) \geq CA_*^t x^0,\ t \in \mathbb{N}_0, \\ &\Rightarrow \quad 0 = \limsup_{t\to\infty} y(t) \geq \limsup_{t\to\infty} CA_*^t x^0. \end{aligned}$$

The combined properties that C and x^0 are positive and that A_* is irreducible with $r(A_*) = 1$ yield the contradiction

$$0 = \limsup_{t\to\infty} CA_*^t x^0 > 0.$$

We deduce that (b) holds, and so there exists $\tau \in N$ such that

$$u(t) \geq u(\tau) > u_*, \quad t \in \mathbb{N},\ t \geq \tau, \tag{2.15}$$

as u is a nondecreasing sequence. Invoking (2.10) again, it follows from (2.15) that

$$A - B\Phi(u(t))F \leq A - B\Phi(u(\tau))F < A - B\Phi(u_*)F = A_*, \quad t \in \mathbb{N},\ t \geq \tau, \tag{2.16}$$

and as $A - B\Phi(u(\tau))F$ is irreducible, (2.11) implies that

$$r(A - B\Phi(u(\tau))F) < r(A_*) = 1\,. \tag{2.17}$$

Hence, for t ∈ N, t>τ, we estimate

$$x(t) = \prod_{j=0}^{t}(A - B\Phi(u(j))F)x^0 = \prod_{j=\tau}^{t}(A - B\Phi(u(j))F)\prod_{j=0}^{\tau-1}(A - B\Phi(u(j))F)x^0$$
$$\leq (A - B\Phi(u(\tau))F)^{t-\tau}A^{\tau}x^0\,,$$

from which the bound (2.7) readily follows. To see that u converges, we prove that u is bounded. For T ∈ N we have that

$$u(T) = u^0 + \sum_{j=0}^{T-1}(u(j+1) - u(j))$$
$$= u^0 + \sum_{j=0}^{T-1}\psi(\|Cx(j)\|) \leq u^0 + \sum_{j=0}^{T-1}\psi(\gamma^j M\|C\|\|x^0\|) < \infty\,,$$

the final inequality being independent of T ∈ N, and where we have used the condition (2.8). Finally, the conjunction of (2.15)–(2.17) necessitates that (2.9) holds, as required. The next section discusses extensions motivated by Theorem 2.3, before which we remark on the above theorem.

Remark 2.4.

(i) For each fixed x^0, the inequality (2.7) ensures that the x(t) component of the solution (x, u) of (2.6) converges to zero exponentially. However, the positive constants M and γ that appear in (2.7) both depend on the initial conditions x^0 and u^0. Therefore, (2.7) does not yet imply that (2.6) is globally exponentially stable.

(ii) The condition (2.8) is a growth rate property of ψ ∈ K at zero; in other words, it states that the action of ψ respects the decay of geometric series. It does not seem overly restrictive. For example, (2.8) holds if ψ ∈ K is H¨older continuous, with any positive exponent, which includes Lipschitz continuous functions and power laws $\mathbb{R}_+ \ni x \mapsto x^\nu,\ \nu > 0.$. Elementary analysis demonstrates that for all $c_1, c_2 > 0,\ \mathbb{R}_+ \ni x \mapsto c_1\ln(1 + c_2x)$ also satisfies (2.8).

(iii) The conclusions of Theorem 2.3 need not hold if the assumption that A − BΦ(ω)F is irreducible for every ω ∈ [0, ∞) in (A2) is relaxed. For example, consider

$$A = \begin{bmatrix} f_1 & 0 \\ 0 & f_2 \end{bmatrix},\quad B = \begin{bmatrix} 1 \\ 0 \end{bmatrix},\quad F = \begin{bmatrix} f_1 & 0 \end{bmatrix},\quad C = \begin{bmatrix} 0 & 1 \end{bmatrix},\quad v \mapsto \psi(v) = v,$$

where $f_1 > 1$ and $f_2 \in (0, 1)$. Here A − δBF is reducible for every δ ∈ [0, 1). Furthermore, $y(t) = f_2^t x_2^0 \to 0$ exponentially as t → ∞, and thus

$$u(t) = u^0 + \sum_{j=0}^{t-1}\|y(t)\| \to u^0 + \frac{x_2^0}{1 - f_2} =: u_\infty < \infty \quad \text{as } t \to \infty.$$

However, for all $x_1^0 > 0,\ u^0, x_2^0 \geq 0,$ and $f_2 \in (0, 1)$, there exists Φ : R+ → [0, 1), Φ ∈ K, such that $f_1(1 - \Phi(u\infty)) > 1$ and thus

$$x_1(t+1) = f_1(1 - \Phi(u(t)))x_1(t) \geq x_1(t) \not\to 0 \quad \text{as } t \to \infty.$$

(iv) If $x^0 = 0$, then x(t) = 0 for all t ∈ N_0, and hence (2.7) trivially holds. In this case, u(t) = u0 for all t ∈ N, and so (0, u0) is an equilibrium of (2.6) for all u^0 ∈ R+. Note that u∞ = u^0 need not satisfy (2.9). However, one striking property is that for all nonzero x^0, and any $u^0 \geq 0$ the limiting control effort u∞ is stabilizing.

2.1. Extensions. Here we present directions in which the simple adaptive control model (2.6) may be extended. The motivation for doing so is, first, to show how the model (2.6) may be tailored to accommodate some of the nuanced scenarios likely to be encountered in pest management (Propositions 2.5 and 2.7) and, second, to demonstrate how the ideas behind (2.6) may be extended to different classes of model (Propositions 2.11 and 2.14).

A disadvantage of the scheme (2.6) is that the limiting control effort u_∞ ensured by Theorem 2.3 may be impractically large, becoming prohibitively expensive or risking environmental damage. Therefore, the first two extensions proposed are alternative strategies that seek to reduce a pest population with a smaller control effort.

A rate-limited adaptive control scheme. Consider first the adaptive control scheme

$$t \in \mathbb{N}_0 \quad \begin{cases} x(t+1) = (A - B\Phi(u(t))F)x(t)\,, & x(0) = x^0, \\ y(t) = Cx(t)\,, & \\ u(t+1) = \begin{cases} u(t) + \psi(\|y(t)\|)\,, & u(t) \leq \theta(t)\,, \\ u(t)\,, & u(t) > \theta(t)\,, \end{cases} & u(0) = u^0, \end{cases}$$

(2.18a) and (2.18b)

where the strictly increasing, unbounded sequence θ is an additional design parameter. The terms in (2.18a) and their interpretation are the same as those in (2.6). The scheme (2.18) differs from (2.6) by the inclusion of θ in the update law for u in (2.18b). The interpretation of (2.18b) is that θ acts as a bound on the size of permitted control actions. While u(t) is no greater than θ(t), the control effort u(t) of the adaptive scheme (2.18) is updated in the same manner as that in (2.6). When u(t) is larger than θ(t), then u(t+1) = u(t), that is, u(t) is maintained at its current level. In effect, the inclusion of θ further limits the rate at which u may increase and in applications may reflect economic, capacity, or legislative constraints. Simulations of the rate-limited adaptive control scheme (2.18) are contained in Figure 3, where it is demonstrated that, although the exponential stabilization may take longer than in (2.6), the rate of increase is further controlled and the limiting control effort u_∞ may be smaller than that produced by (2.6).

Proposition 2.5. Given the rate-limited adaptive control system (2.18), assume that Φ satisfies (A1), $(A, B, F) \in \mathbb{R}_+^{n\times n} \times \mathbb{R}_+^{n\times m} \times \mathbb{R}_+^{m\times n}$ *and* Φ satisfy (A2) ψ ∈ K, θ is a strictly increasing, unbounded sequence, and $C \in \mathbb{R}_+^{p\times n},\ C \neq 0$.Then, for every positive x^0 and nonnegative u^0, the conclusions of Theorem 2.3 hold. Proof. The proof is very similar to that of Theorem 2.3, and so we provide only an outline. Letting $u_* > 0$ be such that $A_* := A - B\Phi(u^*)F$ has $r(A_*) = 1$, we again seek a contradiction and assume that $u(t) \leq u_*$ for all t ∈ N_0. Note that u defined by (2.18b) is still a non-decreasing sequence, and the assumption that it is bounded implies that it converges, say with limit $z \leq u_*$. The assumptions that θ is unbounded and strictly increasing imply that there exists s ∈ N such that z< θ(t) for all t ∈ N, t>s. Therefore, for such t, u(t) ≤ z ≤ θ(t), and hence the sequence $\psi(\|y\|)$ is

$$\sum_{j=s}^{T}\psi(\|y(j)\|) = \sum_{j=s}^{T}[u(j+1) - u(j)] = u(T+1) - u(s) \leq z - u(s) < \infty$$

for all T ∈ N. The proof now follows that of Theorem 2.3, joining at the convergence in (2.12).

Remark 2.6. The conclusions of Proposition 2.5 do not hold in general if the increasing threshold θ for u is bounded or replaced by a constant. Although θ may increase arbitrarily slowly, loosely speaking, the smaller θ is and the slower it increases, the larger we might expect transient growth of the state x to be.

A neutral-zone adaptive control scheme. The second extension is

$$t \in \mathbb{N}_0 \quad \begin{cases} x(t+1) = (A - B\Phi(u(t))F)x(t), & x(0) = x^0, \\ y(t) = Cx(t), & \\ u(t+1) = \begin{cases} u(t) + \psi(\|y(t)\|), & L_1 \leq \|y(t)\|, \\ u(t), & L_2 \leq \|y(t)\| < L_1, \\ \mu u(t), & \|y(t)\| < L_2, \end{cases} & u(0) = u^0, \end{cases} \tag{2.19a}$$

and (2.19b)

where $L_1 \geq L_2 > 0$ and $\mu \in (0, 1)$ are additional design parameters. The terms in (2.19a) and their interpretation are the same as those in (2.6). The interpretation of the update law for u in (2.19b) is that when the output y(t) is larger than some prescribed level, meaning $L_1 \leq \|y(t)\|$, the control effort u increases as before. When the output is deemed small enough—smaller than the threshold L_2—then the control effort decreases geometrically. Between L_1 and L_2 a constant control effort is maintained: there is no adaptation of u, hence the nomenclature "neutral-zone." In control engineering the terms "dead-zone" or "deadband" are also sometimes used. The motivation for (2.19) is to trade off the cost of increasing the control effort u against the possible increase in observed pest population y.

Simulations of the neutral-zone adaptive control scheme (2.19) are contained in Figure 4. As suggested there, when $r(A) \geq 1$, we should not expect the scheme (2.19) to result in convergence of x and u. Instead we obtain the following result. Proposition 2.7. Given the neutral-zone adaptive control system (2.19), assume that Φ satisfies (A1), $(A, B, F) \in R^{n\times n}_+ \times R^{n\times m}_+ \times R^{m\times n}_+$ and Φ satisfy (A2), $\psi \in K$ and satisfies (2.8), and $C \in R^{p\times n}_+$, $C \neq 0$. For $(x^0, u^0) \in R^n_+ \times R+$, $L_1 \geq L_2 > 0$, and $\mu \in (0, 1)$, the solution (x, u) of (2.19) is bounded. The proof of Proposition 2.7 is elementary, but tedious, and hence is relegated to the appendix.

Remark 2.8. The proof of Proposition 2.7 is in fact independent of how u in (2.19b) is defined to decay when $\|y(t)\|$ < L2. Consequently, the exponential decay of u proposed in (2.19b) could be replaced by any scheme with u(t) → 0 when $\|y(t)\| \leq L_2$. Furthermore, Proposition 2.7 remains true in the case that the update law u(t + 1) = u(t) is omitted, by taking $L_1 = L_2$. Although the above result guarantees boundedness of the state and control effort, it does not give any indication of what these bounds are and how they depend on L_1, L_2, or γ. Again, for practical applications, one would need to trade off the costs of eradicating a pest against the cost of not (sufficiently) reducing pest abundance. For example, when the lower threshold L_2 in (2.19b) is small, then control u is only reduced once a pest abundance less than L_2 is observed, and until that point u is either increasing or kept at a fixed level, which may be costly.

Input-to-state stability. The models considered thus far have overlooked external signals acting on the state or output. In the context of pest management the former could denote previously unmodelled immigration, and the latter measurement or sampling error. To incorporate such a situation the model (2.6) becomes

$$\left.\begin{aligned} x(t+1) &= (A - B\Phi(u(t))F)x(t) + d_1(t), & x(0) &= x^0, \\ y(t) &= Cx(t) + d_2(t), & & \\ u(t+1) &= u(t) + \psi(\|y(t)\|), & u(0) &= u^0, \end{aligned}\right\} \quad t \in \mathbb{N}_0, \tag{2.20}$$

where $u^0 \geq 0$ and $\psi \in K$ are design parameters and d_1 and d_2 are external signals. We shall assume that d_1 and d_2 are (locally) bounded and respect the nonnegativity constraints of state and output. Without further assumptions on the class of disturbances, however, we should not expect convergence of the state x or control effort u of (2.20). The following result demonstrates that the state x of (2.20) admits an input-to-state stability (ISS)-type estimate. We refer the reader to [46] for more background and information on ISS.

Proposition 2.9. Given the disturbed adaptive control system (2.20), assume that Φ satisfies (A1), (A, B, F) ∈ Rn×n + × Rn×m + × Rm×n + and Φ satisfy (A2), $\psi \in K$, and $C \in R^{p\times n}_+$, $C \neq 0$. Then, for all positive x^0 and nonnegative u^0 and all disturbance signals d1 $\in {}^\infty{}_{loc}(N_0; R_n +)$ and $d_2 \in {}^\infty{}_{loc}(N_0; R^p)$ with the property that for some $\delta > 0$

$$y(t) \geq \delta Cx(t), \quad t \in \mathbb{N}_0, \tag{2.21}$$

there exist $M_1, M_2 > 0$ and $\gamma \in (0, 1)$ such that the solution x of (2.20) satisfies

$$\|x(t)\| \leq M_1\gamma^t\|x^0\| + M_2\|d_1\|_{\ell^\infty(0,t-1)}, \quad t \in \mathbb{N}_0. \tag{2.22}$$

The constant M_2 may be chosen to depend only on x^0 and δ.

Proof. We restrict to the case that $r(A) \geq 1$ (the result is immediate otherwise). Let $u_* > 0$ be such that $A_* := A - B\Phi(u_*)F$ has $r(A*) = 1$. While $u(t) \leq u_*$, we may estimate for $\tau \geq 1$

$$u(\tau) = u^0 + \sum_{j=0}^{\tau-1}\psi(\|y(j)\|) \geq \sum_{j=0}^{\tau-1}\psi(\delta\|Cx(j)\|)$$

$$\geq \sum_{j=0}^{\tau-1}\psi\left(\delta\left\|CA_*^jx^0 + \sum_{k=0}^{j-1}CA_*^{j-1-k}d_1(k)\right\|\right) \geq \sum_{j=0}^{\tau-1}\psi\left(\delta\|CA_*^jx^0\|\right). \tag{2.23}$$

Since $r(A*) = 1$, A∗ is irreducible, $\delta > 0$, and $C \neq 0$, the summation on the righthand side of (2.23) is greater than u∗ at some time τ, independently of u^0, d_1, and d_2. Then, since u is nondecreasing,

$$u(t) \geq u(\tau) > u_* \quad \forall t \in \mathbb{N}, \ t \geq \tau.$$

Arguing as in the proof of Theorem 2.3, writing $A_\tau := A - B\Phi(u(\tau))F$, we obtain the estimate

$$x(t+1) \leq A_\tau x(t) + d_1(t), \quad t \in \mathbb{N}_0, \ t \geq \tau,$$

and thus for $t \geq \tau + 1$

$$x(t) \leq A_\tau^{t-\tau}x(\tau) + \sum_{j=\tau}^{t-1}A_\tau^{t-1-j}d_1(j). \tag{2.24}$$

Since $\gamma := r(A_\tau) < 1$, the inequality (2.24) readily yields for $t \geq \tau + 1$

$$\|x(t)\| \leq N_1 r(A_\tau)^{t-\tau}\|x(\tau)\| + N_2\|d_1\|_{\ell^\infty(\tau,t-1)} \leq N_3\gamma^t\|x(0)\| + N_2\|d_1\|_{\ell^\infty(0,t-1)} \tag{2.25}$$

for positive constants N_1, N_2, N_3 depending only on A and A_τ. For $t \in \{1, 2,...,\tau\}$ we may estimate

$$x(t) \leq \gamma^t(\gamma^{-t}A^t)x^0 + \sum_{j=0}^{t-1} A^{t-1-j}d_1(j)\,. \tag{2.26}$$

Bounding the right-hand side of (2.26) yields that for $t \in \{1, 2,...,\tau\}$

$$\|x(t)\| \leq \left(\max_{\substack{k\in\mathbb{N}_0\\0\leq k\leq\tau}} \|\gamma^{-k}A^k\|\right)\gamma^t\|x^0\| + \left(\sum_{j=0}^{\tau}\|A^j\|\right)\|d_1\|_{\ell^\infty(0,t-1)}\,. \tag{2.27}$$

Combining (2.25) and (2.27) establishes the estimate (2.22).

Remark 2.10.

- The estimate (2.22) is not a "true" ISS estimate, as the M_1 and γ terms appearing there depend on both x^0 and d_1 in general. Similarly, M_2 in (2.22) depends on x^0. These dependencies are a consequence of the nonlinear feedback $-B\Phi(u)F\,x$. We have included the result to demonstrate that the inclusion of nonzero, bounded additive perturbations d_1 to (2.6) still results in boundedness of the state, and the bound depends linearly on $\|d_1\|_\infty$.
- The assumption that the output disturbance d2 satisfies (2.21) is certainly satisfied if d2 is a proportional output measurement error, meaning that

$$d_2(t) = \varepsilon(t)Cx(t), \quad t \in \mathbb{N}_0\,,$$

where ε is an unknown sequence of measurement errors with the property that $\varepsilon(t) > \varepsilon_0 > -1$ for some constant ε_0. In this case, the condition (2.21) holds with $\delta = 1 + \varepsilon_0 > 0$.

- The term robustness is used in control to quantify how much uncertainty (in some sense) a control scheme may tolerate and still function qualitatively as intended. The ISS estimate (2.22) is a robustness property with respect to the additive disturbances as in (2.20). There are other notions of robustness in adaptive control, and we refer the reader to [19], in which a framework is constructed for describing and analyzing robustness of adaptive control schemes (in the continuous-time case).

Infinite-dimensional state space. Here we demonstrate that in certain circumstances the assumption of a finite-dimensional state space may be relaxed. A key ingredient in the proofs so far is strict monotonicity of the spectral radius with respect to monotonicity of operators; that is,

$$0 \leq A_2 < A_1 \quad \Rightarrow \quad r(A_2) < r(A_1) \tag{2.28}$$

holds. When $A_1, A_2: \mathbb{R}^n_+ \to \mathbb{R}^n_+$ and $\mathbb{R}^n$ is equipped with the usual partial ordering of $0 \leq x \iff x \in \mathbb{R}^n_+$, then irreducibility of A_2 is sufficient for (2.28) (indeed, see (2.11)). The monograph [38] contains nonequivalent sufficient conditions for (2.28) depending on what one is prepared to assume about the (possibly infinite-dimensional) Banach space X, its partial ordering, and the operators $A_1, A_2 : X \to X$. We appeal to the results and terminology of [38], noting that the authors there use the term positive when we would instead use nonnegative.

Let X denote an ordered real Banach space, so that X is equipped with a partial order ≤ (also ≥) that respects vector space addition and multiplication by nonnegative scalars. The positive cone C induced by (X, ≥) is the set of $x \in X$ such that $x \geq 0$ and is a closed, convex set (so that if $x, y \in C$ and $\delta \geq 0$, then $x + y, \delta x \in C$) with the property that $-x \in C$ implies that $x = 0$. The cone C is called reproducing if $X = C \ominus C$, meaning that every $z \in X$ may be written as $z = x_1 - x_2$ with $x_1, x_2 \in C$. For real Banach spaces X_1, X_2 with respective positive cones C_1, C_2, a bounded linear operator $T : X_1 \to X_2$ is called positive if $T\,C_1 \subseteq C_2$. A bounded, positive operator A is called ν-bounded [38, pp. 90, 91] for $\nu \in C$, $\nu \neq 0$, if for all $x \in C$ there exist functions $0 < \eta_1(x) \leq \eta_2(x)$ for $x \neq 0$

$$\eta_1(x)\nu \leq Ax \leq \eta_2(x)\nu\,.$$

Consider now the adaptive control system (2.6), where

$$A: \mathcal{X} \to \mathcal{X}, \quad B: \mathbb{R}^m \to \mathcal{X}, \quad F: \mathcal{X} \to \mathbb{R}^m, \quad C: \mathcal{X} \to \mathbb{R}^p \tag{2.29}$$

are bounded, positive operators. For Φ satisfying (A1), we record the following assumption:

(A2)′ For all $\omega \in [0,\infty)$, $A - B\Phi(\omega)F$ is positive, compact, and νω-bounded with $r(A - B\Phi_\infty F) < 1$, where $\Phi_\infty = \lim_{\omega\to\infty}\Phi(\omega)$.

Proposition 2.11. Let (X, ≥) denote a real ordered Banach space with reproducing positive cone C, and let A, B, F, C as in (2.29) denote bounded, positive operators, with C nonzero. Assume that Φ satisfies (A1), $\psi \in K$, and that (A, B, F) and Φ satisfy (A2)'. Then, the conclusions of Theorem 2.3 apply to the adaptive control system (2.6) specified by (A, B, F, C) in (2.29), Φ, and ψ.

Proof. The proof is identical to that of Theorem 2.3 once (2.17) is established. Under our assumptions, this follows from [38, Theorems 16.2 and 16.3] with A in those results taken as $A_\tau := A - B\Phi(u(\tau))F$, $\delta = 1$, and $y_0 = w$, the positive eigenvector of $A_* = A - B\Phi(u^*)F$ corresponding to $r(A*) = 1$, the existence of which is ensured by [38, Theorem 11.5]. A careful argument using $A\tau \neq A*$ shows that $A\tau w \neq w$. Remark 2.12. The assumptions that we have made in formulating Proposition 2.11 are motivated by the class of Integral Projection Models (IPMs) where we wish the results to apply. An IPM is a discrete-time linear difference equation on the natural state space $L^1(\Omega)$ specified by integral operator

$$A: L^1(\Omega) \to L^1(\Omega), \quad (Av)(\xi) = \int_\Omega k(\zeta,\xi)v(\zeta)\,d\zeta\,, \quad v \in L^1(\Omega)\,, \text{ almost all } \xi \in \Omega,$$

for some nonnegative-valued kernel

$$\Omega \times \Omega \ni (\zeta,\xi) \mapsto k(\zeta,\xi) \geq 0\,, \tag{2.31}$$

where, for simplicity say, Ω is the closure of some bounded set in $\mathbb{R}^n$, $n \in \mathbb{N}$. IPMs were introduced by [47] (see also [48,

49, 50]) as a tool for population modelling where the discrete age- or stage-class variable of a matrix model is replaced by a continuous variable, such as stem width of a plant species. When $L^1(\Omega)$ is equipped with the usual partial order $f \leq g$ if $f(\xi) \leq g(\xi)$ for almost all $\xi \in \Omega$, then the nonnegativity assumption in (2.31) implies that A in (2.30) is a positive operator. Moreover, by [38, Theorem 2.1], (2.31) is sufficient to infer that A in (2.30) is bounded. Integral operators are known to be compact under quite general assumptions on k (for example, $k \in L^2(\Omega \times \Omega)$). It follows from the H older inequality that if there exist $\eta 1, \eta 2 \in L\infty(\Omega)$ with $0 \leq \eta 1, \eta 2 \neq 0$ and if there also exists $\nu \in L^1, 0 \leq \nu \neq 0,$, with

$$\eta_1(\zeta)\nu(\xi) \leq k(\zeta,\xi) \leq \eta_2(\zeta)\nu(\xi)\,, \quad \text{almost all } \zeta,\xi \in \Omega\,,$$

then A in (2.30) is ν-bounded (see [38, p. 96]).

Continuous-time. Here we demonstrate that the continuous-time counterpart of Theorem 2.3 holds. In continuous-time the adaptive control system (2.6) becomes

$$\left.\begin{aligned} \dot{x}(t) &= (A - B\Phi(u(t))F)x(t)\,, & x(0) &= x^0,\\ y(t) &= Cx(t)\,, & &\\ \dot{u}(t) &= \psi(\|y(t)\|)\,, & u(0) &= u^0,\end{aligned}\right\} \quad t \in \mathbb{R}_+\,,$$

where $\dot{x} = \frac{d}{dt}x$, $A \in \mathbb{R}^{n\times n}$ n is Metzler (with the definition recalled from the notation section in the introduction), $B \in \mathbb{R}_+^{n\times m}$, $F \in \mathbb{R}_+^{m\times n}$, and $C \in \mathbb{R}_+^{p\times n}$ for m, n, p ∈ N. To motivate our insistence that A in (2.32) is Metzler, we recall that for $M \in \mathbb{R}^{n\times n}$ the continuous-time linear system of differential equations

$$\dot{z}(t) = Mz(t)\,, \quad t \in \mathbb{R}_+\,, \quad z(0) = z^0 \tag{2.33}$$

has unique solution $z(t) = e^{Mt}z^0$ for $t \in \mathbb{R}_+$, where e^{Mt} denotes the matrix exponential of M t. It is well known that the solution z of the differential equation (2.33) is (componentwise) nonnegative for all $z^0 \geq 0$ if and only if M is a Metzler matrix. In the applied context of pest management, for a meaningful model we naturally require that the x component of a solution of (2.32) is nonnegative.

Returning to (2.32), as before the function ψ ∈ K and initial control effort $u^0 \geq 0$ are design parameters. For continuous-time models, $\Phi : R+ \rightarrow R^{m\times m}{}_+$ is still assumed to satisfy (A1), and the analogous assumption to (A2) on (A, B, F) and Φ is the following:

(A3) For all $\omega \in [0,\infty)$, the matrix $A - B\Phi(\omega)F$ is Metzler and irreducible and $\alpha(A - B\Phi_\infty F) < 0$, where $\Phi_\infty := \lim_{\omega\to\infty}\Phi(\omega)$.

The interpretation of the continuous-time pest model and control action in (2.32) is the same as that in (2.6). Unlike difference equations, when considering differential equations, existence of (global) solutions must be addressed. So as to not detract from the purpose and flow of the present section, the proof of the next lemma is postponed to the appendix. The main result of this section is Proposition 2.14, which mirrors Theorem 2.3.

Lemma 2.13. Under assumption (A3), for every $(x^0, u^0) \in \mathbb{R}_+^n \times \mathbb{R}_+$, there exists a solution (x, u) : $[0,\infty) \rightarrow \mathbb{R}_+^n \times \mathbb{R}_+$ *of* (2.32)

Proposition2.14.Giventheadaptivecontrolsystem(2.32)assume that Φ: $\mathbb{R} \rightarrow \mathbb{R}_+^{m\times m}$ *satisfies* (A1), $(A,B,F) \in \mathbb{R}^{n\times n} \times \mathbb{R}_+^{n\times m} \times \mathbb{R}_+^{m\times n}$ *and* Φ *satisfy* (A3), $\psi \in \mathcal{K}$, *and* $C \in \mathbb{R}_+^{p\times n}$, $C \neq 0$. *For positive* x^0 *and nonnegative* u^0, *let* (x,u) *denote a solution of* (2.32) *defined on* $[0,\infty)$. *Then, there exist* $M,\gamma > 0$ *such that* (2.32) *satisfies*

$$\|x(t)\| \leq Me^{-\gamma t}\|x^0\|\,, \quad t \in \mathbb{R}_+\,. \tag{2.34}$$

If additionally ψ has the property that

$$\forall\,\delta,\lambda > 0: \int_{t\in\mathbb{R}_+} \psi(\delta e^{-\lambda t})\,dt < \infty\,, \tag{2.35}$$

then u converges and its limit $u_\infty \geq 0$ (which depends on x^0, u^0, and ψ) is stabilizing, that is,

$$\alpha(A - B\Phi(u_\infty)F) < 0\,. \tag{2.36}$$

The proof of Proposition 2.14 is very similar to that of Theorem 2.3, mutatis mutandis, and hence is relegated to the appendix. Here we instead provide two remarks. The first contains further comments on adaptive control scheme (2.32) and compares and contrasts it with other simple adaptive controllers in the literature. The second contains motivation for our presentation of the continuous-time result.

Remark 2.15.

- We do not claim that solutions of (2.32) are unique. However, if ψ is locally Lipschitz continuous, then by, for example, [43, Theorem 4.18], (2.32) has a unique solution on [0, ∞), and furthermore, (2.35) is satisfied. The condition (2.35) on ψ ∈ K is the continuous analogue of the discrete condition (2.8) and is also satisfied by Holder continuous functions.
- The conclusions of Proposition 2.14 are similar to those of [31, Lemma 3.4] or [26, Theorems 3.5 and 3.11]. That said, there are two key differences, which means that in the context of positive state systems, Proposition 2.14 is (to the best of our knowledge) novel. First, in the aforementioned results it is required that BF (in our notation) has simple null structure (meaning that R^n = im (BF) ⊕ ker (BF)), which by [26, Proposition 3.9] is equivalent to F B being invertible and is an assumption that cannot be dropped in general [26, Example 3.10]. When B = b, F = f ∈ R^n, bf^T having simple null structure is equivalent to the triple (A, b, f^T) having relative degree one. This assumption is not required for Proposition 2.14 and is replaced by the irreducibility assumption in (A3). For example, the triple

$$A = \begin{bmatrix} -a_1 & a_2 \\ a_3 & -a_4 \end{bmatrix}, \quad B = \begin{bmatrix} 1 \\ 0 \end{bmatrix}, \quad F = \begin{bmatrix} 0 & a_2 \end{bmatrix},$$

where $a_i > 0$ for each i ∈ {1, 2, 3, 4}, and $\mathbb{R}_+ \ni x \mapsto \Phi(x) = 1 - e^{-x}$ together satisfy (A3), but BF does not have simple null structure because

$$\mathrm{im}(BF) = \ker(BF) = \left\{ \begin{bmatrix} \gamma \\ 0 \end{bmatrix} : \gamma \in \mathbb{R} \right\} \neq \mathbb{R}^2 .$$

Second, in Proposition 2.14, the measured variable y = C_x and the output feedback F_x used in (2.32) are not equal in general. This comment applies to the discrete-time system (2.6) as well.

- If $x^0 = 0$, then x(t) = 0 for all t ∈ R_+, and hence (2.34) trivially holds, but u(t) = u^0 for all t ∈ R+ and thus $u\infty = u^0$ need not satisfy (2.36). However, if $x^0 \neq 0$, then the limiting control effort u∞ satisfies (2.36) for every $u^0 \geq 0$. See also Remark 2.4(iv) It is known that a (continuous-time) high-gain adaptive controller applied to a linear model converges to a limit v∞, and v∞ is itself generically stabilizing [51, 52]. Still, even this assertion admits the case that there are infinitely many x^0 for which v∞ is not stabilizing. For nonlinear systems the limit v_∞ need not be stabilizing for an open set of initial conditions [53].
- High-gain adaptive controllers for componentwise nonnegative, continuoustime linear systems have been considered in [36, Chap. 15], particularly in a context of models for general anesthesia. Proposition 2.14 above is comparable to [36, Theorem 15.2]—the key difference being that in the latter the whole state is assumed available to inform the control strategy, so that (2.2) is replaced by the state-feedback $v(t) = -K(t)x(t)$, where K where K is to be determined adaptively, and C = I in (2.32). There the rate of change of adaptation ψ is not included, although the authors do consider adaptive stabilization to a nonzero state equilibrium [36, Theorem 15.1].

Remark 2.16. The motivation for presenting the continuous-time analogue (2.32) of (2.6) is twofold. The first is for mathematical completeness, to demonstrate that the corresponding continuous-time version of Theorem 2.3 holds. The second is that continuous-time models are used for population modelling as well as discrete-time models, and we refer the reader to [54, Chap. 1] or [55, Chap. 2] for examples. We comment that many continuous-time population models are in fact specified by partial or delay differential equations, such as the McKendrick–von Foerster partial differential equation [56, 57] or models for blowfly [58], respectively. These models fall beyond the scope of (2.32) and require more sophisticated mathematical treatment, but their semidiscretizations (leading to systems of ordinary differential equations) may result in models of the form (2.32). Thus (2.32) may be seen as a starting point for considering simple adaptive control for continuous-time models that arise in pest management.

A switching adaptive control mechanism for pest management. The approach taken to pest management so far in this paper is predicated on the assumption that the single management strategy $v = -B\Phi(u)F$ in (2.1) is capable of eradicating a pest once the control effort u is sufficiently large, formulated as assumption (A2). A high-gain–like adaptive controller for increasing u based on the measured output y is shown to achieve this task in Theorem 2.3. Here we consider the realistic situation wherein the control law v = −BΦ(u) F may not eradicate the pest, irrespective of how large the control effort u is: a biological possibility illustrated in the example below.

Example 3.1. A general 3 × 3 Lefkovitch [59] matrix has the structure

$$A = \begin{bmatrix} s_1 + f_1 & f_2 & f_3 \\ g_1 & s_2 & 0 \\ 0 & g_2 & s_3 \end{bmatrix} \geq 0 , \tag{3.1}$$

where f_i are recruitment rates into the population and s_i and g_i are rates of stasis within a stage-class and growth to the next stage-class, respectively. Suppose that a pest management strategy targets the third stage-class exclusively, so that $[0 \;\; 0 \;\; 1]^T$ (2.2). If

$$r\left(\begin{bmatrix} s_1 + f_1 & f_2 \\ g_1 & s_2 \end{bmatrix}\right) = r\left(\begin{bmatrix} s_1 + f_1 & f_2 & f_3 \\ g_1 & s_2 & 0 \\ 0 & 0 & 0 \end{bmatrix}\right) \geq 1 ,$$

then assumption (A2) fails for any F ≥ 0 and Φ satisfying (A1). In other words, controlling the third stage-class alone here cannot eradicate the pest. Motivated by the above example, in response we assume that access is available to several different pest management strategies, for example, targeting distinct life-cycle stages of a pest. Owing to the problematic uncertainty in both the pest dynamics and the efficacy of any given control strategy, we propose a so-called switching adaptive control scheme (also known as a universal adaptive control scheme). The design is in the spirit of M˚artensson [26, 27] or Helmke, Pr¨atzel-Wolters, and Schmid [60] and, as we proceed to demonstrate, cycles through potentially (a priori infinitely) many controllers (that is, courses of management action) before converging in finite-time to a stabilizing controller. We consider the switching system

$$t \in \mathbb{N}_0 \quad \begin{cases} x(t+1) = (A - B\mathscr{K}(s(t))F)x(t) , & x(0) = x^0 , \\ y(t) = Cx(t) , & \\ s(t+1) = s(t) + \|y(t)\| , & s(0) = s^0 . \end{cases}$$

(3.2a)

(3.2b)

(3.2c)

The function K in (3.2a) depends on two chosen sequences: the matrix sequence $K = (K(j))_{j \in \mathbb{N}} \subseteq \mathbb{R}^{m \times m}$ and the strictly increasing, unbounded sequence of positive numbers $\tau = (\tau(j))_{j \in \mathbb{N}_0} \subseteq [0, \infty)$ with $\tau(0) = 0$. The terms K, τ, and s^0 are all design parameters, and, once chosen, K is defined as

$$\mathbb{R}_+ \ni z \mapsto \mathscr{K}(z) := \begin{cases} K(1), & z = 0 , \\ K(j), & z \in (\tau(j-1), \tau(j)], \; j \in \mathbb{N} . \end{cases}$$

(3.2d)

We note that the assumptions τ(0) := 0, τ is strictly increasing, and τ is unbounded together imply that K (z) is well defined for all z ≥ 0. Given a solution (x, s) of (3.2a)–(3.2d), it is convenient to introduce the "counting" function i: $N_0 \rightarrow N$, defined as

$$\mathbb{N}_0 \ni t \mapsto i(t) = j \in \mathbb{N} \quad \text{such that} \quad s(t) \in (\tau(j-1), \tau(j)] , \tag{3.2e}$$

that records the controller used in (3.2a) at time t, that is,

$$\mathcal{K}(s(t)) = K(i(t)) \quad \forall t \in \mathbb{N}_0 .$$

It is readily seen that i is a nondecreasing function

Example 3.2.

(i) In an applied pest management context, the interpretations of x(t) in (3.2a) and y(t) in (3.2b) are as defined in (2.1) or (2.4)—they denote the pest population and some measured portion of the population, respectively. The scalar variable s(t) in (3.2a) and (3.2c) is an increasing switching variable, as it determines which control law (that is, management strategy) is applied at timestep t ∈ N, through the matrix function K and the sequence of switches τ.

(ii) To connect (3.2) with the single control strategy (2.6) proposed in section 2, here we might assume that

$$K(j) = \operatorname{diag}\{\phi_1(g_1(j)), \ldots, \phi_m(g_m(j))\}, \quad j \in \mathbb{N},$$

Where $\phi_j \in \mathcal{K}$ maps $\mathbb{R}_+ \to [0,1]$ and $g \subseteq \mathbb{R}_+^m$ is a sequence of control efforts with the property that

$$\limsup_{j\to\infty} g_k(j) = \infty \quad \forall k \in \{1, 2, \ldots, m\} . \tag{3.3}$$

For example, if three strategies are available that target stages one, two, and n > 2 of an n-stage model, then denoting the jth row of A by a^T_j, we have

$$B = [e_1 \ \ e_2 \ \ e_n], \ K(j) = \begin{bmatrix} \phi_1(g_1(j)) & 0 & 0 \\ 0 & \phi_2(g_2(j)) & 0 \\ 0 & 0 & \phi_3(g_3(j)) \end{bmatrix}, \ F = \begin{bmatrix} a_1^T \\ a_2^T \\ a_n^T \end{bmatrix}$$

Informally, the control scheme (3.2) tries control effort g(1) via the matrix K(1) for some length of time, determined by τ and the switching signal s. The control effort g(1) may be sufficient to stabilize the pest population. If not, then switching signal s must increase, and the control scheme (3.2) switches to g(2) and g(3) (although possibly bypassing some g(k) for k ∈ N altogether) and so on. The aim is to provide conditions on (3.2) and, in this example, the progressively "more effective" sequence g such that a stabilizing controller k_* ∈ N is reached but not switched from, meaning that $r(A - BK(k_*)F) < 1$ and s(t) is never greater than $\tau(k_*)$

Of course, as highlighted by Martensson [26, Remark 5, p. 41], cycling through infinitely many controllers is physically impossible. The interpretation of (3.2a), highlighted by Example 3.2, is that there are in fact only finitely many courses of management action, parameterized by an (infinite) sequence of control efforts that are "eventually increasing," in a sense such as in (3.3). As we demonstrate in our main result of this section, Theorem 3.5, under reasonable assumptions on (3.2), only finitely many control actions are required to stabilize the state to zero (in other words, eradicate the pest). As one might expect when the model parameters are unknown, the theorem does not say which control effort shall stabilize (3.2), only that one shall, and in finite-time. To that end, the first assumption is an underlying structural assumption: (B1) For each $j \in \mathbb{N}$, $A_j := A - BK(j)F$ is nonnegative and primitive with an index of primitivity ∈ N that is independent of j.

Intuitively, the interpretation of a common index of primitivity in (B1) is that the location of the zero/nonzero entries of A_j is independent of j (of course, the nonzero entries of A_j are expected to have values that do depend on j). The next result is a technical lemma that is a consequence of assumption (B1), and its proof is relegated to the appendix. Lemma 3.3. Given the switching system (3.2), assume that (A, B, F) and sequence K satisfy (B1)and that C, F ≠ 0, and let (x, s) denote the solution of (3.2). Then, there exists a positive, nondecreasing sequence β such that

$$\|Fx(t)\| \leq \beta(i(t))\|Cx(t)\| \quad \forall t \geq \ell . \tag{3.4}$$

Remark 3.4. The sequence β in (3.4) is not unique and may be bounded. For example, if m = p = 1 and

$$F = [f_1 \ \ \ldots \ \ f_n] \quad \text{and} \quad C = [c_1 \ \ \ldots \ \ c_n]$$

are both strictly positive, then evidently the constant sequence

$$\beta(j) := \frac{\max_k f_k}{\min_k c_k}, \quad j \in \mathbb{N}_0 ,$$

satisfies (3.4) for all t ∈ $\mathbb{N}_0$.

The second assumption pertains to the existence of stabilizing controllers: (B2) There exists R ∈ $R^{n\times n}_+$ with r(R) < 1 and a subsequence ρ ⊆ N such that $A_{\rho(j)} \leq R$ for all j ∈ N

The terms of the subsequence ρ enumerate the stabilizing controllers, and assumption (B2) implies that there are infinitely many such control schemes. The third assumption relates the C operator in (3.2c) to the triple (A, B, F) and sequence K and enables the conclusions of Theorem 3.5 below to hold under a weaker assumption on τ:

(B3) There exist $q \in \mathbb{N}$, $S \in \mathbb{R}_+^{n\times n}$ with $r(S) < 1$ and a bounded sequence $D \subseteq \mathbb{R}_+^{n\times p}$ such that for every $j \in \mathbb{N}$, $j \geq q$, $0 \leq A_j - D_jC \leq S$.

In other words, (B3) means that for every control strategy j after the qth there is a matrix D_j such that the output-feedback $-D_jy = -D_jCx$ stabilizes A_j.

Theorem 3.5. Given the switching adaptive control system (3.2), assume that (A, B, F) ∈ $R^{n\times n}_+ \times R^{n\times m}_+ \times R^{m\times n}_+$ and K satisfy (B1) and (B2), and that C ∈ $R^{p\times n}_+$ is nonzero. For a given strictly increasing, unbounded sequence τ, let (x, s) denote the solution of (3.2). For all positive x^0 and nonnegative s^0, if τ has the property that

$$\forall \lambda > 0, \quad \exists j_* \in \mathbb{N} : \frac{\tau(j)}{\tau(j-1)} \geq \lambda\beta(j-1) \quad \forall j \geq j_* , \tag{3.5}$$

where β is defined as in Lemma 3.3, then x(t) → 0 as t → ∞ and s converges in an interval (τ(r − 1), τ(r)] for some r ∈ N with $r(A_r) < 1$. If (B3) holds and τ has the property that

$$\frac{\tau(j)}{\tau(j-1)} \to \infty \quad as\ j \to \infty, \tag{3.6}$$

then the same conclusions hold. Proof. An argument similar to that used in the proof of Theorem 2.3 proves that s cannot converge in an interval $(\tau(j-1), \tau(j)]$ where $j \in \mathbb{N}$ is such that $r(A_j) \geq 1$.

If s converges in an interval $(\tau(\rho(j)-1), \tau(\rho(j))]$ for some $j \in \mathbb{N}$ where, we recall, the subsequence ρ is from assumption (B2), then there is nothing to prove. Therefore, it suffices to consider the case that $r \in \mathbb{N}$ and $T \in \mathbb{N}$ are such that

$$s(T-1) \leq \tau(\rho(r)-1) < s(T), \tag{3.7}$$

with $T \in \mathbb{N}$, $T = T(r) \geq l+2$. In other words, (3.7) means that $T \in \mathbb{N}$ is the first time-step when (3.2) has entered the stabilizing control scheme ρ(r). We rewrite the dynamics in (3.2a) as

$$\begin{aligned} x(t+1) &= (A - B\mathscr{K}(s(t))F)x(t) \\ &= (A - BK(\rho(r))F)x(t) + B[K(\rho(r)) - \mathscr{K}(s(t))]Fx(t) \\ &= A_{\rho(r)}x(t) + B[K(\rho(r)) - K(i(t))]Fx(t), \quad x(0) = x^0, \quad t \in \mathbb{N}_0. \end{aligned} \tag{3.8}$$

While $t \in \mathbb{N}$ is such that $s(T+t) \leq \tau(\rho(r))$, then $i(t) = \rho(r)$, and thus from (3.8)

$$0 \leq x(T+t) = A^t_{\rho(r)} A_{\rho(r)-1} x(T-1) \leq R^t A x(T-1), \tag{3.9}$$

by assumption (B2) and as $A_{\rho(r)-1} \leq A$. A telescoping series argument shows that

$$\begin{aligned} s(T+t) = s(T+t) - s(T-1) + s(T-1) &= \sum_{j=T-1}^{T+t-1} [s(j+1) - s(j)] + s(T-1) \\ &= \sum_{j=T-1}^{T+t-1} \|Cx(j)\| + s(T-1) \\ &\leq \sum_{j=0}^{t} \|Cx(T-1+j)\| + \tau(\rho(r)-1), \quad \text{by (3.7)}, \\ &\leq M_0\|x(T-1)\| + \tau(\rho(r)-1), \quad \text{by (3.9)}, \end{aligned} \tag{3.10}$$

for some constant $M_0 > 0$ that is independent of r and T . The variation of parameters formula applied to (3.8) yields that

$$x(T-1) = A^{T-1-\ell}_{\rho(r)} x(\ell) + \sum_{j=\ell}^{T-2} A^{T-2-j}_{\rho(r)} B[K(\rho(r)) - K(i(j))]Fx(j), \tag{3.11}$$

and hence

$$\|x(T-1)\| \leq \|A^{T-1-\ell}_{\rho(r)} x(\ell)\| + \left\| \sum_{j=\ell}^{T-2} A^{T-2-j}_{\rho(r)} B[K(\rho(r)) - K(i(j))]Fx(j) \right\|. \tag{3.12}$$

Assumption (B1) implies that for every $k_1, k_2 \in \mathbb{N}$, the operator $B(K(k_1)-K(k_2))|_{\mathrm{im}F}$ is bounded since

$$-Ax \leq B[K(k_1) - K(k_2)]Fx \leq Ax \quad \forall\, x \in \mathbb{R}^n_+$$

and as each $z \in \mathbb{R}^n$ may be expressed as $z = x_1 - x_2$ with $x_1, x_2 \in \mathbb{R}^n_+$. Consequently, the sequence with terms

$$A^t_{\rho(r)} B[K(\rho(r)) - K(i(t))]|_{\mathrm{im}F}, \quad t \in \mathbb{N}_0,$$

belongs to l^∞, and thus the second term on the right-hand side of (3.12) may be estimated by the Holder inequality, with complementary exponents $p = 1$ for the $F x(j)$ term and $q = \infty$ for the other. We infer that

$$\begin{aligned} \|x(T-1)\| &\leq c_1 + c_2\|Fx\|_{\ell^1(\ell,T-2)} \leq c_1 + c_2\beta(i(T-2))\|Cx\|_{\ell^1(\ell,T-2)} \\ &\leq c_1 + c_2\beta(\rho(r)-1)\|Cx\|_{\ell^1(\ell,T-2)} \end{aligned} \tag{3.13}$$

for some constants $c_1, c_2 > 0$, where to obtain the estimate (3.13) we have used (3.7) to note that

$$s(T-2) \leq s(T-1) \leq \tau(\rho(r)-1) \quad \Rightarrow \quad i(T-2) \leq \rho(r) - 1,$$

and the property that β is nondecreasing. Another telescoping sequence argument for s applied to (3.13) demonstrates that

$$\|x(T-1)\| \le c_1 + c_2\beta(\rho(r)-1)\big[s(T-1)-s(\ell)\big] \le c_1 + c_2\beta(\rho(r)-1)s(T-1)$$

$$\le c_1 + c_2\beta(\rho(r)-1)\tau(\rho(r)-1), \quad \text{by} \tag{3.14}$$

Substituting the estimate (3.14) into (3.10) yields constants d_1, $d_2 > 0$ that are independent of r and T such that

$$s(T+t) \le d_1 + d_2\beta(\rho(r)-1)\tau(\rho(r)-1)\,, \tag{3.15}$$

valid, we recall, for all $t \in N$ such that $s(T+t) < \tau(\rho(r))$. Dividing both sides of (3.15) by $\tau(\rho(r))$ yields that

$$\frac{s(T+t)}{\tau(\rho(r))} \le \frac{d_1}{\tau(\rho(r))} + \frac{d_2\beta(\rho(r)-1)\tau(\rho(r)-1)}{\tau(\rho(r))}, \tag{3.16}$$

and thus, by assumption (3.5), there exists $r \in N$ such that the right-hand side of (3.16) is no greater than one. Therefore, $s(T+t) \le \tau(\rho(r))$ for all $t \in N_0$, and hence s is bounded and thus convergent (in a stabilizing interval), as required. It follows now from (3.9) that $x(t) \to 0$ as $t \to \infty$. Now suppose that (B3) holds. Recapping, we know that s cannot converge in an interval $(\tau(j-1), \tau(j)]$ where $j \in N$ is such that $r(A_j) \ge 1$. If s converges in $(\tau(j-1), \tau(j)]$ with $j \le q$, then there is nothing to prove. Therefore, it suffices to consider $r, T \in N$ such that (3.7) and, additionally, that

$$s(T-2) > \tau(q)\,. \tag{3.17}$$

The inequality (3.17) means that at each time-step $t \ge T-2$, a control law greater than the qth is being applied, and thus assumption (B3) holds. We fix $t_* \in N$, $t_* < T-2$, as the first time-step at which the qth control law was applied, so that

$$s(t_*-1) \le \tau(q-1) < s(t_*) < \tau(q)\,,$$

and note that t_* is independent of r and T. The argument deriving (3.10) is as before and uses (3.7). Letting $A_j := A_j - D_jC \ge 0$ for $j \ge q$, we note that for $t \ge t_*$

$$s(t_*-1) \le \tau(q-1) < s(t_*) < \tau(q)\,,$$

and hence

$$x(T-1) \le S^{T-1-t_*}x(t_*) + \sum_{j=t_*}^{T-2} S^{T-2-j}D_{i(j)}Cx(j)\,. \tag{3.18}$$

Since the sequence D is assumed bounded, $r(S) < 1$, and $t_* \in N$ is fixed, we estimate (3.18) in a similar manner to (3.11)–(3.14) to obtain

$$\|x(T-1)\| \le d_3 + d_4\|Cx\|_{\ell^1(\ell,T-2)} \le d_3 + d_4 s(T-1) \le d_3 + d_4\tau(\rho(r)-1) \tag{3.19}$$

for constants d_3, $d_4 > 0$. Combining (3.10) and (3.19), it follows that

$$s(T+t) \le d_5 + d_6\tau(\rho(r)-1)\,, \tag{3.20}$$

valid for all $t \in N_0$ such that $s(T+t) < \tau(\rho(r))$. Dividing both sides of (3.20) by $\tau(\rho(r))$ yields that

$$\frac{s(T+t)}{\tau(\rho(r))} \le \frac{d_5}{\tau(\rho(r))} + d_6\frac{\tau(\rho(r)-1)}{\tau(\rho(r))}\,. \tag{3.21}$$

The growth property (3.6) of τ implies that the right-hand side of (3.21) is no greater than one for $r \in N_0$ sufficiently large. Therefore, $s(T+t) \le \tau(\rho(r))$ for all $t \in N_0$, and hence s is bounded and thus convergent (in a stabilizing interval), as required. The proof concludes in the same manner as when (B2) holds.

Remark 3.6.

- In section 2, the adaptation law for u in (2.6) included a K function ψ. The role of ψ is to control the rate of increase of adaptation, which corresponds to the rate of increase of pest control effort. The switching variable s in (3.2c) does not contain such a function. Instead, the rate of switching between strategies is determined by the sequence τ—the faster τ grows, in principle, the longer each control strategy is persisted with before switching.
- The key difference between the switching adaptive controllers considered here and those elsewhere in the literature (such as [26, 27] or [60]) is that in general $C \neq F$ here, and so the output F x(t) driving the state in (3.2a), via the feedback $-K_i F\, x(t)$, is not equal to the measured output $C_x(t)$, the variable which is driving the switching signal s(t). However, as we have demonstrated, a consequence of the assumed (and realistically nonrestrictive) primitivity is that there is a coupling between F x(t) and $C_x(t)$ which is sufficient to stabilize (3.2), provided that τ grows sufficiently fast, that is, (3.5) holds

Example. We illustrate the present results using a stage-structured model from [61] of the economically important citrus pest Diaprepes root weevil (DRW; Diaprepes abbreviatus). The species is native to Caribbean islands and arrived in Florida, U.S., in the 1960s [62]. Females deposit egg clusters on the leaves of hostplants. Upon hatching, the larvae drop to the soil surface and burrow underground, where they feed on roots for several months until they pupate. The root damage caused by the larvae can lead to plant death or decline to an unproductive state. In the field it is challenging to study developmental rate and survival of DRW since larvae and pupae live underground, and eggs are hard to find because of their small size (1.2 mm long and 0.4 mm wide), making the estimates of model parameters highly uncertain. Populations are generally monitored via traps that capture adults emerging from the soil. Recommended insecticide treatments are targeted for specific insect stages.[2] For instance, adults are killed with foliar insecticide applications (e.g., fenpropathrin), egg hatching can be prevented with diflubenzuron applied with oil, and larvae attempting to burrow into the soil are killed with a chemical soil barrier (imidacloprid or thiamethoxam). An alternative control strategy are the biocontrol nematodes (Heterorhabditis indica, Steinernema riobrave) that attack and kill larvae underground. However, nematodes are not persistent because as soon as they kill most of the larvae in an area, they die from a lack of food. Hence, they are similar to insecticide application in the sense that they need to be reapplied and nematode dynamics need not be considered here.

A matrix model for DRW is given in [61], with six stage-classes denoting, in increasing order, eggs, two larval stages, pupae, and two adult stages. The timesteps in this model denote weeks. We refer the reader to [61] for more details. The projection matrix A [61, Table 2] has uncertain entries, but each belongs to the intervals in [61, Table 1]. We let A denote the set of all DRW matrices. Every $A \in A$ satisfies $r(A) > 1$. In the following simulations, we numerically generate $A \in A$ and initial condition x^0 with $\|x^0\| = 4000$. We assume that total adult DRW abundance is measured, so that

$$C := \begin{bmatrix} 0 & 0 & 0 & 0 & 1 & 1 \end{bmatrix}$$

Pest management strategies for DRW include the chemical pesticides described above, with a candidate substance for the egg, larval, and adult stages, as well as entomopathogenic nematodes, which attack the larval stage-classes. Let e_j and a^T_j denote the jth standard basis vector in R^6 and the jth row of A, respectively. The effect of the pest control targeting egg, larval, and adult stages may be respectively modelled by (2.4) with

$$B_1 = e_1, \quad B_2 = \begin{bmatrix} e_2 & e_3 \end{bmatrix}, \quad B_3 = \begin{bmatrix} e_5 & e_6 \end{bmatrix}$$
$$\text{and} \quad F_1 = a_1^T, \quad F_2 = \begin{bmatrix} a_2^T \\ a_3^T \end{bmatrix}, \quad F_3 = \begin{bmatrix} a_5^T \\ a_6^T \end{bmatrix}. \tag{4.1}$$

The efficacy functions Φ in (2.4) are in reality unknown but, as always, are assumed to satisfy (A1). In modelling Φ there is a trade-off between biologically motivated explanations and phenomenological solutions. For instance, in the present setting, if we assume that locally DRW larvae are distributed according to a Poisson distribution, and death via nematodes or insecticide application is random and independent, then $e^{-k_1 u}$ is the probability of a DRW larva surviving nematode attack or insecticide application. In the case of nematodes, $k_1 > 0$ is the search efficiency (or "area of discovery"), and u is the number of nematodes applied. In the case of insecticides, the probability of surviving a chemical application is also $e^{-k_1 u}$, in which case u is the chemical concentration, and $k_1 > 0$ is some lethality term. In light of the above discussion, we shall assume that the efficacy function Φ is given by

$$\mathbb{R}_+ \ni u \mapsto \Phi(u) = 1 - e^{-0.6u}. \tag{4.2}$$

More generally, when $\Phi: R_+ \to R^{m\times m}_+$ is such that $\lim_{\omega \to \infty} \Phi(\omega) = I$, the $m \times m$ identity matrix, then it is straightforward to verify that the triple (A, B_j, F_j) satisfies assumption (A2) for each $A \in A$ and $j \in \{1, 2, 3\}$. We first consider control targeting the egg stage-class, so that $B = B_1$ and $F = F_1$. We consider three different choices $\psi_1, \psi_2, \psi_3 \in K$ given by

$$y \mapsto \psi_1(y) = 0.05y^{\frac{2}{3}}, \quad y \mapsto \psi_2(y) = \frac{y}{1+y}, \quad y \mapsto \psi_3(y) = \ln(1+y). \tag{4.3}$$

Recall that the ψ_j functions control the rate of increase of the control effort u(t) in (2.6). Here ψ_1 is a power law with the small scaling parameter 0.05 chosen for ease

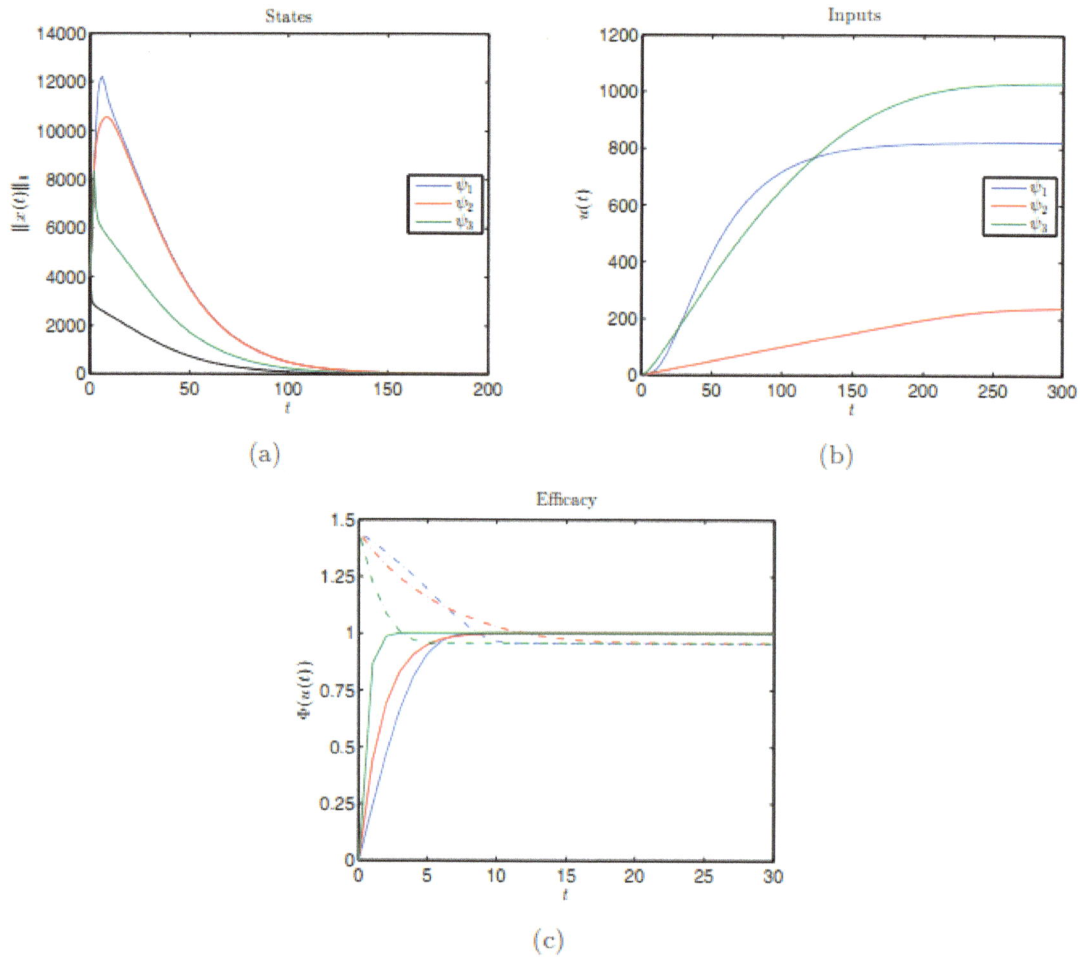

Fig. 1. Simulations of the adaptive control scheme (2.1) applied to the DRW projection model for different choices of ψ in (4.3).

of comparison with ψ_2 and ψ_3. The function ψ_2 (sometimes called a Beverton–Holt or Holling type II nonlinearity) is a bounded function, and ψ_3 is a logarithm, which is unbounded but grows rather slowly. Figure 1 contains simulations of the adaptive control scheme (2.6) with $u^0 = 0$. Recall that in this simulation, and throughout, each time-step denotes one week. So 100 time-steps is roughly two years. As expected from Theorem 2.3, for each ψ_j the state x converges to zero, as seen in Figure 1(a). The control efforts u(t) are plotted in Figure 1(b), and we see that they also converge to a finite limit. Of more relevance is Figure 1(c), which shows the efficacy $\Phi(u(t))$ (solid lines) and the growth rates $r(A - B\Phi(u(t))F)$ (dashed-dotted lines). It is observed that for each ψ_j, u(t) rapidly reaches a level whereby $\Phi(u(t)) \approx 1$, but that the limiting growth rate $r(A - B\Phi(u(t))F) \approx 0.95 < 1$, which provides a lower bound for the rate of decline. By way of comparison, the black line in Figure 1(a) denotes the solution of

$$x(t+1) = (A - B_1F_1)x(t), \quad x(0) = x^0, \quad t \in \mathbb{N}_0,$$

which provides a lower bound in this example for every solution of (2.6) and would be achieved if from the outbreak of infestation the first stage-class was targeted with a strategy that was "100% efficacious." Therefore, in this example ψ_1 and ψ_3 evidently use too great a control effort. We comment, however, that these conclusions may be made only retrospectively (and with knowledge of the to-be-controlled model). Furthermore, the limiting rate of decline ≈ 0.95 of the pest may be deemed too large and motivate pest management strategies that target other stage-classes. The power of Theorem 2.3 is not fully exhibited in Figure 1, as there the matrix A is fixed (but random). To demonstrate the important robustness of the adaptive control scheme (2.6) to model uncertainty, Figure 2 contains simulations of the DRW projection model for 30 pseudorandomly generated projection matrices $A \in \mathcal{A}$ and the efficacy parameter 0.6 from (4.2) replaced by a pseudorandomly drawn number from the interval [0.6, 0.8]. Biologically, these variations may be a consequence of variable environmental factors, such as temperature, sunlight, or rainfall, varying soil quality, or citrus tree health. In each simulation $u_0 = 10$ and $\psi = \psi_2$, and it is observed that the same control strategy achieves stabilization of the state (Figure 2(a)). We reiterate that assumption (A2) holding is crucial for the success of (2.6), yet (A2) does not require explicit knowledge of A, B, F, or Φ. Note that across the simulations the control effort increases approximately linearly and is approximately equal for the first 150–180 time-steps. This is a consequence of the choice of $\psi = \psi_2$ which has two properties: (a) ψ_2 is bounded by one, and (b) $\psi_2(100) = 0.9901 \approx 1$. As such, the choice of ψ_2 does not particularly distinguish large arguments, and so while y(t) ≥ 100—the first 150 time-steps or so—each control effort u grows similarly. Figure 3 contains simulations of the adaptive control scheme with rate limitation (2.18). For variation here we simulated the effects of targeting the larval stageclass, so that $B = B_2$ and $F = F_2$ in (4.1). We have chosen the same efficacy function Φ as in (4.2) but with the parameter 0.6 replaced by 0.05 (so that the control effort is

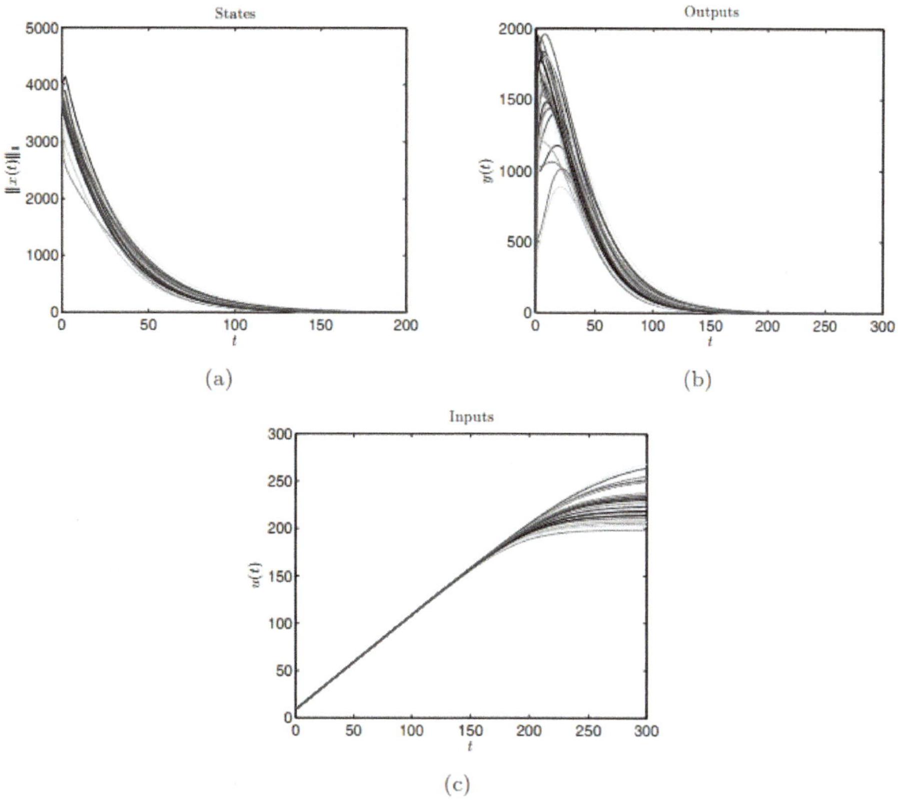

Fig. 2. Simulations of the adaptive control scheme (2.1) applied to the DRW model for 30 DRW projection matrices.

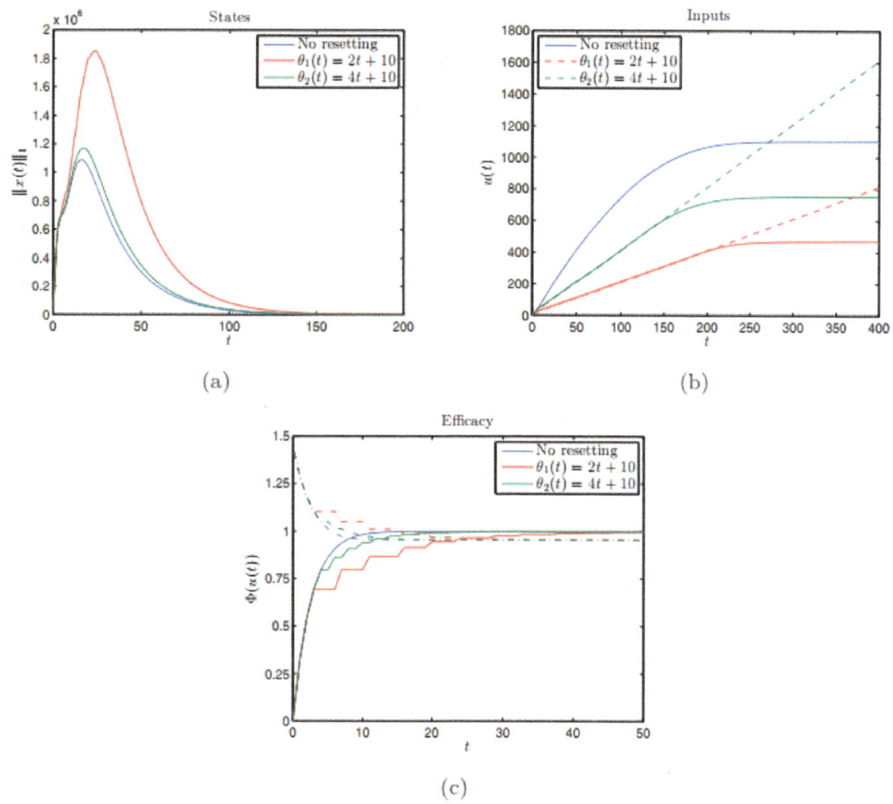

Fig. 3. Simulations of the rate-limited adaptive control scheme (2.18) applied to the DRW projection model for different choices of θ.

hardly efficacious). Our somewhat carefree approach to parameter selection is in part to illustrate the argument that they are not essential for our main results to hold. Here we have chosen u0 = 10, fixed $\psi = \psi_3$ from (4.3), and compared the results of the adaptive control scheme (2.6) (blue curve) with those of the rate-limited adaptive control system (2.18) for two different choices of θ of the form θ(x) = ax + b; see figure legend (red and green curves). Both of the rate-limited adaptive control schemes achieve stabilization of the state (Figure 3(a)) at lower asymptotic control effort than that of (2.6) (Figure 3(b)). We cautiously comment that this need not hold in general, however. Figure 3(c) plots the efficacy $\Phi(u(t))$ (solid lines) and growth rates $r(A-B\Phi(u(t))F)$ (dashed-dotted lines). We see in Figure 3(a) that the choice $\theta = \theta_1$ leads to the largest transient growth, in part, we suspect, as the resulting growth rate $r(A-B\Phi(u(t))F)$ in Figure 3(c) is greater than one for longer periods than with θ_2 or with no θ. However, the difference in performance between θ_2 and the unconstrained model (2.6) is negligible, which would motivate using the saturation rate θ_2 in this application. The examples considered thus far have all invoked control schemes that result in asymptotic eradication of the pest. A consequence of the model (2.1), with state x denoting the structured pest population, is that the pest abundance is predicted to grow again once the control action is stopped. This is a limitation with assuming an unstable linear model for the pest where the basin of attraction of the zero equilibrium is zero alone. In applications one might expect (or hope) that once pest abundance is

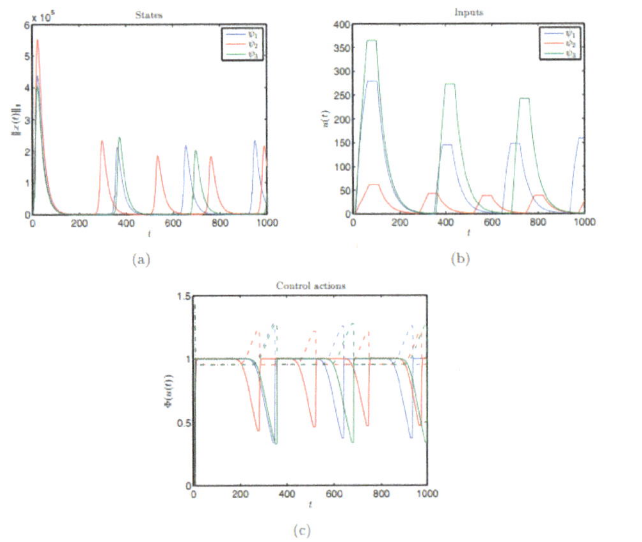

Fig. 4. Simulations of the neutral-zone adaptive control scheme (2.19) applied to the DRW projection model.

reduced to a sufficiently low level, then it would not recover, which would be captured by, say, a nonlinear model with zero as an asymptotically stable equilibrium. Allee effects (see, for example, [63]) provide a biological explanation for why sufficiently small populations may not survive. Simple adaptive controllers for nonlinear pest models is beyond the scope of the present contribution, and instead Figure 4 contains simulations of the neutral-zone adaptive control scheme (2.19), which, we recall, seeks to reduce the (cumulative) control effort by reducing u(t) when the observed pest abundance y(t) is low. Specifically, for the simulation in Figure 4 we choose in (2.19)

$$B = B_2, \quad F = F_2, \quad L_1 = 300, \quad L_2 = 50, \quad \gamma = 0.975, \quad u^0 = 0,$$

and efficacy function Φ given by (4.2) (with parameter 0.6). Recall that the choices of the thresholds L_1, L_2 and decay-rate in control effort γ are the choices of the modeller, and in practice would be influenced by ecological and economic criteria. In this example, we observe in Figure 4(a) that each state trajectory x is bounded, as expected from Proposition 2.7. To illustrate the dynamics of the control scheme (2.19), Figure 4(b) plots u(t) against t, while Figure 4(c) plots $\Phi(u(t))$ (solid lines) and $r(A-B\Phi(u(t))F)$ (dashed-dotted lines) both against t.

As u(t) increases, so does $\Phi(u(t))$, and thus $r(A-B\Phi(u(t))F)$—the asymptotic growth rate of the pest—decreases. However, once y(t) is sufficiently small, u(t) decreases and the process reverses. Broadly speaking, when $r(A - B\Phi(u(t))F) > 1$, the pest abundance increases, and the cycle repeats. From Figure 4(b) we see that the bounded function ψ_2 requires considerably less

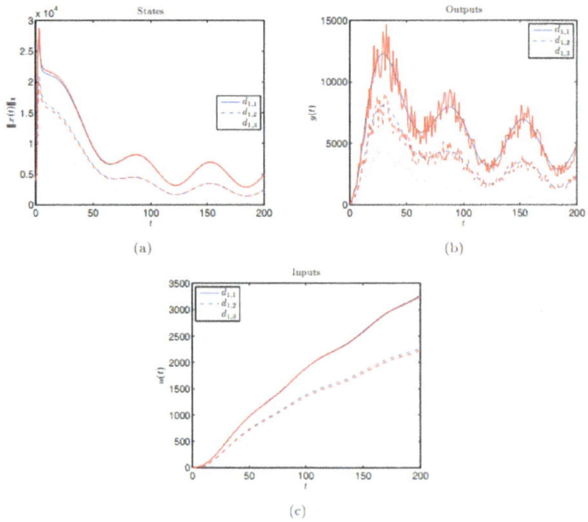

Fig. 5. Simulations of the adaptive control scheme with additive disturbances (2.20) applied to the DRW projection model.

cumulative effort than the other two strategies yet maintains total pest abundance comparable to that of the other choices of ψ. The simulations in Figure 5 illustrate the performance of the adaptive control scheme with additive disturbances (2.20). The motivation for such a model, we recall, is to explore the effect of previously unmodelled dynamics on (2.6), which could denote immigration, emigration when the state is disturbed, or sampling error when the output is disturbed. We choose $\psi = \psi_1$ in (4.3), and Φ is still given by (4.2). We assume that d_1 denotes periodic immigration into the adult stage-classes, with three different amplitudes, so that

$$d_{1,j}(t) = k_j \begin{bmatrix} 0 & 0 & 0 & 0 & 1 & 1 \end{bmatrix}^T (1 + \sin(0.1t)), \quad t \in \mathbb{N}_0, \quad k_j \in \{100, 50, 10\}. \tag{4.4}$$

The blue curves in Figure 5 are generated by (2.20) with $d_2 = 0$ (no measurement error) and $d_1 = d_{1,k}$ as above. The red curves are generated by (2.20) with d1 as in (4.4) and

$$d_2 = \varepsilon(t)Cx(t), \quad t \in \mathbb{N}_0,$$

with $\varepsilon(t)$ (pseudo)randomly drawn from a truncated uniform distribution between −0.2 and 0.2; that is, a proportional measurement error of up to 20% is made at each time-step. In Figure 5(a) the ISS estimate (2.22) is exhibited: the norm of the state is bounded and decreases, asymptotically linearly, with decreasing $\|d_1\|_\infty$. The trajectories in Figures 5(a) and 5(c) may be compared to those in Figure 1 (blue curves), where the same model is projected with no disturbances. We note that, apart from on the outputs themselves, the effects of the output errors d_2 (red curves) are seemingly much smaller than those of d_1 (different line styles). As an illustration of the theory, our final example considers cycling between control strategies that target different stage-classes, with an increasing effort over time, and thus uses the switching adaptive control scheme (3.2). We set

$$B := \begin{bmatrix} B_1 & B_2 & B_3 \end{bmatrix}, \; F := \begin{bmatrix} F_1^T & F_2^T & F_3^T \end{bmatrix}^T,$$
$$K(j) := \operatorname{diag}\{\phi_1(g_1(j)), \phi_2(g_2(j)), \phi_2(g_2(j)), \phi_3(g_3(j)), \phi_3(g_3(j))\}, \quad j \in \mathbb{N}, \tag{4.5a}$$

where φ_i is the efficacy function of control action j ∈ {1, 2, 3} and $g \subseteq R^3_+$ is the predetermined sequence of control efforts

$$g(3j) = \begin{bmatrix} g_1(3j) \\ g_2(3j) \\ g_3(3j) \end{bmatrix} = 1.2 \begin{bmatrix} j \\ 0 \\ 0 \end{bmatrix}, \quad g(3j+1) = 1.2 \begin{bmatrix} j \\ j \\ 0 \end{bmatrix}, \quad g(3j+2) = 1.2 \begin{bmatrix} j \\ j \\ j \end{bmatrix}, \quad j \in \mathbb{N}_0$$

In other words, if at a given starting point the egg stage-class is targeted (g(3j) in (4.5b)), then this course of action is persevered with until the switching variable s(t) reaches the next switch τ(3j). In this case, the management action switches to target both the egg and larval stage-classes (g(3j + 1) in (4.5b)). This course of management action is persisted with until the switching variable reaches the next switch τ(3j+1). As before, if this occurs, then the management action switches to the third strategy in which egg, larval, and adult stage-classes are all targeted (g(3j + 2) in (4.5b)). Should the switching variable s(t) reach the next switch τ(3j + 2), then the above cycle repeats, but with a subsequently larger control effort. The assumptions (B1) and (B2) hold, by construction, for (A, B, F) and K for every A ∈ A and B,F, and K given by (4.5). Moreover, assumption (B3) holds with $D = (D_j)_{j \in N}$ given by

$$0 \le D_j \le \begin{bmatrix} a_{1,5} & 0 & 0 & 0 & 0 & a_{6,5} \end{bmatrix}^T, \quad j \in \mathbb{N},$$

such that $0 \le A - BK_jF - D_jC$ has (6, 5)th entry equal to zero. For the following numerical simulation we take $\psi = \psi_3$ from (4.3) and φ_1, φ_2, and φ_3 in (4.5a) given by (4.5c)

$$x \mapsto \phi_1(x) = 1 - e^{-0.5x}, \quad x \mapsto \phi_2(x) = \tanh(0.6x), \quad x \mapsto \phi_3(x) = \frac{0.8x}{\sqrt{1 + (0.8x)^2}},$$

which, we comment, have been chosen somewhat arbitrarily to illustrate the theory. Theorem 3.5 holds for the model (4.5), and hence we expect eradication of the pest for any sequence τ that satisfies (3.6). Figure 6 contains simulations of the switching adaptive control system (3.2), with model data (4.5), for five random initial states x^0, $s^0 = 1/2$, and τ given by

$$\tau(t+1) := \ln(1+t)\tau(t), \quad t \in \mathbb{N}, \quad t(0) = 0, \; \tau(1) = 4,$$

which evidently satisfies (3.6) as

$$\frac{\tau(t+1)}{\tau(t)} = \ln(1+t) \to \infty \quad \text{as } t \to \infty.$$

Recall that the sequence τ determines the rate at which the control strategies are changed or, in other words, how long a control strategy is persevered with before changing. Figure 6(a) demonstrates that each state trajectory converges to zero over time— much less time than in Figure 1(a), for example, where only the egg stage-class is targeted. Figure 6(b) plots the switching signal s(t) against t. We see that each signal s converges and, moreover, converges in either (τ(7), τ(8)] or (τ(10), τ(11)]. As expected, these intervals are stabilizing inasmuch as $A_8 := A - BK(8)F$ has $r(A_8)=0.7720 < 1$ and $r(A_{11})=0.7719 < 1$.

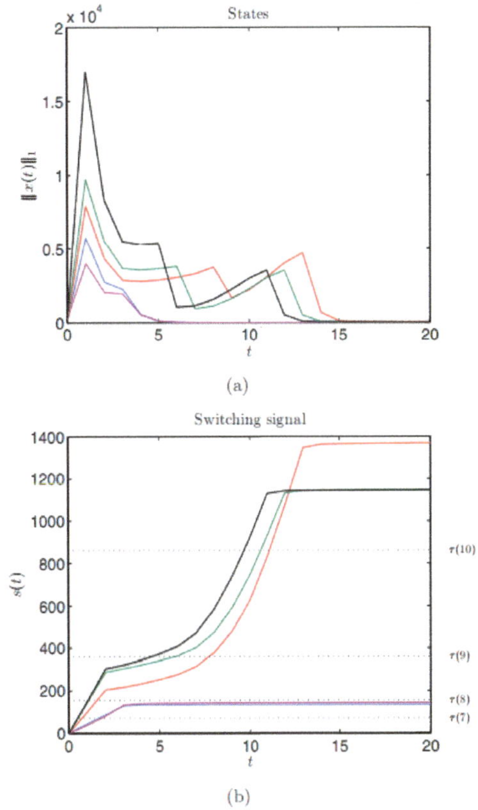

Fig. 6. Simulations of the switching adaptive control scheme (3.2) applied to the DRW projection model.

DISCUSSION

Simple adaptive control has been considered for discrete-time positive state linear systems, motivated by potential applications in pest management. Pest management is a timely and hugely important societal challenge that shall significantly contribute to the future success of economically and environmentally viable agriculture and horticulture. Several different adaptive control schemes have been proposed as strategies for reducing pest abundance. We believe that adaptive controllers are ideally suited for such a task for two reasons. Foremost is the (theoretical) ability, and guarantee, of adaptive control schemes to achieve their objectives in spite of considerable anticipated uncertainty in the pest dynamics. Recall that (A1) and (A2) are the crucial assumptions in section 2, which cover a wide range of matrix models and do not require specific knowledge of A, B, F, or Φ. To recap, if A, B, F, and Φ are known with certainty, then it is possible to choose (the possibly smallest) ˆu such that (2.5) holds, which would also ensure that the constant control effort u(t)=ˆu for every t ∈ N eradicates the pest population over time. The caveat in the previous sentence being how certain is "known with certainty." The second reason, and still of importance, is the ease of computation of management strategies determined by simple adaptive control schemes. We reiterate that finding optimal controls is not necessarily straightforward, either analytically or computationally. A downside of adaptive control schemes is that they are usually conservative— they are not optimal control strategies— which is the price paid for the additional robustness to model uncertainty. In other words, simple adaptive control as presented here trades off a loss of optimality against a gain in robustness. What this means in practice is that the adaptive control law may use more control effort than is required, which is costlier both economically and environmentally. In practical applications, however, the conservatism of adaptive control schemes must be traded off against the economic and social costs of not sufficiently reducing pest abundance—a possible consequence of a supposed optimal management strategy not functioning as intended owing to a lack of robustness, say to model uncertainty. As we have argued elsewhere [64], when seeking to control uncertain population models, open-loop control strategies, that is, those based on (estimated) model parameters and not on measured variables, may not achieve the control objective, whereas robust feedback controls do. The same is true of certain optimal control strategies; see [64] and the references therein. Although borrowing from the tradition of simple, nonidentifier-based adaptive control [30, 31], as a consequence of the naturally assumed componentwise nonnegativity there are mathematical differences between the material presented here and the situations usually considered (such as in [26] or [31]). The most striking difference is which structural assumptions are sufficient for global asymptotic stability of the closed-loop feedback system. For example, prescribed relative degree, minimum phase, and/or minimality may be replaced by irreducibility of the closed-loop system, specifically assumption (A2) (or (A3) in continuous-time). We have sought to explain these structural differences in more detail in Remarks 2.4, 2.15, and 3.6. Another crucial difference is that in typical adaptive control scenarios, the measured output $y = Cx$ is the variable used in the feedback law (here v in (2.2)) to subsequently control the state. In pest management applications, there is no biological reason to assume that the effect of the pest control action (which must depend on the current state) is proportional to the measured population, meaning that presently the feedback takes the form $v = -B\Phi(u)Fx$, with $F \neq C$ in general. Our first result, Theorem 2.3, was extended in different directions, including a rate-limited adaptive control scheme, Proposition 2.5, and a neutral-zone adaptive control scheme, Proposition 2.7. Both of these schemes have been designed to seek to reduce the control effort which, as we mentioned above, may be unnecessarily conservative. The extension of Theorem 2.3 to a class of infinite-dimensional, discrete-time systems was straightforward for Banach spaces with partial orders and positive operators where monotonicity of the spectral radius (2.28) holds. A class of such operators was described in Remark 2.12. We suspect that the analogous continuous-time result, Proposition 2.14, would extend to classes of infinite-dimensional systems where A in (2.32) denotes the generator of a positive operator semigroup (on an ordered Banach space) and the analogous monotonicity of the growth bound holds. The upshot of this seems to be that the additional technical difficulties (see [65]) encountered in establishing high-gain adaptive control in infinite dimensions may again be replaced by an irreducibility or primitivity assumption (understood appropriately). In the context of pest management, there are several extensions that we seek to address in future work. First, as an alternative to chemical pesticide use, we desire strategies for the adaptive deployment of biocontrol agents [66]. Because they are living organisms, with life-cycles that occupy a variety of different time-scales relative to the pest, the dynamics of biocontrol agents should be modelled, as well as those of the pest. There are also likely to be interactions between competing biocontrol agents [67], as well as between the biocontrol and pest populations. Second, the model (2.1) does not directly incorporate a spatial component, the inclusion of which would model the spread of a pest population spatially as well as its change in abundance temporally. Indeed, it is argued that the stages of ecological invasion are (broadly) arrival (or establishment or colonization), spread, and then community level changes [68, 69]. Much attention has been devoted in the mathematical ecology literature to the spread of an invasive population; see, for example, [70] and the references therein. In effect, here we have considered adaptive control for local pest management, where the in situ pest population is well specified by (2.1). There are many cases where management is focused on local control, typically when organized by individual farmers or stakeholders, and the present description is useful. There is also interest in how local adaptive control and observation affect the global properties of the pest in a spatial model, such as abundance and dispersal. In closing, it is our hope that techniques from adaptive control theory shall be added to the toolbox of available pest management strategies and motivate further research from both mathematicians and ecologists. The joint efforts of both groups shall undoubtedly be required in the future modelling of effective pest control strategies to help tackle the crucial issue of food security. As we have sought to explain, there are many possible extensions to the work presented here: the present paper is a starting point.

Appendix.

Proof of Proposition 2.7. We assume that $r(A) \geq 1$; else the result is trivially true. The proof of Theorem 2.3 yields the following property for (2.19):

$$\forall T \in \mathbb{N},\ \exists t_1, t_2 \in \mathbb{N},\ t_1, t_2 \geq T \quad \text{such that} \quad u(t_1) < u_* \quad \text{and} \quad u(t_2) > u_*, \tag{A.1}$$

Where

$$u_* > 0 \quad \text{is such that} \quad r(A - B\Phi(u_*)F) = 1. \tag{A.2}$$

In other words, u cannot eventually be greater than or less than $u*$ as in these regions the state (and thus the output) is exponentially decreasing or increasing, respectively. Arguing in a similar vein,

$$\forall\, T \in \mathbb{N}\,,\ \exists\, t_3, t_4 \in \mathbb{N}\,,\ t_3, t_4 \geq T \quad \text{such that} \quad \|y(t_3)\| < L_1 \quad \text{and} \quad \|y(t_4)\| \geq L_1\,. \tag{A.3}$$

Seeking a contradiction, suppose that u is unbounded. Then, there exists a subsequence $(\tau_t)_{t\in\mathbb{N}_0}$ such that $\tau_t \nearrow \infty$ as $t \to \infty$ and

$$u(\tau_t) \geq t\,, \quad t \in \mathbb{N}_0\,. \tag{A.4}$$

In light of (A.1) and (A.4), for $t \in \mathbb{N}$, let $U_t := \{p_t, \dots, q_t\}$, sets of successive integers, be such that

(i) $\tau_t \in U_t$;
(ii) $u > u_*$ on U_t;
(iii) $\bigcup_{k\in\mathbb{N}} U_k \neq \mathbb{N}$.

We record the fact that for $t \in \{0, 1, \dots, \tau_t - p_t\}$

$$x(t + p_t) \leq A_*^t x(p_t) \leq A_*^t A^r x(p_t - r)\,, \quad r \in \mathbb{N}\,. \tag{A.5}$$

For $t \in \mathbb{N}$ we estimate

$$\begin{aligned} t - u_* \leq u(\tau_t) - u(p_t - 1) &= \sum_{j=p_t-1}^{\tau_t-1} u(j+1) - u(j) \leq \sum_{j=p_t-1}^{\tau_t-1} \psi(\|y(j)\|) \\ &\leq \sum_{j=0}^{\tau_t-p_t} \psi(\|C\|\|x(j+p_t-1)\|) \leq \sum_{j=0}^{\tau_t-p_t} \psi(M_*(r)\gamma^j\|x(p_t-r)\|) \\ &\leq \sum_{j\in\mathbb{N}_0} \psi(M_*(r)\gamma^j\|x(p_t-r)\|) \end{aligned} \tag{|A.6}$$

for each $r \in \mathbb{N}$ and some constant $M_*(r)$, where we have used (A.5). However, from the growth condition (2.8) of ψ, it follows from (A.6) that for each $r \in \mathbb{N}$

$$\|x(p_t - r)\| \to \infty \quad \text{as } t \to \infty. \tag{A.7}$$

Letting $A(t) := A - B\Phi(u(t))F$, we note that

$$C\prod_{j=1}^{k} A(j)x(p_t - k - r) = y(p_t - r)$$

$$\Rightarrow \quad \left\| C\prod_{j=1}^{k} A(j)x(p_t - k - r)\right\| = \|y(p_t - r)\|\,, \quad k, r \in \mathbb{N}. \tag{A.8}$$

By the assumption of a common period of A(t), (A.7) implies that k may always be chosen bounded so that the left-hand side of (A.8) is arbitrarily large. However, by (A.3), there exists $r \in \mathbb{N}$ such that the right-hand side of (A.8) is bounded, which is a contradiction.

We conclude that u is bounded (say by $v < \infty$), and thus x satisfies the bounds

$$A_v x(t) := (A - B\Phi(v)F)x(t) \leq x(t+1) \leq Ax(t)\,, \quad t \in \mathbb{N}_0\,. \tag{A.9}$$

Clearly, if x is bounded, then so is y. Conversely, from the irreducibility of A_v and A, and as C ≠ 0, it follows from (A.9) that if x is unbounded, then so is y, and furthermore, for each L > 0 and t ∈ N

$$\|y(t)\| \leq L \quad \Rightarrow \quad \|x(t)\| \leq M(L) < \infty . \tag{A.10}$$

Seeking another contradiction, we now suppose that x is unbounded, so equivalently that y is unbounded. Similarly to our arguments for u, such a supposition implies the existence of $(\rho_t)_{t\in\mathbb{N}_0}$ such that $\rho_t \nearrow \infty$ as $t \to \infty$ and

$$\|y(\rho_t)\| \geq t, \quad t \in \mathbb{N}_0 . \tag{A.11}$$

In light of (A.3) and (A.11), let $Y_t := \{v_t,...,w_t\}$, sets of successive integers, be such that

(i) $\rho_t \in Y_t$;
(ii) $\|y\| > L_1$ on Y_t;
(iii) $\bigcup_{k\in\mathbb{N}} Y_k \neq \mathbb{N}$.

In particular, the construction of Y_t is such that

$$\|y(v_t - 1)\| \leq L_1 \quad \forall t \in \mathbb{N} \tag{A.12}$$

(else Y_t may be enlarged and relabelled). For t ∈ N, $t \geq L_1$ we first estimate

$$\begin{aligned} t - L_1 &\leq \|y(\rho_t)\| - \|y(v_t - 1)\| \leq \|y(\rho_t) - y(v_t - 1)\| \\ &\leq \|C\| \cdot \|A^{\rho_t-(v_t-1)} + I\| \cdot \|x(v_t - 1)\| \leq N(L_1)(r(A))^{\rho_t - v_t} \end{aligned} \tag{A.13}$$

for some constant N, where we have used (A.12), (A.10), and the fact that r(A) > 1. The inequality (A.13) necessitates that

$$\rho_t - v_t \to \infty \quad \text{as } t \to \infty. \tag{A.14}$$

By property (ii) above, u increases on Y_t, and furthermore, for $t \geq v_t$

$$u(t+1) = u(v_t) + \sum_{j=v_t}^{t} \psi(\|Cx(j)\|) \geq (t - v_t)\psi(L_1) . \tag{A.15}$$

As ψ ∈ K and $L_1 > 0$, so that $\psi(L_1) > 0$, there exists k ∈ N such that

$$k\psi(L_1) > u_* , \tag{A.16}$$

where u_* is as defined in (A.2). Combining (A.14)–(A.16), we see that for sufficiently large t ∈ N

$$u(t) > u_\dagger > u_*, \quad t \in \{v_t + k + 1, \ldots, \rho_t\} \subseteq Y_t .$$

However, now we may estimate

$$\begin{aligned} \|x(\rho_t)\| &\leq \|A_\dagger^{(\rho_t-(v_t+k+1))} x(v_t + k + 1)\| \leq \|A_\dagger^{(\rho_t-(v_t+k+1))} A^{k+2}\| \cdot \|x(v_t - 1)\| \\ &\leq P(L_1) r(A_\dagger)^{\rho_t - v_t} \to 0 \quad \text{as } t \to \infty \end{aligned}$$

by (A.14), as $A_\dagger := A - B\Phi(u_\dagger)F \leq A*$ has $r(A_\dagger) < 1$. Therefore, $\|x(\rho_t)\| \to 0$ as $t \to \infty$. which contradicts (A.11). We conclude that y (and thus also x) is bounded. Proof of Lemma 2.13. Existence of solutions of (2.32) on [0, τ) for some τ > 0 follow from standard Peano existence theory. Suppose that a maximal solution (x, u) is defined on [0, ω) for some ω > 0. We proceed to show that $(x([0,\omega)), u([0,\omega)))$ is bounded and thus relatively compact. For $t \in [0,\omega)$, $x(t)$ is given by

$$x(t) = e^{\int_0^t (A - B\Phi(u(s))F)\,ds} x^0 =: e^{H(t)} x^0 ,$$

and as Φ is bounded, a straightforward estimate shows that $\|H(t)\| \leq Mt$ for some M > 0. Hence,

$$\|x(t)\| \leq e^{Mt}\|x^0\| \leq e^{M\omega}\|x^0\| , \quad t \in [0, \omega) ,$$

from which we estimate for t ∈ [0, ω)

$$0 \leq u(t) = u^0 + \int_0^t \psi(\|Cx(s)\|)\,ds \leq u^0 + \omega\psi(e^{M\omega}\|C\| \cdot \|x^0\|) .$$

By [43, Corollary 4.10], we conclude that ω = ∞, as required. Proof of Proposition 2.14. Assuming that α(A) ≥ 0 (else the result is trivial), let u∗ > 0 be such that $A_* := A - B\Phi(u_*)F$ has $\alpha(A_*) = 0$. Such a u_* exists by assumption (A3), (2.10), and as Φ is continuous. Seeking a contradiction, if u is such that u(t) ≤ u∗ for all t ∈ R_+, then as $t \mapsto \frac{d}{dt}u$ is nonnegative, u is nondecreasing and bounded from above, and thus u converges to z ≤ u∗. Consequently, by the fundamental theorem of calculus, for T ≥ 0

$$0 \leq \int_0^T \psi(\|y(s)\|)\,ds = \int_0^T \dot{u}(s)\,ds = u(T) - u(0) \to z - u^0 < \infty \quad \text{as } T \to \infty,$$

and so $t \mapsto \psi(\|y(t)\|) \in L^1.$ On the other hand, $u \leq u_*$ implies that

$$\dot{x}(t) = (A - B\Phi(u(t))F)x(t) \geq (A - B\Phi(u_*)F)x(t) , \quad t \in \mathbb{R}_+ ,$$

and hence

$$\begin{aligned} x(t) \geq e^{A_* t}x^0 , \ t \in \mathbb{R}_+ \quad &\Rightarrow \quad y(t) \geq Ce^{A_* t}x^0 , \ t \in \mathbb{R}_+ \\ &\Rightarrow \quad \liminf_{t\to\infty} y(t) \geq \liminf_{t\to\infty} Ce^{A_* t}x^0 . \end{aligned} \tag{A.17}$$

The combined properties that C and x^0 are nonnegative and nonzero and A_* is irreducible with $\alpha(A_*) = 0$ yield that

$$\liminf_{t\to\infty} Ce^{A_* t}x^0 > 0 ,$$

which, in light of (A.17) and the fact that y is continuous and ψ ∈ K, contradicts $t \mapsto \psi(\|y(t)\|) \in L^1$. We conclude that there exists τ ∈ N such that

$$u(t) \geq u(\tau) > u_* , \quad t \in \mathbb{R}_+ , \ t \geq \tau ,$$

as u is nondecreasing. The proof of the estimate (2.34) now mirrors that of Theorem 2.3, instead noting that for Metzler A_1, A_2 ∈ $R^{n\times n}$, $A_2 \leq A_1$, $A_1 \neq A_2$, and A_2 irreducible imply that $\alpha(A_2) < \alpha(A_1)$. To prove that u converges, we prove that it is bounded. To that end, for t ∈ R+

$$0 \le u(t) = u(0) + \int_0^t \dot{u}(s)\, ds = u^0 + \int_0^t \psi(\|y(s)\|)\, ds \le u^0 + \int_0^t \psi(\|C\| M e^{-\gamma s} \|x^0\|)\, ds$$
$$\le u^0 + \int_{\mathbb{R}_+} \psi(\|C\| M e^{-\gamma s} \|x^0\|)\, ds < \infty$$

by (2.34) and (2.35), completing the proof.

Proof of Lemma 3.3. The integer in assumption (B1) is such that

$$\prod_{k=1}^{\ell} A_{j_k} \gg 0 \quad \forall \{j_1, \dots, j_\ell\} \subseteq \mathbb{N}.$$

Therefore, as

$$F \neq 0, C \neq 0, \text{ it follows that for } t \geq \ell \qquad \text{(A.18)}$$

$$x(t) = \prod_{j=0}^{t} A_{i(j)} x^0 \quad \Rightarrow \quad \|Fx(t)\|, \|Cx(t)\| > 0 \quad \Rightarrow \quad \|Fx(t)\| \le \gamma(i(t)) \|Cx(t)\| .$$

with

$$\gamma(i(t)) := \frac{\|Fx(t)\|}{\|Cx(t)\|} > 0 , \quad t \geq \ell .$$

Defining

$$\beta : \mathbb{N} \to \mathbb{R}, \quad \beta(j) := \sup_{k \le j} \gamma(k) ,$$

it follows from (A.18) that (3.4) holds and β is evidently non-decreasing.

REFERENCES

1. H. C. J. Godfray and T. Garnett, Food security and sustainable intensification, Philos. Trans. R. Soc. Lond. B Biol. Sci., 369 (2014), 20120273.
2. U.S. EPA, Pesticides Industry Sales and Usage 2006 and 2007 Market Estimates, http://www. epa.gov/pesticides/pesticides-industry-sales-and-usage-2006-and-2007-market-estimates.
3. D. Pimentel, H. Acquay, M. Biltonen, P. Rice, M. Silva, J. Nelson, V. Lipner, S. Giordano, A. Horowitz, and M. D'amore, Environmental and economic costs of pesticide use, BioScience, 42 (1992), pp. 750–760.
4. E.-C. Oerke, Crop losses to pests, J. Agr. Sci., 144 (2006), pp. 31–43.
5. D. Pimentel, Pest management in agriculture, in Techniques for Reducing Pesticide Use: Economic and Environmental Benefits, D. Pimentel, ed., John Wiley & Sons, Chichester, UK, 1997, pp. 1–11.
6. H. J. Barclay, Models for pest control using predator release, habitat management and pesticide release in combination, J. Appl. Ecol., 19 (1982), pp. 337–348.
7. M. Hassell and R. May, Generalist and specialist natural enemies in insect predator-prey interactions, J. Anim. Ecol., 55 (1986), pp. 923–940.
8. Hastings, R. J. Hall, and C. M. Taylor, A simple approach to optimal control of invasive species, Theor. Popul. Biol., 70 (2006), pp. 431–435.
9. J. Blackwood, A. Hastings, and C. Costello, Cost-effective management of invasive species using linear-quadratic control, Ecol. Econ., 69 (2010), pp. 519–527.
10. K. Shea and H. P. Possingham, Optimal release strategies for biological control agents: An application of stochastic dynamic programming to population management, J. Appl. Ecol., 37 (2000), pp. 77–86.
11. R. G. Haight and S. Polasky, Optimal control of an invasive species with imperfect information about the level of infestation, Resour. Energy Econ., 32 (2010), pp. 519–533.
12. Chad`es, T. G. Martin, S. Nicol, M. A. Burgman, H. P. Possingham, and Y. M. Buckley, General rules for managing and surveying networks of pests, diseases, and endangered species, Proc. Natl. Acad. Sci. USA, 108 (2011), pp. 8323–8328.
13. R. Bellman, Dynamic Programming, Princeton University Press, Princeton, NJ, 1957.
14. W. B. Powell, Approximate Dynamic Programming: Solving the Curses of Dimensionality, Wiley Ser. Probab. Stat. 703, John Wiley & Sons, Hoboken, NJ, 2007.
15. H. M. Regan, M. Colyvan, and M. A. Burgman, A taxonomy and treatment of uncertainty for ecology and conservation biology, Ecol. Appl., 12 (2002), pp. 618–628.
16. B. K. Williams, Uncertainty, learning, and the optimal management of wildlife, Environ. Ecol. Stat., 8 (2001), pp. 269–288.
17. J. Doyle, Guaranteed margins for LQG regulators, IEEE Trans. Automat. Control, 23 (1978), pp. 756–757.
18. G. Zames, Feedback and optimal sensitivity: Model reference transformations, multiplicative seminorms, and approximate inverses, IEEE Trans. Automat. Control, 26 (1981), pp. 301– 320.
19. M. French, Adaptive control and robustness in the gap metric, IEEE Trans. Automat. Control, 53 (2008), pp. 461–478.
20. Y. D. Landau, Adaptive Control: The Model Reference Approach, Control Systems Theory 8, Marcel Dekker, New York, 1979.
21. V. G. Sragovich, Mathematical Theory of Adaptive Control, Interdiscip. Math. Sci. 4, World Scientific, Hackensack, NJ, 2006.
22. Astolfi, D. Karagiannis, and R. Ortega, Nonlinear and Adaptive Control with Applications, Comm. Control Engrg. Ser., Springer-Verlag, London, 2008.
23. I. Byrnes and J. C. Willems, Adaptive stabilization of multivariable linear systems, in Proceedings of the 23rd

IEEE Conference on Decision and Control, 1984, pp. 1574–1577.

24. J. C. Willems and C. I. Byrnes, Global adaptive stabilization in the absence of information on the sign of the high frequency gain, in Analysis and Optimization of Systems, Part 1 (Nice, 1984), Lecture Notes in Control and Inform. Sci. 62, Springer, Berlin, 1984, pp. 49–57.
25. Mareels, A simple self-tuning controller for stably invertible systems, Systems Control Lett., 4 (1984), pp. 5–16.
26. B. M˚artensson, Adaptive Stabilization, Ph.D. thesis, Lund Institute of Technology, Lund, Sweden, 1986.
27. B. M˚artensson, Adaptive stabilization of multivariable linear systems, in Differential Geometry: The Interface between Pure and Applied Mathematics, Contemp. Math. 68, AMS, Providence, RI, 1987, pp. 191–225.
28. S. Morse, Recent problems in parameter adaptive control, in Mathematical Tools and Models for Control, Systems Analysis and Signal Processing, Vol. 3 (Toulouse/Paris, 1981/1982), Travaux Rech. Coop. Programme 567, CNRS, Paris, 1983, pp. 733–740.
29. R. D. Nussbaum, Some remarks on a conjecture in parameter adaptive control, Systems Control Lett., 3 (1983), pp. 243–246.
30. Ilchmann, Non-Identifier-Based High-Gain Adaptive Control, Lecture Notes in Control and Inform. Sci. 189, Springer-Verlag, London, 1993.
31. Ilchmann, Non-identifier-based adaptive control of dynamical systems: A survey, IMA J. Math. Control Inform., 8 (1991), pp. 321–366.
32. S. Holling, ed., Adaptive Environmental Assessment and Management, John Wiley & Sons, Chichester, UK, 1978.
33. Walters, Adaptive Management of Renewable Resources, Macmillan, New York, 1986.
34. K. Williams, Adaptive management of natural resources—Framework and issues, J. Environ. Manage., 92 (2011), pp. 1346–1353.
35. H. R. Heinimann, A concept in adaptive ecosystem management—An engineering perspective, Forest Ecol. Manag., 259 (2010), pp. 848–856.
36. W. M. Haddad, V. Chellaboina, and Q. Hui, Nonnegative and Compartmental Dynamical Systems, Princeton University Press, Princeton, NJ, 2010.
37. Berman, M. Neumann, and R. J. Stern, Nonnegative Matrices in Dynamic Systems, John Wiley & Sons, New York, 1989.
38. M. A. Krasnoselskij, J. A. Lifshits, and A. V. Sobolev, Positive Linear Systems: The Method of Positive Operators, Heldermann Verlag, Berlin, 1989.
39. Berman and R. J. Plemmons, Nonnegative Matrices in the Mathematical Sciences, SIAM, Philadelphia, 1994.
40. L. Farina and S. Rinaldi, Positive Linear Systems: Theory and Applications, WileyInterscience, New York, 2000.
41. Guiver, D. Hodgson, and S. Townley, Positive state controllability of positive linear systems, Systems Control Lett., 65 (2014), pp. 23–29.
42. Pimentel, Environmental and economic costs of the application of pesticides primarily in the United States, Environ. Dev. Sustain., 7 (2005), pp. 229–252.
43. H. Logemann and E. P. Ryan, Ordinary Differential Equations. Analysis, Qualitative Theory and Control, Springer, London, 2014.
44. H. Caswell, Matrix Population Models: Construction, Analysis, and Interpretation, Sinauer Associates, Sunderland, MA, 2001.
45. Stott, S. Townley, D. Carslake, and D. Hodgson, On reducibility and ergodicity of population projection matrix models, Methods Ecol. Evol., 1 (2010), pp. 242–252.
46. D. Sontag, Input to state stability: Basic concepts and results, in Nonlinear and Optimal Control Theory, Lecture Notes in Math. 1932, Springer, Berlin, 2008, pp. 163–220.

CHAPTER 13

SURVEY OF NORTHEASTERN HOP ARTHROPOD PESTS AND THEIR NATURAL ENEMIES

Lilian B. Calderwood, Scott A. Lewins, and Heather M. Darby

Keywords: *Humulus lupulus*, hop aphid, potato leafhopper, two spotted spider mite, integrated pest management

Commercial hop (*Humulus lupulus* L.) production has not occurred in the northeast region of the United States for 150 yr. Vermont production peaked in 1860 when the state produced 289,690 kg of dried hops (**Kennedy 1860**). A combination of the spread of hop downy mildew (*Pseudoperonospora humuli*) (Miyabe and Takah G. W. Wilson), the expansion of production in western states, and the passing of prohibition laws later in the 1920s contributed to the decline of the 19th-century northeast hop industry. Today, the Pacific Northwestern states of Washington, Oregon, and Idaho remain the dominant hop production states. However, hop production in nontraditional regions is growing and in 2013 accounted for 2% of the total U.S. hop acreage (**Hop Growers of America 2014**). Over the past 5 yr, this specialty crop has resurged in the northeast, fueled by the local food movement. In Vermont alone, >40 breweries are seeking local ingredients including but not limited to hops. As acreage increases, regionally specific pest challenges have become apparent and growers look to extension services for relevant scouting and management strategies.

Integrated pest management (IPM) programs are crop, pest, and region specific. Arthropod communities are a reflection of climate, landscape, and management practices (**Schweiger et al. 2005**). There are clear differences in climate and landscape between the traditional Pacific Northwestern hop-growing region and the rejuvenated northeast. Therefore, it is important that northeast hop IPM is developed to respond to regional arthropod community abundance and phenology. Twospotted spider mite (*Tetranychus urticae* Koch) (Acari: Tetranychidae) and hop aphid (*Phorodon humuli* (Schrank)) (Hemiptera: Aphididae) have been documented and researched extensively as economically damaging arthropod pests in the Pacific Northwest and European hop-growing regions (**James 2003a,b**; **Mahaffee et al. 2009**; **Woods et al. 2014**). To our knowledge, arthropod communities in hop yards were last reported for the northeast region in the 1940s as reports from the Cornell Agricultural Experiment Station that document hop aphid, twospotted spider mite, and potato leafhopper (*Empoasca fabae* (Harris)) (Hemiptera: Cicadellidae) as major arthropod pests in New York hop yards (**Magie 1944**).

Communities of predators and parasitoids can contribute to pest management in agricultural landscapes (**Symondson et al. 2002**). Assemblages of generalist and specialist natural enemy arthropods reduce pest abundance in hop yards (**Campbell and Cone 1994**; Gardiner et al. 2003; **James 2003a,b**; **Gent et al. 2009a,b**; **Grasswitz and James 2009, 2011**; **Woods et al. 2014**).

In an effort to provide northeast hop growers and stakeholders with current pest management strategies, it was important to first identify predominant pests and subsequent natural enemies in Vermont hop yards. The objectives of this publication were to 1) present current hop pest biology and management tactics, 2) report the phenology of arthropod pests observed over three growing seasons (2012–2014), 3) report abundance and peak date for each pest, and 4) document natural enemy abundance.

Citation: Lilian B. Calderwood, Scott A. Lewins, and Heather M. Darby, Survey of Northeastern Hop Arthropod Pests and Their Natural Enemies, DOI: 10.1093/jipm/pmv017.

BIOLOGY AND MANAGEMENT OF MAJOR NORTHEASTERN HOP PESTS

Twospotted Spider Mite

Twospotted spider mites overwinter in the crown of hop plants, woody debris, and trellis pole crevasses as diapausing adult females. As soon as temperatures warm in the spring, females emerge, migrate to new growth, feed, and lay up to 16 eggs per day. Five to eight generations of spider mite are observed during the hop-growing season in the Pacific Northwest (**Mahaffee et al. 2009**). Twospotted spider mites are a midlate season pest that thrives in hot and dry environmental conditions. Spider mites develop from egg to adult in 7–10 d at 28–30°C (**Mahaffee et al. 2009**). Scouting for spider mites on the underside of leaves is an important practice given the potential for rapid population increase in the right conditions (**Weihrauch 2004**).

Symptoms

Twospotted spider mites are small and translucent. Eggs, larvae, and nymphs are white to clear in color while adults can appear white to yellow in color with two dark spots on the back (**Fig. 1**). Females are the largest life stage of this mite with a length of ~0.5 mm. Males are ~0.2 mm in length. This pest feeds with piercing–sucking mouthparts on leaf and cone mesophyll cells. The first sign of damage is "stippling," pin-prick sized brown spots on leaves (**Fig. 2**). Adults and eggs first appear on the underside of leaves in the space between leaf veins. As the population grows, webbing will appear and brown stippling will expand, eventually turning whole leaves brown as a result of desiccation. Foliar injury reduces plant vigor while spider mite feeding on cones can drastically reduce the marketability of the hop product via desiccation which causes cones to shatter (**Mahaffee et al. 2009**).

Fig. 1. Twospotted spider mite, *Tetranychus urticae* Koch, adults, nymphs, and eggs.

Fig. 2. Characteristic twospotted spider mite "stippling" damage at a moderate mite infestation level.

Management

Pesticide application is the traditional management tactic used to control twospotted spider mite populations below economically damaging levels in hop production. Spray applications are based on economic thresholds ranging from 1–2 mites per leaf in June and 5–10 mites per leaf in July in Washington State (**Strong and Croft 1995**) to 60 mites per leaf in German hop yards (**Wright and Cone 1999**, **Weihrauch 2004**). Economically damaging levels of twospotted spider mite can be a repercussion of spraying broad-spectrum insecticides and fungicides (**James 2002, 2003a,b**; **Gent et al. 2009a,b**; **Woods et al. 2012**). Twospotted spider mite is a pest of hops whether broad-spectrum insecticides are sprayed or not. However, economic damage to hop yield and quality from this pest is seen with the application of pesticides aimed at reducing other arthropod pests. For example, spring application of imidacloprid is a successful method of reducing hop aphid populations, but natural enemies of hop aphid and twospotted spider mite are also eliminated. This provides an opportunity for secondary pests (twospotted spider mite) to reach uncontrollable outbreak levels (**James and Price 2002**; **James 2003a,b**). Hormoligosis, increased female fecundity, has been observed in spider mites when exposed to imidacloprid insecticides (**Sclar et al. 1998**, **James and Price 2002**, **Ako et al. 2004**). Resident natural enemy communities are able to regulate spider mite populations in the absence of pesticides on hops (**Huffaker et al. 1969**, **James et al. 2001**, **Woods et al. 2014**). Due to twospotted spider mite product resistance, research on spider mite pest management in commercial production is evolving toward the conservation of antagonistic arthropods and narrow-spectrum acaricides (**James 2003a,b**).

Hop cultivars have different susceptibility levels to twospotted spider mites and high farnesol concentration in leaves has been suggested as a possible mechanism behind their preference (**Regev and Cone 1975**). **Peters and Berry (1980)** report variation in the density of leaf trichomes on different hop cultivars. It was observed that twospotted spider mite development time was higher on leaves with more dense leaf trichomes (**Peters and Berry 1980**).

Spider mite populations increase in dry, dusty conditions. Ground cover between hop rows increases humidity, reduces dusty conditions, and therefore reduces favorable twospotted spider mite habitat. Flowering vegetation increases moisture and provides shelter and alternative food resources for predators of twospotted spider mite (**Grasswitz and James 2009**, **Lu et al. 2014**). Spider mites can be controlled by an assemblage of natural enemy arthropods, viruses, and pathogens (**Jeppson et al. 1975**); *Stethorus punctum* spp. (Coleoptera: Coccinellidae) (**James 2003a**), *Orius* spp. (Hemiptera: Anthocoridae), and predatory mites (Mesostigmata: Phytoseiidae) have been identified as effective predators of this pest commonly found in hop yards (**Grasswitz and James 2009**).

Damson-Hop Aphid

Cool, wet growing seasons are favorable for the damson-hop aphid. Hop aphid alternates between hop as the summer host and *Prunus* spp. (Rosales: Prunaceae) as winter hosts. Hop aphids have a holocyclic life cycle where adult, winged females (alate) give birth to live nymphs which develop into wingless (apterous) males and females in addition to alate female reproductive adults. Hop aphids overwinter as eggs on *Prunus* spp. and adult, wingless females hatch in the spring, laying up to four generations of apterous females and alates (**Mahaffee et al. 2009**). Recently hatched alatae fly from*Prunus* spp. to hop in early spring when temperatures reach 13°C. Up to 10 generations of apterous hop aphid on hop per season have been observed in the Pacific Northwest depending on weather conditions and management practices. Toward the end of the season, apterous females produce alates which return to *Prunus* spp. in the fall. They lay alate males and females which subsequently lay overwintering eggs (**Wright et al. 2005**, **Mahaffee et al. 2009**).

In Spain, peak hop aphid date is typically in mid-July (**Lorenzana et al. 2013**), while in Washington State peak populations are observed in late August and continue into early September (**Campbell and Cone 1994**). The abundance of alternate *Prunus* spp. hosts in European farm landscapes has been observed to increase the number of migrant female aphids into hop yards in comparison to migrant populations in Washington state where there are fewer surrounding *Prunus* spp. (**Campbell and Cone 1994**). While foliar feeding can reduce plant productivity at high population levels, the moist habitat that hop aphids provide for sooty mold to grow in cones, reduces hop yield and quality.

Symptoms

These soft bodied, pear-shaped insects are found on the underside of hop leaves and range in color from white to light green. Hop aphids are often found on the upper and lower leaf surfaces of new hop growth (**Fig. 3**). Hop aphids have piercing–sucking mouthparts which are used to feed on the phloem of hop plants. On hop, aphids are smaller than on *Prunus* spp, the winter host plant. Hop aphids secrete a sugary substance commonly referred to as "honeydew" that provides the perfect medium for growth of sooty mold, an Ascomycete fungus, to grow in hop cones (**Wright et al. 2005**, **Mahaffee et al. 2009**,**Lorenzana and Hermoso 2010**). With high aphid population levels during cone development and at harvest time, fuzzy looking grey-black colored mold can be found when hop cones are pulled apart. Sooty mold is the indirect impact that aphids have on hop quality (**Fig. 4**).

Fig. 3. Hop aphids on new hop growth.

Fig. 4. Hop cone infested with hop aphids and sooty mold.

Management

Hop aphid population ecology has been studied, and economic thresholds have been determined (**Campbell 1977**, **Wright et al. 2005**, **Lorenzana et al. 2009**). A study conducted in Spain recommends holding off insecticides application against hop aphid until an economic threshold of 8–10 hop aphids per leaf is reached (**Lorenzana et al. 2009**). An economic threshold specific to the northeast region for hop aphids does not exist. Biological control of hop aphid including the release of parasitoids (**Wright and James 2001**) and predators (**Campbell and Cone 1994**) has been described as an effective management tool. Hop aphids are reported to thrive on plant parts highest in nitrogen and in hop yards with ample nitrogen application (**Gent et al. 2009a**). Hop plants are fast growing and reach a height of 5 m in one growing season (Neve 1991). The large flush of new vegetative growth in hop yards serves as a food source for hop pests and aphids in particular.

Hop cultivars vary in their susceptibility to hop aphid (**Campbell 1983**, **Dorschner and Baird 1988**,**Weihrauch and Moreth 2005**). 'Cascade' has been reported as highly susceptible to hop aphid in the Pacific Northwest (**Dorschner and Baird 1988**). The mechanisms behind the variation in hop aphids by cultivar are likely a combination of plant nutrition and leaf chemistry. Cultivars with high essential oil content are more attractive to hop aphids. In particular, high levels of cariophyllene and farnesene have been highlighted as attractive essential oils to hop aphids (Kralj et al. 1998).

Potato Leafhopper

Potato leafhopper has not been reported or studied on hops since the 1940s (**Magie 1944**). A native to the northeast, this pest has been researched in other northeastern production systems including snap bean (*Phaseolus vulgaris*) (**Nault et al. 2004**), alfalfa (*Medicago sativa*) (**Lamp et al. 2007**), maple (*Acer* spp.) (Frank et al. 2007), and grape (*Vitis vinifera*) (**Lenz et al. 2012**). Potato leafhopper is a polyphagous, migratory insect with an appetite for >200 broad leaf plants (**DeLong 1971**). Adult females can arrive to northern Vermont anytime between May and August, making it an unpredictable pest. Adults overwinter in southeastern states as far north as Virginia (**Taylor and Shields 1995**). Spring wind currents carry this species of leafhoppers north. Typically, adult females arrive to Midwestern states first and northeastern states last (**Sidumo et al. 2005**). Upon arrival, females feed and lay eggs in hop leaf and stem tissue. Potato leafhoppers can develop at temperatures between 10 and 24°C (**Simonet and Pienkowski**

1980, Sher and Shields 1991). On alfalfa, nymphs hatch between 3 and 10 d after oviposition. Wingless nymphs go through five instars over the course of 10–14 d before finally molting into an winged adult. In another 7–10 d, females will begin oviposition (**DeLong 1971**). On an average, it takes 3 wk for an egg to develop into an adult (Hogg 1985). Depending on spring arrival time and temperature, potato leafhoppers will have two or three generations per season at northern latitudes.

Symptoms

Potato leafhoppers are light green, wedge-shaped insects. Adults are 3.0 mm in length while first-instar nymphs can be 0.5 mm long (**Fig. 5**). Injury from this pest is called "hopperburn" and described as "V"-shaped leaf chlorosis where the outer edges and tip turn yellow and can develop further into brown leaf necrosis (**Fig. 6**). Potato leafhoppers feed with piercing–sucking mouthparts on mesophyll cells or phloem depending on the plant host (**DeLong 1971**). Either the leafhopper or the plant immune system restricts phloem flow to the leaf edges resulting in leaf edge yellowing and curling. Visual hopperburn is not present until after significant leafhopper feeding has occurred. It is documented that these symptoms are the result of host plant physiological response to potato leafhopper feeding as exhibited in alfalfa (**Lamp et al. 2007**) and grape (**Lamp et al. 2011**, **Lenz et al. 2012**). Adult potato leafhopper feeding on first-year hop leaves was shown to reduce net photosynthesis of eight hop cultivars tested in 2014 (L. B. Calderwood, unpublished data).

Fig. 5. Potato leafhopper, *Empoasca fabae* (Harris), nymphs of different first, third, fourth, and fifth instars on the underside of a hop leaf. Small yellow spots on leaves are lupulin glands, not arthropods.

Fig. 6. Leaf chlorosis and necrosis damage caused by potato leafhopper feeding called "hopper burn."

Management

Potato leafhopper IPM programs involve monitoring of the population weekly. In alfalfa, sweep netting and sticky traps are used to assess leafhopper populations (**Degooyer et al. 1998**). In hops, scouting the underside of several leaves from each cultivar is recommended. Potato leafhoppers are visible with the naked eye and have a signature side-to-side scuttling movement. There is evidence that copper sulfate fungicides (Bordeaux mixture), primarily used for hop downy mildew control, were indirectly used to manage potato leafhopper populations in the 1800s (**Magie 1944**). Copper hydroxide-based sprays are currently the most widely used products to manage hop downy mildew in the northeast. As mentioned above, the impact of fungicides on two spotted spider mite, hop aphid, and natural enemies has been investigated in the Pacific Northwest (**James 2002, 2003a,b**; **Gent et al. 2009a,b**; **Woods et al. 2012**). The natural enemy assemblage for potato leafhopper is similar to that of hop aphid (**Table 1**), and spraying for potato leafhopper, another soft-bodied insect, is likely to have similar secondary outbreak repercussions.

Table 1. Major hop pests and their natural enemy groups

Pest common name	Pest species	Natural enemy common name	Natural enemy taxon
Twospotted spider mite	*Tetranychus urticae*	Predatory mites	Phytoseiidae
		Spider mite destroyer	*Stethorus punctum* spp.
		Minute pirate bug	Anthocoridae
Hop aphid	*Phorodon humuli*	Minute pirate bug, bigeyed bugs, and damsel bugs	Anthocoridae, Geocoridae, Nabidae
		Green lacewing and brown lacewing	Chrysopidae and Hemerobiidae
		Ladybird beetles	Coccinellidae
		Parasitoid wasps	Parasitica
		Syrphid fly	Syrphidae
Potato leafhopper	*Empoasca fabae*	Minute pirate bug, bigeyed bugs, and damsel bugs	Anthocoridae, Geocoridae, Nabidae

		Green lacewing and brown lacewing	Chrysopidae and Hemerobiidae
		Ladybird beetles	Coccinellidae
		Parasitoid wasps	Parasitica
		Spiders	Aranea

Potato leafhoppers are known to be repelled by glandular trichomes produced at different densities on alfalfa (Shockley and Backus 2002), potato (**Kaplan et al. 2008**), and dry bean (**Gonzales et al. 2004**) cultivars. Preliminary data indicated that some hop cultivars were more susceptible to potato leafhopper feeding than others. Additionally, first-year hop plants have been observed to be particularly susceptible to potato leafhopper feeding injury (unpublished data). Further research should identify which hop cultivars are more and less susceptible to this emerging pest and explore mechanisms behind cultivar susceptibility.

Hop Downy Mildew

Although hop downy mildew incidence was not included in this arthropod survey, the biology and management of this pathogen are of great importance, as it is a major challenge in northeastern hop yards. The pathogen has been positively confirmed in numerous yards in NY, MA, and VT by the UVM diagnostic lab and at times by Dr. David Gent of Oregon State University. Hop downy mildew is caused by the oomycete pathogen *Pseudoperonospora humuli*. Spring in the northeast is often cool and wet providing a perfect habitat for the spread of hop downy mildew. The pathogen overwinters as mycelium in ground leaf litter and on hop crowns (**Skotland and Johnson 1983**).

Fig. 7. Hop downy mildew primary basal spike with short internodes, chlorosis, and necrosis.

Research on eastern downy mildew strains is of particular need. It is unclear weather oospores are an overwintering stage of the disease cycle in eastern states (**Magie 1942**, **Skotland and Johnson 1983**). Given moist conditions in the spring, sporangia are carried by wind and rain containing zoospores, the primary inoculum. Zoospores enter hop leaves through leaf stomata, germinate, and produce more sporangia, which release secondary inoculum zoospores that infect additional plants. Hop downy mildew zoospores can arrive to a hop yard via wind currents or in already infected rootstock. The crowns that harbor zoospores and oospores over the winter and give rise to infected shoots in the spring, have systemic downy mildew. These shoots are called primary basal spikes (**Fig. 7**) because they release the first inoculum of the season. When downy mildew is systemic, the pathogen lives in the hop yard year round and will continue to spread unless meticulously managed.

Symptoms

Early spring (March–May) pale green-yellow shoots with short internodes are primary basal spikes. Secondary inoculum infects already trained bines in late spring forming areal spikes. Both basal and areal spikes have a "Christmas tree" appearance in comparison to healthy hop bines (**Figs. 7** and **8**). Chlorotic, stunted bines that fall away from strings are also a symptom of secondary inoculum. Foliar lesions are present on the leaves of both basal and areal spikes. Lesions are sections of leaf cells that form angled brown spots on the underside of leaves (**Fig. 9**). On severely infected spikes, the entire underside of leaves will be covered in brown sporangia producing more zoospores (**Mahaffee et al. 2009**). This hop-specific disease reduces hop moisture content and impacts the appearance of hop cones which directly reduces the quality of hops delivered to brewers. Hop cones infected with downy mildew are prone to early ripening, browning, uneven drying in the oast, a shorter shelf life, and less desirable brewing characteristics, all of which affect the marketability of the product (**Skotland and Johnson 1983**, **Mahaffee et al. 2009**). Fungicide sprays can cause phytotoxicity evidenced by spotting on leaves (**Fig. 10**) in hot and humid weather conditions. Severe fungicide burn can result in plant desiccation. Scouting observations indicate that certain hop cultivars are more susceptible to copper-based fungicides than others.

Fig. 8. Hop downy mildew areal spike with short internodes and chlorosis.

Fig. 9. Hop downy mildew foliar lesions on the underside of a hop leaf.

Fig. 10. Phytotoxicity from application of copper-based fungicide.

Management

It is difficult to reduce disease in an already infested hop yard. A combination of resistant cultivars, hop yard sanitation, and fungicides are used to reduce infestation levels of hop downy mildew. No hop cultivar is currently 100% resistant. When the more susceptible cultivars contract the disease, the resistant cultivars are exposed and remain subject to economic losses from reduced yield and cone quality. The most resistant cultivars include Fuggle, Newport, Perle, Spalter, Wye Challenger, Hallertauer Gold, Hallertauer Magnum, and Hallertauer Tradition (**Gent et al. 2009b**).

"Crowning" or spring pruning is an important spring management practice where late winter or early spring crown growth is mechanically or chemically removed. In the Pacific Northwest, **Gent et al. (2012)**provides evidence for the reduction in hop downy mildew infection. Hop yards that were pruned twice showed the lowest disease severity and growers were able to apply one fewer fungicide application. Crowning date and number influenced yield of 'Willamette' in this study (**Gent et al. 2012**). Research is needed on the impact of crowning on harvest date and yield for the shorter growing season of the northeast region. Later maturing cultivars in Europe have reduced yield when cut back at later dates (**Goenia and Micibski 1972**). After crowning, scouting for basal spikes should take place. If basal spikes are present they should be removed from the hop yard immediately. Sanitation of hop yard clippers and pruning tools is a critical and easy way to reduce the risk of disease spread (**Skotland and Johnson 1983**, **Gent et al. 2012**).

Pruning must be paired with fungicide applications for hop downy mildew management (**Hunger and Horner 1982**, **Mahaffee et al. 2009**). A variety of products and application times are currently exercised in northeastern hop yards in hopes of controlling this disease. According to our 2014 pesticide survey, which included grower responses from Vermont, New York, and Massachusetts, the most commonly sprayed fungicides are copper based. A mean of four fungicide applications are sprayed per season with some farms 12 applications of a copper-based fungicide per season. In Europe, 10–16 fungicide applications are common practice. It is clear from research in Washington and Oregon that making the timing of fungicide sprays more precise can reduce the number of fungicide applications per season. Using a forecasting system to calculate emergence of the first systemically infected spikes of downy mildew reduced the amount of fungicide sprayed in the Pacific Northwest without increasing disease severity (**Gent et al. 2010**).

NATURAL ENEMY ARTHROPODS OF IMPORTANCE

Parasitoid Wasps

Parasitic hymenopterans are very small wasps abundant in diverse agricultural landscapes. Several species are known parasitoids of potato leafhopper and hop aphid and their presence in hop yards has been documented (**Wright and James 2001**, **Grasswitz and James 2009**). *Anagrus* spp. are known for their parasitism of potato leafhopper eggs (**Lovinger et al. 2000**) while a suite of additional parasitoid species play a role in landscape biocontrol. Assemblages of several parasitoid species have been shown to keep aphid populations in check (**Sigsgaard 2002**). The effectiveness of parasitoids as biological control agents in hops has not been studied (**Grasswitz and James 2009**). However, **Wright and James (2001)** reared 802 parasitoids from 83 hop aphids collected on *Prunus* spp. (alternative hop aphid host) in Washington State. The most abundant species reared was *Lysiphlebus testaceipes*(Cresson). **Grasswitz and James (2011)** identified nine generalist parasitoid species in Pacific Northwest hop yards when assessing parasitism on hop looper (*Hypena humuli* Harris), a minor lepidopteran pest of hops. The parasitoids found in this study were not identified to species.

Spiders and Predatory Mites

There are over 30,000 species of spider. This diverse group of arthropods primarily feeds on other arthropods, making them effective members of predatory guilds (**Riechert and Lockley 1984**). A diversity of these generalist predators are present in healthy agricultural ecosystems. In alfalfa, spiders are highlighted as predators of potato leafhopper (**Harwood and Obrycki 2007**). Predatory mites (Acari: Phytoseiidae) including *Phytoseiulus persimilis* Athias-Henriot, *Neoseiulus californicus* McGregor,*Neoseiulus fallacis* Garman, *Amblyseius andersoni* (Chant), *Galendromus occidentalis* Nesbitt have controlled twospotted spider mite populations in Pacific Northwest (**James 2002, James et al. 2003**) and United Kingdom (**Barbour et al. 2003**) hop yards. *N. fallacis* has been shown to manage spider mites in eastern orchard systems (**Metzger 2001**). Similar to spiders, predatory mites and spider mites can be seen with the naked eye. Additional steps should be taken to measure predatory mite presence and the potential impact they may have on spider mite populations in nontraditional hop-growing regions.

Predatory True Bugs and Flies

In the minute pirate bug family (Hemiptera: Anthocoridae), the *Orius* genus is reported as an effective generalist predator of hop aphids (**Lorenzana and Hermoso 2010**). According to studies in alfalfa, minute pirate bugs are effective predators of potato leafhopper (**Wieser Erlandson and Obrycki 2010**). Damsel bugs (Nabidae) and big eyed bugs (Geocoridae) are present in Pacific Northwest hop yards (**Grasswitz and James 2011**) and are observed occasionally in northeastern hop yards. Spined soldier bugs (Pentatomidae: *Podisus* spp.) have been spotted in northeastern hop yards and have been shown to be effective generalist biocontrol agents against Lepidopteran pests of cotton (**Lopez et al. 1976**) and tomato (**De Clercq et al. 1998**). Syrphid fly larvae are specifically known as effective biocontrol agents against aphids (Diptera: Syrphidae) (**Bugg 1992**). Finally, species of predatory midge, *Aphidoletes* spp. (Diptera: Cecidomyiidae), are shown to be effective members of the aphid biological control community in temperate apple orchards. Aphid midges thrive in cool, moist conditions and are potentially abundant in northeastern hop yards (**Wyss et al. 1999**, **Minarro et al. 2005**).

Lacewings

Green lacewings are effective consumers of soft-bodied insects in Pacific Northwest hop yards (**James 2006**, **Lorenzana et al. 2013**). **James (2006)** demonstrated that goldeneyed lacewing (*Chrysopa oculata* Say) populations were increased with the placement of methyl salicylate-baited sticky cards. Lacewing larvae were observed to feed on four potato leafhoppers per day at high pest densities under lab conditions (**Weiser Erlandson and Obrycki 2010**).

Lady Beetles

The Coccinellidae family of beetles includes many generalist predators. *Hippodamia convergens* Guérin-Méneville, *Harmonia axyridis* Pallas, and the native *Coleomegilla maculata* De Geer are three predatory lady beetle species found in hop yards (**James 2003a**, **Campbell and Cone 1994**). Lady beetles are well known for their consumption of soft-bodied arthropod pests, and certain species are mass reared for release in some farm landscapes. Lady beetle larvae have a voracious appetite for aphids and will consume more individual pests than the adult life stage (Koch 2003). *H. axyridis* has been reported to feed on twospotted spider mites and hop aphids in hop yards (**James 2003b**). *C. maculata* are specifically reported to feed on potato leafhoppers in alfalfa systems. The *Stethorus punctum picipes*(Casey), commonly known as the spider mite destroyer, is a small lady beetle that feeds on spider mites in the west (**James 2003a**). *Stethorus punctum punctum* (LeConte) is the spider mite destroyer of the eastern United States and is conserved for biological control in apple production (**Felland and Hull 1996**).

Arthropod Scouting Methodology

Seven hop yards in Vermont were scouted every other week June–August for 3 yr (2012–2014; **Fig. 11**). In 2012, there were few hop yards in Vermont and those selected for this survey were growers willing to participate. Hop yards varied in acreage, cultivar diversity, and management practices. As more farmers started growing hops in Vermont the number of hop yards scouted increased in 2013 and 2014 (**Table 2**).

Fig. 11. Hop yard scouting sites in Vermont. The "North" micro region includes Alburgh and North Hero yards, "West" includes Addison and Ferrisburgh yards, and "East" includes Calais, Berlin A, and Berlin B yards.

Table 2. Characteristics of hop yards scouted in Vermont

Farm	Years scouted	No. of cultivars	Farm size[a]	Micro region	Management
Addison	2012, 2013, 2014	7	Large	West	Organic[b]
Alburgh	2012, 2013, 2014	24	Large	North	Organic
Berlin A	2012, 2013, 2014	3	Medium	East	Nonorganic[c]
Calais	2013, 2014	1	Small	East	Organic practice[d]
North Hero	2013, 2014	4	Small	North	Organic practice
Berlin B	2014	5	Medium	East	Nonorganic
Ferrisburgh	2014	5	Large	West	Nonorganic

[a] Small yards had <60 hills, medium yards had 60–200 hills, and large yards had >200 hills.

[b] Hop yards certified organic by Vermont Organic Farmers, LLC, Richmond, VT were considered "Organic."

[c] Hop yards that were not certified organic were considered "Nonorganic."

[d] Hop yards that were not certified but follow organic management were considered "Organic Practice."

Scouting took place every other week for a total of six collection dates at each location annually. A group of five to seven plants was considered a plot. Three leaves on one plant in each plot were sampled during each visit between ground level and 2 m. Both top and bottom leaf surfaces were visually examined with Optivisor lenses (Donegan Optical Company Inc., Lenexa, KS). Arthropods were identified and counted in the field. Pests were identified to species level, while natural enemies were identified to the level at which an ecological role could be assigned (**Table 1**). Parasitoid wasps and predatory mites were not reported. Statistical comparisons were not presented because it was the goal to show community variation by location and management practices. The data presented are a representation of the arthropod community found in northeastern hop yards. Season sums and peak date means were calculated for arthropod taxons.

Beyond farm location, variables impacting arthropod presence including farm size, management practice, and microclimate region were assigned to each farm. Small, medium, or large size was also assigned to each farm. Small yards had <60 hills (<120 plants), medium yards had 60–200 hills (120–400 plants), and large yards had >200 hills (>400 plants). "Organic" yards were certified organic (Vermont Organic Farmers, LLC, Richmond, VT) while "Nonorganic" yards practiced a range of management tactics between organic but not certified, and conventional and were not certified organic. The "North" micro region included Alburgh and North Hero farms, "West" included Addison and Ferrisburgh farms, and "East" includes Calais, Berlin A, and Berlin B farms.

Scouting Results

Larger yards had a higher total number of pest and natural enemy individuals. However, there were hot spots of pests present on small- and medium-sized yards. With higher numbers of pests, the natural enemy community increased in abundance. Aphid and potato leafhopper natural enemy assemblages were similarly composed of generalist predators. The "East" micro region had fewer growing degree-days each year of the study. Locations with higher elevations included Calais (338 m) and Berlin (268 m). Alburgh, North Hero, Addison, and Ferrisburgh are at elevations between 27 m and 32 m along the large, freshwater Lake Champlain. Overall, Calais and Berlin yards had lower pest levels, yet they are small- and medium-sized yards.

Twospotted Spider Mite

Twospotted spider mite populations were highest in 2012 and 2014. The number of spider mite generations cannot be concluded from these data, but there is an overall late season population increase in all hop yards. In 2012, the spider mite peak date was in early August in all locations. North and West locations had a similar early August peak date in 2014 while East locations exhibited twospotted spider mite peaks in late August (**Table 3**). Peak dates in **Table 6** indicate that spider mite destroyer populations mirrored or lagged behind twospotted spider mite populations. This pattern was especially clear in large yards and in years with high twospotted spider mite populations (**Table 4**). In Addison, the decreased ratio of twospotted spider mites to spider mite destroyers from 2012 to 2014 indicates that this predator population can become established. Without considering other yard variables, spider mite destroyers appear to have reduced the 2014 twospotted spider mite population in that location (**Table 4, Fig. 12**). Addison, being a large yard, had fewer twospotted spider mites than Berlin B, a medium-sized yard. Ferrisburgh had the highest total number of twospotted spider mites and low natural enemy abundance in 2014.

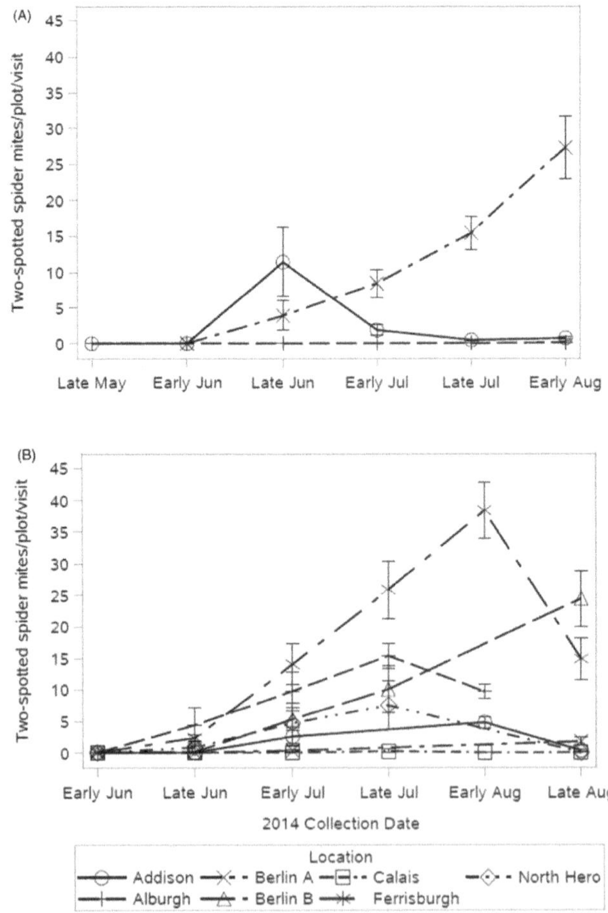

Fig. 12. Mean number of twospotted spider mites counted per plot, per yard visit in 2012 (A) and 2014 (B). Three farms were scouted in 2012 and seven in 2014. No spider mites were observed in 2014.

Table 3. Twospotted spider mite (TSSM) season sum, peak date, and peak date mean (± SEM) by year and farm

TSSM	Farm	Farm size	Micro region	Season sum	Peak date	Peak date mean (± SEM)
2012	Alburgh	Large	North	3,207	8-Aug.	27.32 (± 4.38)
	Addison	Large	West	809	6-Aug.	0.70 (± 0.17)
	Berlin A	Medium	East	4	7-Aug.	0.13 (± 0.13)
2013	Alburgh	Large	North	145	12-Aug.	2.00 (± 0.34)
	North Hero	Small	North	0	N/A	0.00
	Addison	Large	West	172	15-July	0.91 (± 0.23)
	Calais	Small	East	0	N/A	0.00
	Berlin A	Medium	East	0	N/A	0.00
2014	Alburgh	Large	North	3,206	4-Aug.	14.58 (± 1.82)
	North Hero	Small	North	209	21-July	7.50 (± 3.85)
	Addison	Large	West	623	6-Aug.	4.77 (± 0.90)
	Ferrisburgh	Large	West	3,609	6-Aug.	38.39 (± 4.34)
	Calais	Small	East	2	25-July	0.25 (± 0.25)
	Berlin A	Medium	East	63	22-Aug.	1.80 (± 0.69)
	Berlin B	Medium	East	1,046	22-Aug.	24.38 (± 4.50)

Table 4. Twospotted spider mite (TSSM) and spider mite destroyer (SMD) adult and larvae combined season totals and peak dates

Year	Farm	Season total		Peak date	
		TSSM	SMD	TSSM	SMD
2012	Alburgh	3,207	121	8-Aug.	8-Aug.
	Addison	809	21	6-Aug.	24-July
	Berlin A	4	1	7-Aug.	24-July
2013	Alburgh	145	2	12-Aug.	24-July
	North Hero	0	0	N/A	N/A
	Addison	172	28	15-July	15-July, 29-July[a]
	Calais	0	0	N/A	N/A
	Berlin A	0	2	N/A	N/A
2014	Alburgh	3,206	163	4-Aug.	11-Aug.
	North Hero	209	35	21-July	18-Aug.
	Addison	623	52	6-Aug.	6-Aug.
	Ferrisburgh	3,609	36	6-Aug.	6-Aug.
	Calais	2	0	25-July	N/A
	Berlin A	63	0	22-Aug.	N/A
	Berlin B	1,046	0	22-Aug.	N/A

[a] SMD was equally high on both dates in Addison.

Hop Aphid

Hop aphid populations were highest in 2014. The wet spring of 2013 and continual precipitation throughout 2014 allowed hop aphids to flourish (**Table 5**, **Fig. 13**). Two hop aphid cycles were observed in 2014. In years of high population levels, hop aphid peak date ranged from early to late August depending on geographic location. As expected, locations with high total pest numbers had the highest peak date means. Natural enemy counts were highest in locations and years with high hop aphid populations. This indicates that increased natural enemy presence could aid in pest population management. Where predators became established there were lower hop aphid season sums. Specifically in Alburgh, the hop aphid population almost doubled from 2013 to 2014. In both 2013 and 2014, Addison had high aphid populations. In contrast, Ferrisburgh had the highest hop aphid population of all yards with low natural enemy abundance (**Table 5**). Again, it is difficult to compare hop aphid to potato leafhopper control by natural enemies because these assemblages are very similar (**Table 1**). Although spiders are not included in the hop aphid natural enemy assemblage, it is possible that these generalist predators feed on hop aphids when other prey is not available. Alburgh had the highest spider count of all locations in 2012 and 2014 when the highest number of aphids was observed.

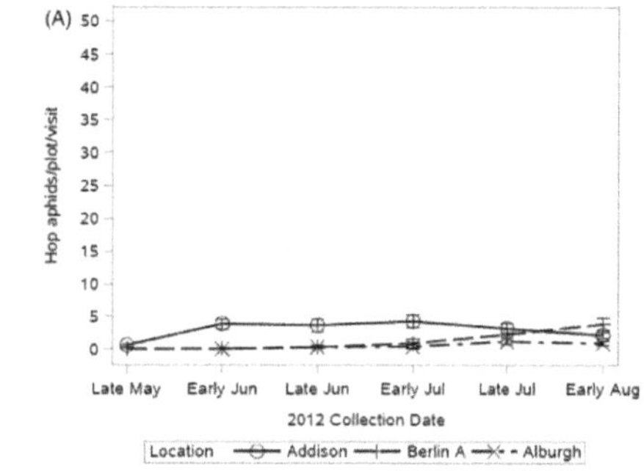

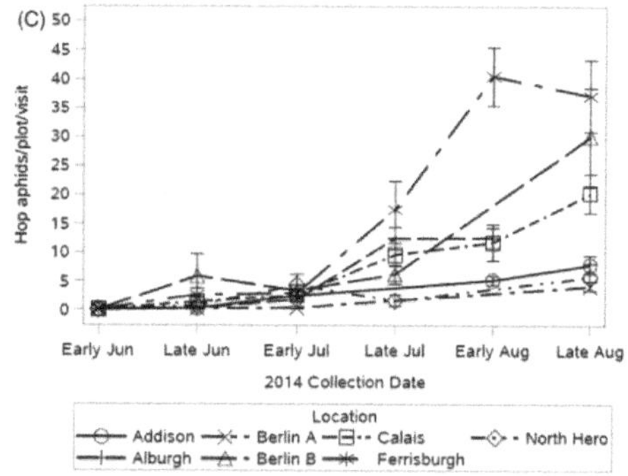

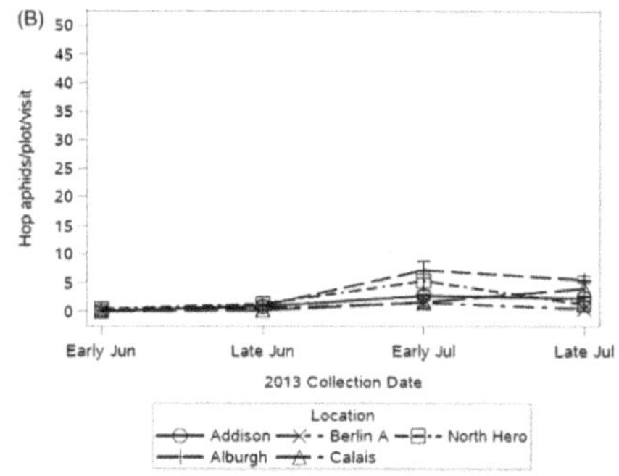

Fig. 13. Mean number of hop aphids counted per plot, per yard visit in 2012 (A), 2013 (B), and 2014 (C). Three farms were scouted in 2012, five in 2013, and seven in 2014.

Table 5. Hop aphid (HA) season sum, peak date, and peak date mean (± SEM) by year and farm

HA	Farm	Farm size	Micro region	Season sum	Peak date	Peak date mean (± SEM)
2012	Alburgh	Large	North	172	25-July	1.22 (± 0.55)
	Addison	Large	West	984	24-July	3.18 (± 0.74)
	Berlin A	Medium	East	224	7-Aug.	3.86 (± 0.95)
2013	Alburgh	Large	North	1562	12-Aug.	12.93 (± 1.60)
	North Hero	Small	North	204	9-July	5.35 (± 1.73)
	Addison	Large	West	522	29-July	3.00 (± 0.54)
	Calais	Small	East	93	6-Aug.	5.88 (± 2.13)
	Berlin A	Medium	East	97	6-Aug.	0.80 (± 0.28)
2014	Alburgh	Large	North	2864	4-Aug.	16.93 (± 3.20)
	North Hero	Small	North	201	18-Aug.	5.81 (± 2.30)
	Addison	Large	West	1080	27-Aug.	7.96 (± 1.51)
	Ferrisburgh	Large	West	3708	6-Aug.	40.53 (± 5.06)
	Calais	Small	East	365	24-Aug.	20.25 (± 3.39)
	Berlin A	Medium	East	149	22-Aug.	4.27 (± 0.74)
	Berlin B	Medium	East	1131	22-Aug.	30.04 (± 8.50)

Table 6. Potato leafhopper (PLH) season sum, peak date, and peak date mean (± SEM) by year and farm

PLH	Farm	Farm size	Micro region	Season sum	Peak date	Peak date mean (± SEM)
2012	Alburgh	Large	North	730	29-June	6.01 (± 0.60)
	Addison	Large	West	588	28-June	6.50 (± 0.97)
	Berlin A	Medium	East	573	30-June	14.70 (± 1.33)
2013	Alburgh	Large	North	349	24-June	3.11 (± 0.65)
	North Hero	Small	North	291	24-June	6.75 (± 1.73)
	Addison	Large	West	945	18-June	5.65 (± 0.79)
	Calais	Small	East	69	9-July	4.38 (± 1.44)
	Berlin A	Medium	East	284	9-July	5.67 (± 0.84)

2014	Alburgh	Large	North	113	8-July	0.59 (± 0.14)
	North Hero	Small	North	34	21-July	1.12 (± 0.34)
	Addison	Large	West	183	11-July	0.83 (± 0.17)
	Ferrisburgh	Large	West	62	9-July	0.65 (± 0.18)
	Calais	Small	East	3	25-July	0.38 (± 0.26)
	Berlin A	Medium	East	6	11-July	0.17 (± 0.10)
	Berlin B	Medium	East	19	25-July	0.33 (± 0.12)

Potato Leafhopper

Potato leafhopper populations were highest in 2012 and 2013 across all locations with peak dates in late June and early July (**Fig. 14**). Two generations of potato leafhopper were observed in 2012 and 2013 on multiple farms. This migratory pest arrived late to the northeast in 2014. A low, later population was therefore observed in 2014 with peak dates throughout July. Eastern sites had very few individuals in the 2014 season. Addison was the location with the highest potato leafhopper pressure, yet it is also the location with the highest natural enemy abundance. Medium-sized Berlin A had similar potato leafhopper abundance to Addison in 2012 (**Table 6**). Both locations exhibited natural enemy presence. In 2013, Berlin A potato leafhopper pest abundance dropped and natural enemy abundance increased, potentially indicating the impact of natural enemies. Spiders were a prevalent potato leafhopper natural enemy in all locations. Potato leafhopper was typically an early season pest, attacking new growth. However, given the unpredictable migration of this pest, females could arrive before natural enemy populations become established in the spring. Potato leafhopper population numbers were less variable by location or farm size than hop aphids or twospotted spider mites.

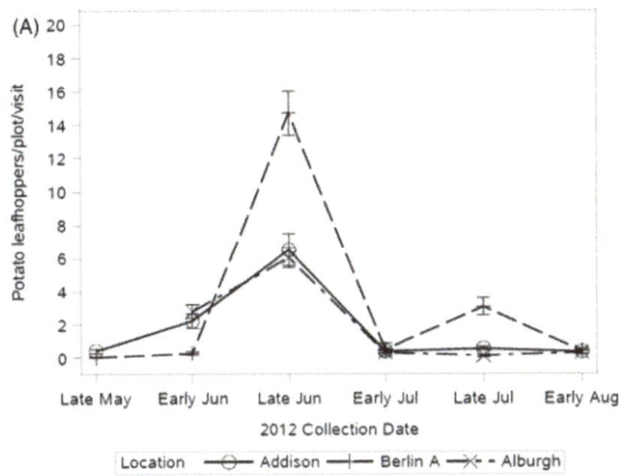

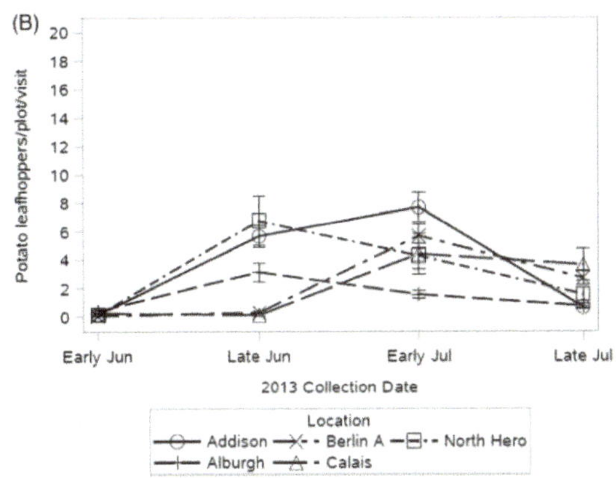

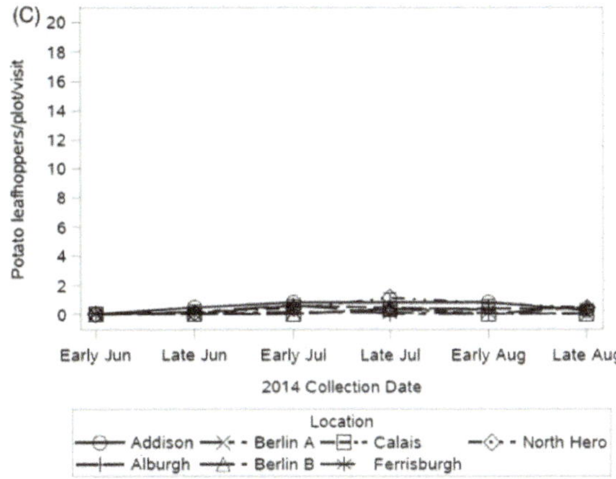

Fig. 14. Mean number of potato leafhoppers counted per plot, per yard visit in 2012 (A), 2013 (B), and 2014 (C). Three farms were scouted in 2012, five in 2013, and seven in 2014.

Minor Pests

Eastern comma, *Polygonia comma* Harris (Lepidoptera: Nymphalidae), adults and larva (**Fig. 15**) were abundant only in 2012 and have not been reported as a problem since. Japanese beetle, *Popillia japonica* Newman (Coleoptera: Scarabaeidae) (**Fig. 16**), was reported as a problem on hops in southern parts of the northeast region. Low numbers of this pest were observed in 2012. A Japanese beetle hot spot was observed at the Berlin A location in late July through early August of 2013. Hop looper was spotted on all scouted hop yards during at least 1 yr of the survey at low abundance (**Table 7**). The foliar chewing damage that Japanese beetle and eastern comma larvae incur to hop plants is visually shocking. However, because Japanese beetles are foliar feeders and become a problem in late July or early August, their damage is unlikely to reduce hop yield.

Fig. 15. Eastern comma larva, a minor hop pest.

Fig. 16. Japanese beetle adults and defoliation. This is a minor pest in northern hop yards yet a more serious pest in southern locations.

Table 7. Minor pests including Eastern comma (*Polygonia comma* Harris) adult and larvae, hop looper (*Hypena humuli* Harris) larvae, and Japanese beetle (*Popillia japonica* Newman) adult season sums are reported

Year	Farm	Farm size	Micro region	Season sum		
				Eastern comma	Hop looper	Japanese beetle
2012	Alburgh	Large	North	55	7	1
	Addison	Large	West	321	7	5
	Berlin A	Medium	East	100	3	4
2013	Alburgh	Large	North	5	1	1
	North Hero	Small	North	0	0	1
	Addison	Large	West	0	2	7
	Calais	Small	East	0	2	0
	Berlin A	Medium	East	1	7	93
2014	Alburgh	Large	North	0	10	1
	North Hero	Small	North	0	5	0
	Addison	Large	West	0	8	0
	Ferrisburgh	Large	West	0	7	0
	Calais	Small	East	0	5	0
	Berlin A	Medium	East	0	7	0
	Berlin B	Medium	East	0	12	0

Generalist Predators

Because the arthropod community was disturbed as plants were approached, parasitoid wasps were seldom observed and the magnitude of generalist predators may not be represented. Across farms, lady beetles and spiders were the most abundant generalist predators. Lacewings were the next most abundant natural enemy group followed by syrphid flies and finally minute pirate bugs. As expected, large farms had the highest natural enemy abundance. However, over the course of 3 yr the smaller North Hero and Calais farms exhibited an increase in the number of total generalist predators. Hop yards that were sampled all 3 yr (Alburgh, Addison, and Berlin A) had the highest generalist natural enemy abundance in 2012 (**Table 8**). These generalist natural enemy populations coincide with high pest abundance in 2012. Hop aphids and potato leafhoppers totaled 2,201 in 2012, 1,680 in 2013, and 356 in 2014 in the yards scouted for 3 yr.

Table 8. Season sums by farm and year for generalist predators including lady beetle (Coccinellidae) adults and larva, spiders, lacewings (Chrysopidae and Hemerobiidae) adult and larva, syrphid fly (Syrphidae) adults, and minute pirate bug (*Orius* spp.) adults

Year	Farm	Farm size	Micro region	Season sum				
				Lady beetles	Spiders	Lacewings	Syrphid flies	Minute pirate bugs
2012	Alburgh	Large	North	56	39	70	0	0
	Addison	Large	West	109	32	31	2	0
	Berlin A	Medium	East	33	9	9	0	0
2013	Alburgh	Large	North	4	6	3	7	0
	North Hero	Small	North	3	17	5	0	0
	Addison	Large	West	45	43	17	11	1
	Calais	Small	East	0	12	0	0	0
	Berlin A	Medium	East	5	29	8	1	0
2014	Alburgh	Large	North	8	56	1	0	1
	North Hero	Small	North	15	23	0	0	0
	Addison	Large	West	34	26	3	0	0
	Ferrisburgh	Large	West	41	28	7	3	0
	Calais	Small	East	18	9	0	0	0
	Berlin A	Medium	East	6	26	1	0	0
	Berlin B	Medium	East	17	11	1	0	0

Specific Farm Factors

Season totals and graphical phenology of pest and peak date means by location provide insight into northeast hop yard arthropod patterns. However, the effects of cultivar genetic variation, specific management practices, surrounding habitat, previous crop, and hop yard age on the arthropod community were not accounted for in this survey. There are specific management practices that we believe to have influenced certain pest populations. Alburgh exhibited almost the same total number of spider mites in 2012 and 2014. Pyganic, a pyrethrin broad-spectrum insecticide, was sprayed multiple times at this location for management of potato leafhopper in 2012. This likely caused the late season secondary outbreak of twospotted spider mites at this site. Hop quality and yield were negatively affected by this outbreak in 2012. Insecticides have not been applied in the Alburgh hop yard since. The same total number of spider mites in 2014 did not have an impact on hop quality or yield in the unsprayed yard.

Moving Forward

The patterns observed in this survey were the first published documentation of northeastern pests and natural enemies since the 1940s (Magie 1944). We consider the hop arthropod pests of concern in the northeast to be twospotted spider mite, hop aphid, and potato leafhopper. We observed similar pest phenology to other hop-growing regions for twospotted spider mites and hop aphids.

Arthropod communities are known to change based on climate, landscape, and management practices (Schweiger et al. 2005). Microclimate and hop yard size factors reported here observationally influenced pest abundance. As expected, where high pest populations were observed, natural enemies were more abundant. Habitat diversity and surrounding landscape, not reported, are well-known variables that increase natural enemy presence (Landis et al. 2000, Rusch et al. 2010).

Hop cones grow throughout the canopy of plants. Although logistically challenging, we suggest that mid and late season pest monitoring could be improved by sampling above eye level in the hop canopy. Unbaited sticky traps can be pinned to trellis wire to monitor flying pests and natural enemies. Growers should visually inspect the underside of leaves on all cultivars as weekly scouting. Parasitoid wasps, not reported, are important and numerous members of hop aphid and potato leafhopper natural enemy assemblages (unpublished data).

Weather was an indicator of pest abundance in northeastern hops. In a dry, hot year, we expect to see high twospotted spider mite populations, yet with cooler temperatures and early or continued precipitation, we expect to see high aphid populations throughout August. Potato leafhopper population prediction will rely on southern reports of presence on other crops, such as alfalfa, that are routinely monitored.

Economic threshold levels have not been developed for these pests in the northeast. When making pest control decisions, eastern Extension professionals reference Pacific Northwestern and European economic threshold ranges for twospotted spider mites (10–100 mites per leaf; Wright and Cone 1999,Weihrauch 2004) and hop aphid (5–70 aphids per leaf; Lorenzana et al. 2009). The literature suggests that thresholds for these pests vary considerably both locally and regionally (Weihrauch 2004,Lorenzana et al. 2009). The research hop yard in Alburgh, VT, has not exceeded a yard mean of 60 twospotted spider mites per leaf or a yard mean of 10 hop aphids per leaf during peak pest abundance. These levels have, however, been reached in pockets of the Alburgh hop yard.

This survey provided evidence that natural enemies and insecticide applications impact pest populations. Natural enemies were present to varying degrees in northeast hop yards, and once established, they appeared to maintain pest populations. This was particularly clear in the Addison hop yard. Insecticides have the opposite effect, as they kill natural enemies and increase the risk of a twospotted spider mite outbreak. This was observed in Alburgh (2012) and Ferrisburgh (2014) yards.

The work presented was the first step toward developing appropriate IPM tactics for modern day northeastern hop production. As evidenced, it is always important to consider site-specific factors that may influence arthropod populations. Northeastern hop arthropod and disease IPM requires further research. Based on this survey, economic thresholds for the three major arthropod pests should be developed. Although some research on natural enemy populations in hop yards has been conducted in the Pacific Northwest, development of conservation and augmentative biological control protocols would reduce pesticide use and provide management options for organic growers. Finally, research on hop downy mildew and weed management will be required in order to produce high quality northeastern hops.

ACKNOWLEDGMENTS

We continue to appreciate the growers who opened their farms to our scouting efforts and have participated in our grower surveys. We thank the permanent and seasonal employees of the UVM Extension Northwest Crops and Soils Team for their time spent collecting data. This research would not have been possible without financial support from the Northeast Sustainable Agriculture Research and Education (SARE) program (GNE12-033) and the United States Department of Agriculture-National Institute of Food and Agriculture (USDA-NIFA) Northeastern IPM Center (027269).

REFERENCES CITED

1. Ako, M., C. Borgemeister, H. M. Poehling, A. Elbert, and R. Nauen. 2004. Effects of neonicotinoid insecticides on the bionomics of twospotted spider mite (Acari: Tetranychidae). J. Econ. Entomol. 97: 1587–1594.
2. Barbour, A., C. Campbell, H. Crane, R. Lilley, and E. Tregidga. 2003. Biocontrol of two-spotted spider mite Tetranychus urticae on dwarf hops by the Phytoseiid mites Phytoseiulus persimilis and Neoseiulus californicus. Biocontrol Sci. Technol. 13: 275–284.
3. Bugg, R. 1992. Habitat manipulation to enhance the effectiveness of aphidophagous hover flies (Diptera: Syrphidae). Sustain. Agricul. Res. Educ. Prog. 5: 1–9.
4. Campbell, C. 1983. Antibiosis in hop (Humulus lupulus) to the damson-hop aphid, Phorodon humuli. Entomol. Exp. Appl. 33: 57–62.
5. Campbell, C. 1977. Distribution of damson-hop aphid (Phorodon humuli) migrants on hops in relation to hop variety and wind shelter. Ann. Appl. Biol. 87: 315–325.
6. Campbell, C. and W. Cone. 1994. Influence of predators on population development of Phorodon humuli (Homoptera: Aphididae) on hops. J. Environ. Entomol. 23: 1391–1396.
7. De Clercq, P., F. Merlevede, K. Mestdagh, K. Vandendurpel, J. Mohaghegh, and D. Degheele. 1998. Predation on the tomato looper Chrysodeixis chalcites (Esper) (Lep., Noctuidae) by Podisus maculiventris (Say) and Podisus nigrispinus (Dallas) (Het., Pentatomidae). J. Appl. Entomol. 122: 93–98.
8. DeLong, D. 1971. The bionomics of leafhoppers. Annu. Rev. Entomol. 16: 179–210.
9. Degooyer, T., L. Pedigo, and M. Rice. 1998. Evaluation of grower-oriented sampling techniques and proposal of a management program for potato leafhopper (Homoptera: Cicadellidae) in alfalfa. J. Econ. Entomol. 91: 143–149.
10. Dorschner, K. and C. Baird. 1988. Susceptibility of hop to Phorodon humuli. Entomol. Exp. Appl. 49: 245–250.
11. Felland, C. and L. Hull. 1996. Overwintering of Stethorus punctum punctum (Coleoptera: Coccinellidae) in apple orchard ground cover. Journal of Environmental Entomology 25: 972–976.
12. Frank, S., W. Klingeman, S. White, and A. Fulcher. 2013. Biology, injury, and management of maple tree pests in nurseries and urban landscapes. J. Integr. Pest Manage. 4: 1–14.
13. Hop Growers of America. 2014. US hop acreage estimate. pp. 1–2. Yakama, Washington, DC.
14. Gent, D., C. Ocamb, and J. Farnsworth. 2010. Forecasting and Management of Hop Downy Mildew. Plant Disease 94: 425–431.
15. Gent, D. H., M. Nelson, G. Grove, W. Mahaffee, W. W. Turechek, and J. L. Woods. 2012. Association of spring pruning practices with severity of powdery mildew and downy mildew on hop. Plant Dis. 96: 1343–1351.
16. Gent, D. H., D. James, L. Wright, D. Brooks, J. Barbour, A. Dreves, A. G. Fisher, V. Walton. 2009a. Effects of powdery mildew fungicide programs on two-spotted spider mite (Acari: Tetranychidae), hop aphid (Hemiptera:
17. Aphididae), and their natural enemies in hop yards. Hortic. Entomol. 102: 274–286.
18. Gent, D. H., J. Barbour, A. Dreves, D. James, R. Parker, and D. Walsh (Eds). 2009b. Field guide for integrated pest management in hops, vol. 1. Oregon State University, University of Idaho, U.S. Department of Agriculture - Agricultural Research Service, and Washington State University.
19. Gonzales, J., D. Coyne, D. Lindgren, D. Schaaf, and K. Eskridge 2004. Heritability of the resistance to potato leafhopper in dry bean. Hortic. Sci. 39: 1578–1580.
20. Goenia, A. and B. Micibski 1972. The influence of term of rootstock cutting on the spring infestation of hop with Peronospora. Rocz. Nauk Rolniczych 2: 7–22.
21. Grasswitz, T. and D. James 2011. Phenology and impact of natural enemies associated with the hop looper (Hypena humuli) in Washington State, USA. Int. J. Integr. Pest Manage. 57: 329–339.
22. Grasswitz, T. and D. James 2009. Influence of hop yard ground flora on invertebrate pests of hops and their natural enemies. J. Appl. Entomol. 133: 210–221.
23. Harwood, J. and J. Obrycki 2007. Web-site selection strategies of linyphiid spiders in alfalfa: Implications for biological control. BioControl 52: 451–467.
24. Huffaker, C., M. Vrie, and J. McMurty 1969. The ecology of tetranychid mites and their natural control. Annu. Rev. Entomol. 14: 125–174.

25. Hunger, R. and C. Horner 1982. Control of hop downy mildew with systemic fungicides. Plant Dis. 66: 1157–1159.

26. James, D. 2002. Selectivity of the acaricide, befenazate, and aphicide, pymetrozine, to spider mite predators in Washington hops. International J. Acarol. 28: 175–179.

27. James, D. 2003a. Pesticide susceptibility of two Coccinellids (Stethorus punctum picipes and Harmonia axyridis) important in biological control of mites and aphids in Washington hops. Biocontrol Sci. Technol. 13: 253–259.

28. James, D. 2003b. Toxicity of imidacloprid to Galendromus occidentalis, Neoseiulus fallacis and Amblyseius andersoni (Acari: Phytoseiidae) from hops in Washington State, USA. Exp. Appl. Acarol. 31: 275–281.

29. James, D. 2006. Methyl salicylate is a field attractant for the goldeneyed lacewing, Chrysopa oculata. Biocontrol Sci. Technol. 16: 107–110.

30. James, D. and J. Coyle. 2001. Which pesticides are safe to beneficial insects and mites? Agric. Environ. News 178: 12–14.

31. James, D. and T. Price. 2002. Fecundity in two spotted spider mite (Acari: Tetranychidae) is increased by direct and systemic exposure to imidacloprid. J. Econ. Entomol. 95: 729–732.

32. James, D., T. Price, and L. Wright. 2003. Mites and aphids in Washington hops: Candidates for augmentative or conservation biological control, pp. 189–194. In Proceedings, First International Symposium on Biological Control of Arthropods, Honolulu, HI.

33. Jeppson, L., H. Keifer, and E. Baker. 1975. Mites injurious to economic plants. University of California Press, Berkeley CA.

34. Kaplan, I., G. Dively, and R. Denno. 2008. Variation in tolerance and resistance to the leafhopper Empoasca fabae (Hemipetera: Cicadellidae) among potato cultivars: Implications for action thresholds. J. Econ. Entomol. 101: 959–968.

35. Kennedy, J. 1860. Eighth census of agriculture, Vermont. United States Department of Agriculture, Washington, DC.

36. Lamp, W. O., D. Miranda, L. Culler, L. Alexander. 2011. Host suitability and gas exchange response of grapevines to potato leafhopper (Hemiptera: Cicadellidae). J. Econ. Entomol. 104: 1316–1322.

37. Lamp, W. O., L. Alexander, and M. Nguyen 2007. Physiological response of glandular-haired alfalfa to potato leafhopper (Hemiptera: Cicadellidae) injury. J. Environ. Entomol. 36: 195–203.

38. Landis, D., S. Wratten and G. Gurr 2000. Habitat management to conserve natural enemies of arthropod pests in agriculture. Annu. Rev. Entomol. 45: 175–201.

39. Lenz, M., R. Isaacs, J. Flore, and G. Howell 2012. Photosynthetic performance of Pinot gris (Vitis vinifera L.) grapevine leaves in response to potato leafhopper (Empoasca fabae Harris) infestation. Am. J. Enol. Vitic. 63: 357– 366.

40. Lopez, J., R. Ridgway, and R. Pinnell 1976. Comparative efficacy of four insect predators of the bollworm and tobacco budworm. J. Environ. Entomol. 5: 1160–1164.

41. Lorenzana, A., and A. Hermoso 2010. Population development of Phorodon humuli and predators (Orius spp.) within hop cones: Influence of aphid density on hop quality. Crop Prot. 29: 832–837.

42. Lorenzana, A., M. Hermoso-de-Mendoza, M. Seco, and P. Casquero 2013. Population dynamics and integrated control of the damson-hop aphid Phorodon humuli (Schrank) on hops in Spain. Span. J. Agric. Res.11: 505–517.

43. Lorenzana, A., A. Hermoso de Mendoza, and M. Seco 2009. Thresholds for management of Phorodon humuli (Hemiptera: Aphididae) on hops. J. Entomol. Sci. 44: 198–208.

44. Lovinger, A., D. Liewehr, and W. Lamp 2000. Glandular trichomes on alfalfa impede searching behavior of the potato leafhopper parasitoid. Biol. Control 18: 187–192.

45. Lu, Z.-X., P. Zhu, G. Gurr, X. Zheng, D. Read, K. Heong, Y. Yang, and H. Xu 2014. Mechanisms for flowering plants to benefit arthropod natural enemies of insect pests: Prospects for enhanced use in agriculture. Insect Sci. 21: 1–12.

46. Magie, R. O. 1944. Disease and insect control on hops; technical bulletin 708. New York State Agricultural Experiment Station. Cornell University. Geneva, NY.

47. Magie, R. O. 1942. The epidemiology and control of downy mildew on hops; technical bulletin 267. New York State Agricultural Experiment Station, Cornell University. Geneva, NY.

48. Mahaffee, W., S. Pethybridge, and D. Gent (Eds). 2009. Compendium of Hop Diseases and Pests. American Phytopathological Society Press, St. Paul, MN.

49. Metzger, J. 2001. Neoseiulus fallacis (Garman) (Acari: Phytoseiidae) as a potential biological control agent for spider mites (Acari: Tetranychidae) in Virginia vineyards. M.S. thesis, Virginia Polytechnic Insitute and State University, Blacksburg.

50. Minarro, M., J. Hemptinne, E. Dapena. 2005. Colonization of apple orchards by predators of Dysaphis plantaginea: Sequential arrival, response to prey abundance and consequences for biological control. Biological Control 50: 403–414.

51. Nault, B., A. Taylor, M. Urwiler, T. Rabaey, and W. Hutchison. 2004. Neonicotinoid seed treatments for managing potato leafhopper infestations in snap bean. Crop Prot. 23: 147–154.

52. Peters, K., and R. Berry 1980. Effect of hop leaf morphology on two spotted spider mite. J. Econ. Entomol. 73: 235–238.

53. Regev, S., and W. Cone. 1975. Chemical differences in hop varieties vs. susceptibility to the two spotted spider mite. J. Environ. Entomol. 4: 697–700.

54. Riechert, S., and T. Lockley. 1984. Spiders as biological control agents. Annu. Rev. Entomol. 29: 299–320.

55. Schweiger, O., J. Maelfait, W. Wingerden, F. Hendrickx, R. Billeter, M. Speelmans, I. Augenstein, B. Aukema, S.

Aviron, D. Bailey, et al. 2005. Quantifying the impact of environmental factors on arthropod communities in agricultural landscapes across organizational levels and spatial scales. J. Appl. Ecol. 42: 1129–1139.

56. Sclar, D. C., D. Gerace, and W. S. Cranshaw. 1998. Observations of population increases and injury by spider mites (Acari: Tetranychidae) on ornamental plants treated with imidacloprid. J. Econ. Entomol. 91: 250–255.

57. Sher, R., and E. Shields. 1991. Potato leafhopper (Homoptera: Cicadellidae) oviposition and development under cool fluctuating temperatures. J. Environ. Entomol. 20: 1113–1120.

58. Sidumo, A., E. Shields, A. Lembo. 2005. Estimating the potato leafhopper Empoasca fabae (Homoptera: Cicadellidae) overwintering range and spring premigrant development by using geographic information system. J. Econ. Entomol. 98: 757–764.

59. Simonet, D., and R. Pienkowski. 1980. Temperature effect on development and morphometrics of the potato leafhopper. J. Environ. Entomol. 9: 798–800.

60. Sigsgaard, L. 2002. A survey of aphids and aphid parasitoids in cereal fields in Denmark, and the parasitoids' role in biological control. J. Appl. Entomol. 126: 101–107.

61. Skotland, C., and D. Johnson. 1983. Control of downy mildew of hop. American Phytopathological Society 1183–1185.

62. Strong, W., and B. Croft. 1995. Inoculative release of Phytoseiid mites (Acarina: Phytoseiidae) into the rapidly expanding canopy of hops for control of Tetranychus urticae (Acarina: Tetranychidae). J. Environ. Entomol. 24: 446–453.

63. Symondson, W., K. Sunderland, and M. Greenstone. 2002. Can generalist predators be effective biocontrol agents? Annu. Rev. Entomol. 47: 561–594.

64. Taylor, P. S., and E. J. Shields. 1995. Phenology of Empoasca fabae (Harris) (Homoptera: Cicadellidae) in its overwintering area and proposed seasonal phenology. J. Environ. Entomol. 24: 1096–1108.

65. Weihrauch, F. 2004. A new monitoring approach for the two-spotted spider mite Tetranychus urticae Koch (Acari: Tetranychidae) in hop culture. J. Plant Dis. Prot. 111: 197–205.

66. Weihrauch, F. and L. Moreth. 2005. Behavior and population development of Phorodon humuli (Schrank) (Homoptera: Aphididae) on two hop cultivars of different susceptibility. J. Insect Behav. 18: 693–705.

67. Weiser Erlandson, L. and J. Obrycki. 2010. Predation of immature and adult Empoasca fabae (Harris) (Hemiptera: Cicadellidae) by three species of predatory insects. J. Kans. Entomol. Soc. 83: 1–6.

68. Woods, J., A. Dreves, D. James, J. Lee, D. Walsh, and D. H. Gent. 2014. Development of biological control of Tetranychus urticae (Acari: Tetranychidae) and Phorodon humuli (Hemiptera: Aphididae) in Oregon hop yards. J. Econ. Entomol.107: 570–581.

69. Woods, J., A. Dreves, G. Fisher, D. James, L. Wright, and D. Gent. 2012. Population density and phenology of Tetranychus urticae (Acari: Tetranychidae) in hop is linked to the timing of sulfur applications. J. Environ. Entomol. 41: 621–635.

70. Wright, L., and D. James. 2001. Parasitoids of the hop aphid (Homoptera: Aphidae) on Prunus during the spring in Washington state. J. Agric. Urban Entomol. 18: 141–147.

71. Wright, L., W. Cone, and D. James. 2005. Sources of spring and fall hop aphid, Phorodon humuli (Schrank), (Homoptera: Aphidae) migrants in south central Washington. J. Entomol. Soc. B. C. 102: 21–26.

72. Wright, L., and W. Cone. 1999. Binomial sequential sampling plans for adult female two-spotted spider mites (Acari: Tetranychidae) on hops. J. Econ. Entomol. 92: 1335–1343.

73. Wyss, E., M. Villiger, and H. Muller-Scharer. 1999. The potential of three native insect predators to control the rosy apple aphid, Dysaphis plantaginea. Biol. Control 44: 171–182.

CHAPTER 14

ECOLOGY, TAXONOMY, AND PEST MANAGEMENT OF BILLBUGS (COLEOPTERA: CURCULIONIDAE) IN ORCHARDGRASS

William R. Kuhn, Roger R. Youngman, Shaohui Wu, Curtis A. Laub

Keywords: Hunting billbug, bluegrass billbug, Curculionidae, Sphenophorus spp, orchardgrass

Orchardgrass, *Dactylis glomerata* L. (Cyperales: Poaceae), is a cool season bunch-type grass, cultivated as a high-value forage crop in the United States. In Virginia, the production value of orchardgrass in 2011 was estimated US$188,772,000, or ≈35% of the total production value of grass hay for the state (**National Agricultural Statistics Service [NASS] 2012**). The bluegrass billbug,*Sphenophorus parvulus* Gyllenhal, and hunting billbug, *Sphenophorus venatus vestitus* Chittenden, are important pests of orchardgrass in Virginia. A survey conducted in 2005 on over 324 ha (800 ac) of orchardgrass in northern Virginia showed losses of 40—100% attributed to billbugs (R.R.Y., unpublished data). The sheltered feeding habits of these species, combined with the lack of effective systemic insecticides for orchardgrass, make billbug control extremely difficult in orchardgrass. Here, we present an overview of billbug ecology, and control options for these pests in orchardgrass.

ORCHARDGRASS

Orchardgrass is a cool season bunchgrass used for forage throughout the world. Native of Eurasia and northern Africa, it was first described by Linnaeus in 1753 (**Peeters 2004**). Cultivation of orchardgrass began after its introduction into the United States, presumably in the 1750s, and in 1763, an improved variety from Virginia was reimported to England (**Christie and McElroy 1995**). Named for its propensity to grow in shady areas such as orchards, orchardgrass is also called cocksfoot in Europe for the likeness of its inflorescence to the foot of a rooster (**Balasko and Nelson 2003**).

In the United States today, various cultivars of orchardgrass are found in every state, including Alaska and Hawaii, as well as Puerto Rico (Natural Resources Conservation Service [NRCS] 2010); however, it is predominant in the northeastern, north—central, and Pacific Northwest regions of the United States (**Christie and McElroy 1995**). Orchardgrass has a few marked characteristics that make it easy to distinguish from other grasses (**Figure 1**). As a bunch-type grass, orchardgrass plants grow in tight, round clusters (**Smith et al. 1986**). The leaves of these clusters are limp, somewhat curly, and green to blue-green in color. Each leaf is folded longitudinally at its base within the bud and has a cross-sectional "V" -shape (**Christie and McElroy 1995**). Stems can reach 20—120 cm in height (**Peeters 2004**).

Orchardgrass became widely accepted as a forage grass in the United States in the 1940s and is now used for pasture, hay, silage, and green chop (**Miller 1984**, **Christie and McElroy 1995**). Key features of this grass include easy and rapid establishment, early maturation, high productivity, shade tolerance, heat tolerance, drought tolerance, and moderate winter-hardiness (**Miller 1984**). Orchardgrass responds well to nitrogenous fertilizers, by application of animal manure, or by planting with a legume, such as alfalfa, *Medicago sativa* L., or ladino clover, *Trifolium repens* L. (**Miller 1984**, **Christie and McElroy 1995**). The first cutting of an orchardgrass stand is

Citation: William R. Kuhn, Roger R. Youngman, Shaohui Wu, Curtis A. Laub, Ecology, Taxonomy, and Pest Management of Billbugs (Coleoptera: Curculionidae) in Orchardgrass. DOI: http://dx.doi.org/10.1603/IPM12022.

best taken at or just before the appearance of the inflorescence, as nutritive value decreases at the onset of anthesis and seed production (**Smith et al. 1986**). A healthy stand of orchardgrass can produce up to three cuttings annually (**Smith et al. 1986**).

Figure 1. Orchardgrass (*D. glomerata*): mature plants (left), a late-summer inflorescence (center), and three examples of the feeding holes on orchardgrass caused by billbug adults (right).

BILLBUGS (*SPHENOPHORUS* SPP.)

The weevils (Coleoptera: Curculionidae) currently belonging to the genus *Sphenophorus* (known as "billbugs"), have a long and confusing nomenclative history, explained in detail by **Vaurie (1951)** and**O'Brien and Wibmer (1982)**. The genus currently known as *Sphenophorus* was originally described as*Calendra* Clairville and Shellenberg, but was spelled as *Calandra* (with an "a") in one figure (**Schellenberg and Clairville 1798**). This caused years of confusion, as different authors used one spelling over the other (e.g., **Schoenherr 1838**, **Pierce 1925**). In 1959, the matter was finally settled when the International Commission on Zoological Nomenclature (ICZN), following recommendations of specialists around the world, suppressed the names *Calendra* and *Calandra* in favor of the name*Sphenophorus* (**ICZN 1958, 1959**). The last major revision of the genus was made by **Vaurie (1951)**.

Several taxonomic keys have been written for the identification of various life stages of weevils to the genus- and species-level of *Sphenophorus*. **Anderson (1948)** provided a generic key to larvae of the curculionid subfamily, Calendrinae (now Dryophthorinae), which included *Calendra*. It should be noted that no species-level key for the larvae of *Sphenophorus* yet exists; however, **Satterthwait (1931b)** devised a species-level key to the pupae of *Calendra* in the United States. Species-level keys for*Sphenophorus* adults were written by **Blatchley and Leng (1916)** for Rhynchophora (now Curculionoidea) of eastern North America, **Vaurie (1951)** for *Calendra* of the United States, **Johnson—Cicalese et al. (1990)** for those species in this genus that are pests of turfgrass in the United States, and **Downie and Arnett (1994)** for beetles of the United States. **Vaurie (1951)** also included a key to differentiate the five *S. venatus* subspecies.

BLUEGRASS BILLBUG

The bluegrass billbug, *S. parvulus*, was described by Leonard Gyllenhal in 1838 (**Schoenherr 1838**). Since that time, it has been reported as a pest of many grass species throughout the United States, and in particular corn, *Zea mays* L., and Kentucky bluegrass, *Poa pratensis* L. (**Bruner 1890**, **Webster 1892**,**Forbes 1902**, **Blatchley and Leng 1916**, **Tashiro and Personius 1970**, **Lindgren et al. 1981**). Additional hosts include barley, *Hordeum vulgare* L., wheat, *Triticum aestivum* L., rye, *Secale cereale* L., ryegrass, *Lolium* spp., bentgrass, *Agristis* spp., and yellow nutsedge, *Cyperus esculentus* L. (**Satterthwait 1931b**, **Johnson—Cicalese 1988**). Although there are many publications that associate this species with injury to various grasses and plants, few authors have connected the bluegrass billbug to orchardgrass. **Satterthwait (1931b)** included orchardgrass as a potential host plant for the bluegrass billbug; **Turner (1955)** reported it as a pest of orchardgrass in Virginia; and **Kuhn et al. (2013)**confirmed its presence in orchardgrass fields in Virginia. The bluegrass billbug is currently found in nearly every state of the continental United States as well as Ontario, Canada (**Vaurie 1951**, **O'Brien and Wibmer 1982**, **Johnson—Cicalese 1988**, **Watschke et al. 1995**).

Eggs of the bluegrass billbug are off-white and shaped like beans, with a size of ≈1.6 by 0.6 mm (**Vittum et al. 1999**). The larvae are small, with soft white bodies and sclerotized head capsules, which range from yellowish in teneral larvae to dark reddish after maturation (**Figure 2**). They are legless, somewhat curved, although shorter and stockier than scarabaeiform larvae, and grow to be 4—6 mm in length (**Watschke et al. 1995**). Pupae begin off-white in color, darken to a rusty brown, and have a length of 5.5—10 mm (**Satterthwait 1931b**). Bluegrass billbug adults are 5—6.5 mm in length (**Blatchley and Leng 1916**) and range in color from rusty brown, after eclosion, to black in color (**Figs. 2** and **3**). This coloration can also be lightened by mud, which frequently covers the body (**Vittum et al. 1999**). The pronotum is covered with tiny evenly spaced and uniformly sized pits; deep punctate furrows run down the elytra, which resemble parallel chains (**Watschke et al. 1995**). The antennae are geniculate and clavate, and arise near the base of the long slender rostrum (**Blatchley and Leng 1916**).

Figure 2. Two billbug (*Sphenophorus* spp.) larvae (right and bottom) and bluegrass billbug (*S. parvulus*) adult (center) on the head of a penny. Scale bar = 1 mm.

Figure 3. Bluegrass billbug (*S. parvulus*) adult, dorsal (left) and lateral (right) views. Scale bar = 1 mm.

Although most information available on the life history of the bluegrass billbug is based on its study in turfgrass, the life history appears to be similar in orchardgrass according to observations by **Turner (1955)**. The bluegrass billbug undergoes a univoltine life cycle in much of its distribution, where newly eclosed adults overwinter under rocks, boards, sticks, dead leaves, and dead vegetation near its host plants (**Bruner 1890, Forbes 1902, Blatchley and Leng 1916, Watschke et al. 1995, Vittum et al. 1999**). However, **Bruner (1890)** reported a second early fall generation in Nebraska, which overwintered as a pupa. Two or more generations per year may occur in the warmer regions of its distribution.

As the weather warms in the spring and the soil temperature at 2.5 cm below the ground surface reaches 18.3—20.0°C (67—69°F; typically mid-March in Virginia; W.R.K., unpublished data), billbug adults first emerge from their overwintering sites and search for suitable host plants (**Watschke et al. 1995**). After a brief period of feeding, oviposition occurs from late April into June and early July (**Webster 1892, Blatchley and Leng 1916**). The female chews holes in the stems of the host, just above the soil surface, and deposits one to three eggs into or nearby each hole (**Webster 1892, Watschke et al. 1995**). **Smith (1913)** reported one female laying 255 eggs over a 3-mo period. The egg hatches in ≈6—7 d, and the larva begins feeding up and down within the stem, filling it with powdery frass (**Webster 1892, Smith 1913, Satterthwait 1932, Watschke et al. 1995**). Once a larva becomes too large to fit inside the stem, it generally burrows out, drops to the ground, and begins feeding internally and externally on the crown and roots of the plant (**Webster 1892, Satterthwait 1932**). Some larvae, however, can remain inside the plant throughout the larval stage (**Watschke et al. 1995**). The larval period has been reported to last from 23 to 60 d, and is followed by pupation in a soil cell near the host plant (**Webster 1892, Forbes 1902, Watschke et al. 1995**). Adults emerge after 8—10 d and are abundant in late August through September (**Forbes 1902, Blatchley and Leng 1916, Watschke et al. 1995**). Newly eclosed adults feed briefly on nearby host plants, then seek out overwintering sites (**Forbes 1902**). Although some adults are capable of flight, it is rarely seen (**Watschke et al. 1995, Young 2002**).

HUNTING BILLBUG

The hunting billbug, *S. venatus vestitus*, was first described by **Frank Hurlbut Chittenden (1904)** as *S. vestitus*, and later designated as a subspecies of *S. venatus* along with four other subspecies: *confluens* Chittenden, *glyceriae* Chittenden, *reticulaticollis* Boheman, and *venatus* Say (**Vaurie 1951**). These five subspecies are geographically distributed across the Atlantic and Pacific coasts and southern United States with little overlap between them. In the mainland United States, the hunting billbug ranges from Washington, DC, to Florida and as far west as New Mexico and southeastern Kansas (**Chittenden 1904, Vaurie 1951, O'Brien and Wibmer 1982**). In addition, it has been reported in Hawaii, Puerto Rico, Mexico, the Bahamas, the Dominican Republic, Martinique, and Japan (**Marsden 1979, O'Brien and Wibmer 1982, Hatsukade 1997**). Among the subspecies, only *S. v. confluens* and the hunting billbug have been reported in orchardgrass (**Kamm 1969, Kuhn et al. 2013**). The hunting billbug is also called the zoyziagrass billbug because it is a pest of zoysiagrass, *Zoysia* spp. (**Kovitvhadi and Kerr 1968**). Its additional host species include Timothy hay, Bermuda grass, *Cynodon dactylon* (L.), wheat, yellow nutsedge, corn, and sugarcane, *Saccharum officinarum* L. (**Satterthwait 1931b, Satterthwait 1932, Woodruff 1966**).

Although it is not currently possible to differentiate between the eggs or larvae of the hunting or bluegrass billbugs, the pupae and adults of these species may be distinguished from one another. The pupae and adults (**Figure 4**) of the hunting billbug are slightly larger than the bluegrass billbug, ranging in size from ≈9 to 12.75 mm and 6 to 11 mm, respectively (**Satterthwait 1931b, 1932**). The pronotum of the adult has a series of nonuniform pits and callosities, which together resemble the letters "Y" or "V" framed by parentheses (**Watschke et al. 1995**). The hunting billbug also has chain-like furrows on its elytra, but they are less distinct than those of the bluegrass billbug.

Figure 4. Hunting billbug (*S. venatus vestitus*) adult, dorsal (left) and lateral (right) views. Scale bar = 1 mm.

The life history of the hunting billbug appears to be similar to that of the bluegrass billbug. However, **Huang (2008)** found that hunting billbug adults are nocturnal, and that eggs may be deposited in or around feeding "notches" created by the female on grass stems. The hunting billbug may have multiple generations in the southern United States, reproducing continually throughout the year, or may overwinter as a larva or pupa in cooler regions (**Watschke et al. 1995, Young 2002, Huang and Buss 2009**).

BILLBUG INJURY, SCOUTING, NATURAL ENEMIES, AND CONTROL IN ORCHARDGRASS

Billbug feeding can destroy a stand of orchardgrass or go completely unnoticed, depending on the size of the billbug population and weather conditions. In addition to causing direct injury to the plant, feeding can also provide the opportunity for infection by rot-inducing bacteria and fungi (**Hanson et al. 1950**). Adult feeding is evidenced in late spring by the appearance of oblong, typically paired holes (≈1—5 mm in diameter) that occur singly or in sets on the leaves of the host plant (**Figure 1**). Each set of holes begins as a single hole made by an adult feeding at the folded base of a leaf, and then unfurls into a mirrored pair as the leaf grows. These holes are a diagnostic feature of billbug adult feeding, which appears on orchardgrass, corn, and other grass species (**Forbes 1905, Satterthwait 1932, Kamm 1969**, W.R.K., unpublished data), and therefore represent a method to detect billbug presence. Larvae feed within the stems and roots in midsummer and damage appears as irregularly shaped patches of brown grass (**Watschke et al. 1995**). Damage caused by larvae is often more evident in dry than well-watered grass, despite being equally infested (**Bruner 1890**). Later in the growing season, infested stems break off easily and the larval frass, which resembles sawdust, spills out of these hollowed stems (**Tashiro and Personius 1970**). Feeding by larvae in later instars causes the most damage (**Turner 1955**).

Several methods have been devised to sample for billbugs. To extract billbug eggs from turfgrass,**Tashiro (1987)** used a blender to break up the turfgrass material and a series of stacked sieves to filter out the resultant turfgrass fragments, leaving the eggs. Larvae can be found by examining the upper layers of soil cores or square-foot samples (**Watschke et al. 1995**). Adult billbugs may be found in turfgrass by drenching the grass with water (**Vittum et al. 1999**), monitoring sidewalks and pavement for adults (**Watschke et al. 1995, Vittum et al. 1999**), using "suction or vacuum samplers" (**Watschke et al. 1995**), and using pitfall traps (**Johnson—Cicalese et al. 1990, Watschke et al. 1995, Vittum et al. 1999**). In addition, **Doskocil (2010)** found several commercial blends of aggregation pheromones from four non-*Sphenophorus* weevil species to be effective attractants for hunting billbug adults in a laboratory setting.

Pitfall traps, which are often used to capture surface-dwelling arthropods, are the most common method for sampling billbug adults. A basic pitfall trap comprises a container buried in the ground so that its rim is flush with the soil surface. The container is partially filled with a killing solution, such as soapy water, ethanol, or ethylene glycol, to both prevent captive arthropods from escaping and to preserve them from decay until the traps can be checked. Several modifications have been made to the basic trap idea. Linear pitfall traps use a piece of buried PVC pipe with a slit cut dorsally along its length to funnel surface-dwelling arthropods into a collection can at one end (**Lawrence 1982**), while in a barrier pitfall trap (**Hansen and New 2005, Laub et al. 2008**), a barrier wedged into the ground diverts arthropods into pitfall traps on either of its ends. In the latter method, multiple pitfall traps may be connected by several barriers, or multiple barriers may be used around a single pitfall trap, increasing the area covered by the trap (**Hansen and New 2005**).

Although a number of cultural, biological, and chemical control methods have been suggested for billbugs in turfgrass and corn, there has been little focus on control in orchardgrass. **Tashiro (1987)** and**Watschke (1995)** reviewed billbug-resistant varieties of turfgrass, but no billbug-resistant varieties of orchardgrass have been reported.

Predators of billbugs include the adults and larvae of carabid beetles (**Bruner 1890**); a tachinid fly,*Myiophasia metallica* (Townsend); as well as the American toad, alligators, and 26 species of birds (**Satterthwait 1932**). Natural parasites and parasitoids include the entomopathogenic fungus, *Beauveria*sp. (**Kamm 1969**); entomopathogenic nematodes of the genus *Mermis* Dujardin or *Gordius* L. (**Bruner 1890**); mites (Acari) (**Forbes 1902**); a braconid wasp, *Vipio belfragei* (Cresson) (**Satterthwait 1932**); and the mymarid wasp, *Anaphes (Patasson) calendrae* (Gahan) (**Satterthwait 1931a, Beardsley 2000**).

Billbug augmentative biological control has been attempted and has included the introduction of*A. calendrae* into Hawaii in 1928 and 1963. The second introduction was successful, as individuals of *A. calendrae* were found in 1995; however, efficacy against the target weevil species is unknown (**Beardsley 2000**). For controlling billbug adults and larvae, **Watschke et al. (1995)** recommended application of various entomopathogenic nematodes (*Steinernema carpocapsae* Weiser, *Steinernema glaseri* Steiner, and multiple *Heterorhabditis* spp.), and moistening turfgrass in the spring to augment the entomopathogenic, soil-borne fungi, *Beauveria* spp.

Watschke et al. (1995) reported that the most effective means of chemical control for billbugs is to target the egg-laying spring adults. Larvae may also be targeted as they leave the stem and drop to the soil to feed on the roots of the plant; however, some larvae remain within the plant and are therefore never exposed to contact or soil insecticides (**Watschke et al. 1995**). In turfgrass, systemic insecticides are effective for controlling billbug larvae and adults; however, these insecticides are not available for orchardgrass. There are only five insecticides currently registered for grass hay and pasture, which includes orchardgrass: β-cyfluthrin, carbaryl, λ-cyhalothrin, malathion, and ζ-cypermethrin (**Youngman et al. 2013**). Of these, however, only λ-cyhalothrin is currently labeled for use on billbugs in grass hay and pasture in Virginia.

BILLBUGS IN ORCHARDGRASS: A POORLY UNDERSTOOD PROBLEM

Billbugs are a major pest of orchardgrass in Virginia; however, most of our current understanding of them is based on studies of their interactions with turfgrass. Further investigations are warranted to verify the life histories of the bluegrass and hunting billbugs on orchardgrass. In addition, other billbug species have been reported in orchardgrass fields, including the southern corn billbug, *Sphenophorus callosus* (Olivier), lesser billbug, *Sphenophorus minimus* (Hart), and Timothy billbug, *Sphenophorus zeae* (Walsh) (**Kuhn et al. 2013**). Feeding studies are needed to determine whether these species regularly feed on orchardgrass and, if so, how likely they are to become pests. Finally, the current measures available for control of billbugs in orchardgrass are extremely limited. New management tools, insecticides or otherwise, are required to effectively control this pest.

ACKNOWLEDGMENTS

We thank Timothy Mize and Kenner Love from the Virginia Cooperative Extension for their help and support, the farmers of Fauquier and Loudoun Counties, VA, for the use of their orchardgrass fields, and the editor of this journal and two anonymous reviewers for their helpful comments in preparation of this manuscript. Funding for this work was graciously provided in part by a Virginia Cooperative Extension Graduate Student Partnership Grant.

REFERENCES CITED

1. Anderson, W. H. 1948. Larvae of some genera of Calendrinae (Rhynchophorinae) and Stromboscerinae (Coleoptera: Curculionidae). Annals of the Entomology Society of America 41: 413– 437.
2. Balasko, J. A., and C. J. Nelson. 2003. Grasses for northern areas, pp. 125–148. In R. F. Barnes, C. J. Nelson, M. Collins, and K. J. Moore (eds.), Forages: An

Introduction to Grassland Agriculture, 6th ed. Iowa State Press, Ames, IA.

3. Beardsley, J. W. 2000. The introduction and establishment of Anaphes (Patasson) calendrae (Gahan) in Hawaii (Hymenoptera: Mymaridae). Proceedings of the Hawaiian Entomology Society 34: 189 –191.
4. Blatchley, W. S., and C. W. Leng. 1916. Rhynchophora or weevils of north eastern America. The Nature Publishing Company, Indianapolis, IN.
5. Bruner, L. 1890. Report on Nebraska insects, pp. 95–106. In Reports of observations and experiments in practical work of the division. US Department of Agriculture, Washington, DC.
6. Chittenden, F. H. 1904. On the species of Sphenophorus hitherto considered as simplex LeConte. Proceedings of the Entomology Society of Washington 6: 127–130.
7. Christie, B. R., and A. R. McElroy. 1995. Orchardgrass, pp. 325–334. In R. F. Barnes, D. A. Miller, and C. J. Nelson (eds.) Forages: An Introduction to Grassland Agriculture, 5th ed. Iowa State Press, Ames, IA.
8. Doskocil, J. P. 2010. The biology and ecology of hunting billbug in North Carolina turfgrass. Ph.D. dissertation, North Carolina State University, Raleigh, NC.
9. Downie, N. M., and R. H. Arnett. 1994. The beetles of northeastern North America, pp. 1634 –1645. Sandhill Crane Press, Gainesville, FL.
10. Forbes, S. A. 1902. The corn billbugs in Illinois. Illinois Agricultural Experiment Station Bulletin 79:435– 461.
11. Forbes, S. A. 1905. The corn bill-bugs, pp. 52–57. In Twenty-third report of the state entomologist on the noxious and beneficial insects of the state of Illinois. Illinois State Journal Co., State Printers, Springfield, IL.
12. Hansen, J. E., and T. R. New. 2005. Use of barrier pitfall traps to enhance inventory surveys of epigaeic Coleoptera. Journal of Insect Conservation 9: 131–136.
13. Hanson, E. W., H. E. Milliron, and J. J. Christensen. 1950. The relation of the bluegrass billbug, Calendra parvula (Gyllenhal) to the development of the basal stem rot and root rot of cereals and grasses in north-central United States. Phytopathology 40: 527–543.
14. Hatsukade, M. 1997. Biology and control of the hunting billbug, Sphenophorus venatus vestitus Chittenden, on golf courses in Japan. International Turfgrass Society Research Journal 8: 987–996.
15. Huang, T.-I. 2008. Billbug (Sphenophorus spp.) composition, abundance, seasonal activity, development time, cultivar preference, and response to endophytic ryegrass in Florida. M.S. thesis, University of Florida, Gainesville, FL.
16. Huang, T.-I., and E. A. Buss. 2009. Billbug (Coleoptera: Curculionidae) species composition, abundance, seasonal activity, and developmental time in Florida. Journal of Economic Entomology 102: 309 –314.
17. (ICZN) International Commission on Zoological Nomenclature. 1958. Report on the question of the application of the generic name "Calandra" Clairville & Schellenberg, 1798 (Class Insecta, Order Coleoptera) and matters incidental thereto. The Bulletin of Zoological Nomenclature 16: 5– 47.
18. (ICZN) International Commission on Zoological Nomenclature. 1959. Opinion 572. The Bulletin of Zoological Nomenclature 17: 112–116.
19. Johnson–Cicalese, J. M. 1988. Billbug pests of turfgrass: biology, host range, and effects of endophyte-infected grasses. M.S. thesis, Rutgers University, New Brunswick, NJ.
20. Johnson–Cicalese, J. M., G. W. Wolfe, and C. R. Funk. 1990. Biology, distribution, and taxonomy of billbug turf pests (Coleoptera: Curculionidae). Environmental Entomology 19: 1037–1046.
21. Kamm, J. A. 1969. Biology of the billbug Sphenophorus venatus confluens, a new pest of orchardgrass (Dactylus glomeratus). Journal of Economic Entomology 62: 808 – 812.
22. Kovitvhadi, K., and S. H. Kerr. 1968. Artificial diet for the zoysiagrass billbug, Sphenophorus venatus vestitus (Coleoptera: Curculionidae), and notes on its biology. Florida Entomology 51: 247–250.
23. Kuhn, W. R., R. R. Youngman, K. Love, M. Mize, S. Wu, and C. A. Laub. 2013. Billbugs (Coleoptera: Curculionidae) New to Orchardgrass (Dactylis glomerata) grown in Virginia. Entomological News (in press).
24. Laub, C., R. R. Youngman, K. Love, and T. Mize. 2008. Using pitfall traps to monitor insect activity. Virginia Cooperative Extension, Blacksburg, VA. No. 444 – 416.
25. Lawrence, K. O. 1982. A linear pitfall trap for mole crickets and other soil arthropods. Florida Entomology 65: 376 –377.
26. Lindgren, D. T., R. C. Shearman, A. H. Bruneau, and D. M. Schaaf. 1981. Kentucky bluegrass cultivar response to bluegrass billbug, Sphenophorus parvulus Gyllenhal. HortScience 16: 339.
27. Marsden, D. A. 1979. How to control the hunting billbug. University of Hawaii Cooperative Extension Entomology Notes No. 2. Hawaii Cooperative Extension, Manoa, HI.
28. Miller, D. A. 1984. Forage crops, pp. 396 – 409. McGraw–Hill, New York, NY.
29. (NASS) U.S. Department of Agriculture, National Agricultural Statistics Service, Virginia Field Office. 2012. 2011 Virginia Agricultural Statistics Bulletin and Resource Directory. NASS, Richmond, VA.
30. (NRCS) U.S. Department of Agriculture, Natural Resources Conservation Service. 2013. PLANTS profile for Dactylis glomerata (orchardgrass). (http://plants.usda.gov/java/profile?symbolDAGL).
31. O'Brien, C. W., and G. J. Wibmer. 1982. Annotated checklist of the weevils (Curculionidae sensu lato) of North America, Central America, & the West Indies (Coleoptera: Curculionidae). Memoirs of the American Entomological Institute 34. American Entomological Institute, Ann Arbor, MI.
32. Peeters, A. 2004. Wild and sown grasses: profiles of a temperate species selection: ecology, biodiversity and use, pp. 127–134. Food and Agriculture Organization of the United Nations and Blackwell Publishing, Rome,

Italy.

33. Pierce, W. D. 1925. The history of the rynchophorid genera Rhynchophorus, Calendra, Sphenophorus and Sitophilus (Coleoptera). Proceedings of the Entomology Society of Washington 27: 113–114.

34. Satterthwait, A. F. 1931a. Anaphoidea calendrae Gahan, a mymarid parasite of eggs of weevils of the genus Calendra. Journal of the New York Entomological Society 39: 171–190.

35. Satterthwait, A. F. 1931b. Key to known pupae of the genus Calendra, with host-plant and distribution notes. Annals of the Entomology Society of America 24: 143–172.

36. Satterthwait, A. F. 1932. How to control billbugs destructive to cereal and forage crops. U.S. Department of Agriculture Farmers' Bulletin 1003, U.S. Government Printing Office, Washington, DC.

37. Schellenberg, J. R., and J. P. Clairville. 1798. Helvetische Entomologie, Verzeichniss der schweizerischen Insekten nach einer neuen Methode geordnet: mit Beschreibungen und Abbildungen, p. 62, pl. II. Bei Orell, Fu¨ssli und Compagnie, Zu¨ rich, Switzerland.

38. Schoenherr, C. J. 1838. Genera et Species Curculionidum, cum synonymia hujus familae, pp. 874, 961. Roret, Parisiis.

39. Smith, D., R. J. Bula, and R. P. Walgenbach. 1986. Forage management, pp. 173–178. Kendell/Hunt Publishing Co., Dubuque, IA.

40. Smith, R. I. 1913. Biological record of little grass billbug (Sphenophorus parvulus). Annual Report of the North Carolina Agriculture Experiment Station 35: 136 –140.

41. Tashiro, H., and K. E. Personius. 1970. Current status of the bluegrass billbug and its control in western New York home lawns. Journal of Economic Entomology 63: 23–29.

42. Tashiro, H. 1987. Turfgrass insects of the United States and Canada, pp. 220 –228, pl. 44 – 45. Cornell University Press, Ithaca, NY.

43. Turner, E.C.J. 1955. Calendra parvula (Gyllenhal), a pest of orchardgrass. Journal of Economic Entomology 48: 335.

44. Vaurie, P. 1951. Revision of the genus Calendra (formerly Sphenophorus) in the United States and Mexico (Coleoptera: Curculionidae). Bulletin of the American Museum of Natural History 93:33–186.

45. Vittum, P. J., M. G. Villani, and H. Tashiro. 1999. Turfgrass insects of the United States and Canada, pp. 42–53. Comstock Publishing Associates, Ithaca, NY.

46. Watschke, T. L., P. H. Dernoeden, and D. J. Shetlar. 1995. Managing turfgrass pests, pp. 254 –262. CRC, Boca Raton, FL.

47. Webster, F. M. 1892. Insects which burrow in the stems of wheat. Bulletin of the Ohio Agriculture Experiment Station 5:72–73.

48. Woodruff, R. E. 1966. The hunting billbug, Sphenophorus venatus vestitus Chittenden, in Florida (Coleoptera: Curculionidae). Florida Department of Agriculture Entomology Circular 45, University of Florida, Gainesville, FL.

49. Young, B. F. 2002. Seasonal activity and biology of the hunting billbug, Sphenophorus venatus vestitus (Coleoptera: Curculionidae), in northwest Arkansas. M.S. thesis, University of Arkansas, Fayetteville, AR.

50. Youngman, R. R. 2013. Insects: orchardgrass, timothy, and bermudagrass, pp. 4 –13– 4 –16. In A. Thomas (ed.), Pest Management Guide: Field Crops 2013. Virginia Cooperative Extension, Blacksburg. (http://pubs.ext.vt.edu/ 456/456-016/Section_4_Insects-4.pdf).

CHAPTER 15

A DATA-DRIVEN, MACHINE LEARNING FRAMEWORK FOR OPTIMAL PEST MANAGEMENT IN COTTON

Matthew H. Meisner, Jay A. Rosenheim, Ilias Tagkopoulos

Keywords: Agricultural informatics, applied pest management, Markov decision processes, optimal pest management.

INTRODUCTION

Maximizing crop yield in commercial agriculture is highly desirable for several reasons. First, the world's growing population is generating an increased demand for agricultural products (Godfray et al. **2010**). Expanding the land area devoted to agriculture is often not feasible or not desirable, so increasing the yield generated from existing farmland may be the only viable way to meet this demand. Second, agriculture is critical to the global economy. Despite this significance, the profit margins in commercial agriculture are often very small. Increasing crop yield can generate increased revenue for farmers, facilitating the continued economic viability of agriculture.

However, yield maximization is not the only consideration that is relevant to a commercial farmer. Another important consideration for farmers is the cost of various agricultural inputs, including pesticides. Inputs may increase yield, but they are costly, and these costs must be considered by a farmer who wishes to maximize profits. Furthermore, indiscriminate pesticide use has other detrimental effects that are more difficult to quantify. First, excessive pesticide use may accelerate the evolution of pesticide resistance (Mallet **1989**), which reduces our capacity to control pests and necessitates the development of more potent pesticides with unknown environmental impact (Roush **1987**). Second, pesticide applications can adversely affect nontarget species. When pesticide applications depress populations of beneficial species, such as pest predators, these applications can contribute to secondary pest outbreaks and therefore augment future crop damage and pesticide costs (Gross and Rosenheim **2011**). Finally, there is abundant evidence that pesticides are detrimental to both human health and ecosystem health (Rosner and Markowitz **2013**), creating a strong incentive to avoid using these chemicals unless they are absolutely necessary.

In this study, we focus specifically on the immediate costs of pesticide applications (i.e., the amount of money that it costs a farmer to apply a pesticide), and examine the tradeoff between these immediate costs and the costs incurred from yield loss due to pest damage. While the less immediate costs of pesticide use (e.g., secondary pest outbreaks, human health) are important, we have excluded them from our analysis for two reasons. First, these costs are more difficult to quantify objectively than are the immediate costs. Second, costs such as the evolution of pesticide resistance and environmental damage from pesticides are not as immediately relevant to farmers, many of whom may be focused on short-term profit maximization, and we seek to provide management recommendations relevant to farmers' actual decision-making.

Currently, farmers lack the computational tools needed to identify management strategies that optimally navigate the tradeoff between minimizing yield loss due to pest damage and minimizing the cost of pesticide applications. Without access to data-driven pest management recommendations, farmers often rely on intuition and personal experience to guide their

Citation: Matthew H. Meisner, Jay A. Rosenheim and Ilias Tagkopoulos, "A data-driven, machine learning framework for optimal pest management in cotton," DOI: 10.1002/ecs2.1263.

crop management decisions, and this may lead to suboptimal decision-making that reduces yield, increases pesticide costs, and exacerbates the deleterious effects of pesticides on human health and the environment. These problems are especially severe when pesticides are inexpensive relative to the value of the crop, as this incentivizes growers to apply them prophylactically.

In some cases, farmers may base their pest management decisions on "economic injury levels"—pest densities at which a profit-maximizing farmer is supposed to apply a pesticide. These levels are typically derived from experimental studies. However, experimental studies are often unable to resolve small effects of pests on yield, even though these small yield declines can be of substantial economic significance to commercial farmers (Rosenheim et al. **2011**). In addition, growers are often provided with a single economic threshold for the entire growing season, which neglects the possibility that a crop's susceptibility to pest damage may vary throughout the growing season.

Here, we identified the optimal pest management strategies for the pest *Lygus hesperus* in commercial cotton fields. The plant bug *L. hesperus* is one of the most damaging pests of cotton, and a frequent target of insecticide applications (Rosenheim et al. **2006**, Godfrey et al. **2013**). We determined at which pest densities and at which times in the growing season profit-maximizing farmers should apply pesticides to treat *L. hesperus*, and we quantified how close current farmer behavior is to this optimal policy. To explore these questions, we took an "ecoinformatics" approach in which we analyzed historical records of commercial cotton production from 566 fields in California's San Joaquin Valley. By aggregating historical records from hundreds of fields and various growers, we compiled a robust data set with the power to quantify how pest densities and different pest management strategies affect yield. Ecoinformatics analyses generally involve harnessing the power of large, preexisting, observational data sets to address important questions in environmental biology, especially ones that may be difficult to study experimentally (Rosenheim et al. **2011**).

First, we quantified the yield loss associated with particular pest densities at different times in the growing season. We then used a finite time horizon Markov decision process (Bauerle and Rieder **2011**) to identify the optimal pest management strategy for each pest density at each time point. Finally, we assessed how closely aligned the observed grower pest management policies are to the optimal policy.

MATERIALS AND METHODS

Data Set

The data set was constructed by collecting existing crop records from commercial cotton fields in California's San Joaquin Valley. The data were shared by both growers and pest control advisors, professional consultants hired to monitor field conditions and recommend crop management strategies. The data set contains records of 1498 unique field-year instances from 566 unique fields, ranging from 1997 to 2008. The following variables were used in our analyses:

1. Cotton yield. Measured once for each field-year instance, cotton lint yield was measured in bales/acre and recorded for 1240 of the 1498 total records.
2. Cotton species. The database consisted of records of two different cotton species:*Gossypium barbadense* L. ("Pima cotton") and *Gossypium hirsutum* L. ("Acala/upland cotton").
3. *Lygus hesperus* densities. Pest control advisors measured *L. hesperus* densities approximately weekly, primarily during June and July. The pest control advisors' sampling procedure consisted of 50 swings of a sweep net across the top of the plant canopy (for the remainder of the manuscript, we use the term "sweep" to refer to this standard, 50-swing sample). As not all pest control advisors sampled on the same days or at exactly the same intervals for all fields, we transformed successive samples into mean *L. hesperus* density estimates (insects/50-swing sweep sample) by calculating the area under the linear curve of*L. hesperus* density vs. time and dividing by the number of days in the sampling interval.
4. Pest management practices. Every time that a pesticide was applied to a field, the date, chemical, and target pest was recorded.

Markov Decision Process Model

Model Structure

To identify optimal crop management strategies, and to quantify the differences in farmer profit under alternate strategies, we modeled farmer decision-making as a discrete time Markov decision process (MDP) with a finite time horizon (Fig. 1). MDPs provide a framework for modeling a system that can be in different states at each discrete point in time. The transitions between states over time are partially under the control of a decision maker, who can choose different actions at each time point, but are also partially stochastic. Transitions are modeled probabilistically, and the transition probabilities between states depend upon the current state and the action taken. MDPs also involve rewards or costs, which are incurred when a particular action is taken or a particular transition occurs. The MDP framework allows us to determine the action in each state at each time that is expected to maximize the sum of rewards over a specified time horizon (Puterman **2005**, Bauerle and Rieder **2011**).

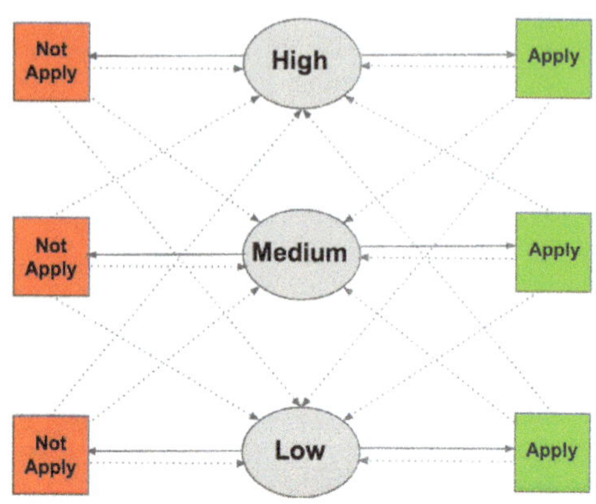

Figure 1. Network of MDP states and actions. In each state (high,

medium, or low pest density), a farmer can choose to apply or not apply a pesticide targeting *L. hesperus*. Solid lines indicate actions, and dashed lines indicate transitions.

For our specific MDP application, modeling farmer decision-making about pest management, we divided the growing season into $T = 8$ week-long intervals spanning 5 June through 30 July. While cotton is typically planted in March or April and harvested in October, consistent measurements of the main cotton pest, *L. hesperus*, are only taken during June and July; consequently, we limited our model to this period of the growing season. We used eight intervals to balance (1) the desire to provide management recommendations with fine temporal resolution (as the optimal pest management strategy may differ throughout the growing season), and (2) the inability to reliably estimate a large number of parameters.

States

At each time point, we classified each field as being in one of three states $s \in S$: low, medium, or high pest density. The exact thresholds for the states were selected so that, aggregated over all eight time points, there was an approximately equal number of observations from each state. The specific boundaries that led to this balance, all in units of insects/sweep, were (0, 0.66) for the low state, (0.66, 1.73) for the medium state, and greater than 1.73 for the high state.

Actions

In our model, we considered two different actions $a \in A$: either applying a pesticide or not applying a pesticide that a farmer can take. For each field at each time point, we determined if a pesticide application for *L. hesperus* had occurred by seeing if an application was recorded for which*L. hesperus* was listed as one of the target pests.

Transition Probabilities

Transition probabilities from state s to state s are denoted as $Q_{a,t}(s, s)$; they depend on the action a, time t, origin state s, and destination state s. Transition probabilities were modeled using a Bayesian ordered logistic regression, so that we could quantify and account for uncertainty in our estimates of these quantities. Ordered logistic regression was selected because the response variable of interest, the identity of the destination state, is a categorical variable with an ordered structure (low, medium, or high) (Agresti **2010**). For every unique combination of origin state, time, and action, we examined the records in the data set which described fields in that state at that time where that action was taken. We then fit an ordered logistic regression with destination state (low, medium, or high) as the ordered categorical response variable. As there are three possible categories, the ordered logistic regression involves two intercept terms, both of which were given uninformative N(0, 100) priors. No predictor variables were included in the model, as the purpose of this regression was only to quantify uncertainty in the estimated transition probabilities, not to draw inferences about what affects these probabilities across fields in the same state where a given action is taken at a given time.

Costs

The final component of the MDP model involves costs, $c_{a,t}(s')$, which depend on the action taken, the time, and the destination state. To calculate these costs, we considered both the cost of pesticide application and the cost due to yield loss from *L. hesperus* damage. First, there is a cost associated with a pesticide application; this cost encompasses both the cost to purchase the product and the fuel and labor costs of applying the product. Using estimates provided by cooperating growers and crop consultants, we estimated the cost of a pesticide application targeting *L. hesperus* to be \$20 per acre. Estimating this cost exactly is challenging for several reasons. First, growers use a variety of different pesticide products to suppress *L. hesperus*, and the prices of these different products vary substantially. Second, different growers may not pay the same price for the same product, as growers often negotiate discounts with chemical suppliers. Finally, some pesticide applications for*L. hesperus* may occur concurrently with other applications to a field; in this case, the additional cost of adding on the treatment for *L. hesperus* may be significantly lower than that of a dedicated application solely for *L. hesperus*. Despite these challenges, we feel that our estimate of \$20 per acre represents a typical cost of a pesticide application targeting *L. hesperus*.

The second component of the cost term involves the cost associated with yield loss due to damage from *L. hesperus*. We used a hierarchical Bayesian linear mixed model (Gelman and Hill **2009**) to estimate the change in yield associated with being in the medium or high pest density state, compared to being in the low pest density state, at each of the eight-time intervals. Our model included cotton yield as the response variable, a random effect for field to control for field-specific differences in yield potential, a random effect for year to control for annual fluctuations in yield, a fixed effect identifying the cotton species to control for variable yield potential between the two species, and fixed effect indicator variables indicating which state each field was in at each time interval. Uninformative N(0, 100) priors were used for all fixed effects, and uninformative InvGamma(0.001, 0.001) priors were used for all variance components. We considered the low state to be a baseline level of insect pressure, and calculated the costs of being in the medium or high state, relative to the low state. As a statistical model can only identifiably estimate two indicator variables for a categorical predictor variable with three levels, we selected one state as a baseline and only considered yield differences between the other states and that baseline state.

We quantified how yield differences between the high and low state, and yield differences between the medium and low state, changed over the growing season using a Bayesian linear regression of the estimated yield differences vs. time. As the intervals are equally spaced, we labeled the time intervals as $t = 1, \ldots, 8$ and regressed the yield differences against these numeric values. We performed a separate regression of yield difference vs. time for the yield differences between the high and low state and the yield differences between the medium and low state. We used uninformative N(0, 100) priors for the intercept and slope, and an uninformative InvGamma(0.001, 0.001) prior for the variance. To account for uncertainty in the estimates of yield differences, we repeated the regressions of yield difference vs. time, 10 000 times, each time using estimated yield differences sampled from the posterior of the mixed model in which these differences were estimated. We obtained 10 000 posterior samples from each individual linear regression, and then analyzed all 10 000 × 10 000 posterior samples in order to perform inference about the slope of the regression.

After calculating, for each of the eight time intervals, the difference in yield between fields in the medium state and the low state, and the difference in yield between fields in the high state and the low state, we converted these differences in yield to differences in farmer revenue, in dollars, using the most recent estimates of the value of cotton per pound published by the UC-Davis Department of Agricultural and Resource Economics: $0.90 per pound for Acala cotton (Hutmacher **2012a**), and $1.30 per pound for Pima cotton (Hutmacher **2012b**). As the difference in value of the two cotton species is substantial, we considered Acala and Pima cotton separately when converting the high vs. low and medium vs. low yield differences into revenue.

The total cost for a particular transition at time t to state s' at $t + 1$, denoted as ca,t (s'), was the sum of the $20 cost of pesticide application, if an application occurred, and the cost of the yield loss due to *L. hesperus* damage if a transition led to either the medium or high state at the next time point. Costs for transitions to the low state only involved the cost of a pesticide application, if it occurred. If we expected yield to be lost in the medium or high states, compared to the low state, the cost was considered to be a negative value.

All statistical models were fit using Markov Chain Monte Carlo, implemented in the Stan probabilistic programming language for Bayesian inference, and accessed through the RStan package in the R language for statistical computing (Stan Development Team, **2013**). For each model, we performed two independent Markov chain simulations of length 10 000, and discarded the first 5000 samples of each as burn-in. We verified that our MCMC simulations had converged by ensuring that $\widehat{R}$, a measure of expected posterior scale reduction if sampling were to be continued indefinitely, was near one (Gelman and Hill **2009**). All code for our MDP analyses was written in the R programming language, without the use of any MDP-related packages.

Solving For Optimal Policies

The MDP modeling framework allows us to identify the optimal policy that a farmer should follow in order to maximize profits over the course of the growing season. For each state at each time, we determined whether or not a profit-maximizing farmer should apply pesticides to suppress*L. hesperus*. The optimal policy, a function of state and time, is given by Eq. 1:

$$\pi_t(s) = \underset{a}{\operatorname{argmax}} \left\{ \sum_{s' \in S} Q_{a,t}(s, s') \left[c_{a,t}(s') + V_{t+1}(s')\right] \right\} t \in [1, 7]$$

And the "value" of each state at each time is given by Eq. 2:

$$V_t(s) = \sum_{s' \in S} Q_{\pi_t(s),t}(s, s') \left[c_{\pi_t(s),t}(s') + V_{t+1}(s')\right] t \in [1, 8]$$

To solve for the optimal policy, one needs estimates of the transition probabilities, costs, and the terminal value, i.e., $V_8(s)\ s \in S$. Using these estimates, we can work backwards using stochastic dynamic programming to solve for the optimal policy in each state at each time (Puterman **2005**). Due to the different values of the two cotton species, we obtained the optimal policy separately for both Acala and Pima cotton.

Parameter Estimation under Uncertainty

As both the transition probabilities and yield declines due to *L. hesperus* are quantities that we had to estimate from the data, there is uncertainty associated with these estimates. To propagate this uncertainty, we obtained, for each time-state combination, 10 000 posterior "samples" of $\sum_{s' \in S} Q_{a,t}(s, s') \left[c_{a,t}(s') + V_{t+1}(s')\right]$ for both possible actions, and selected the action with highest posterior mean as the optimal action. The samples were obtained by considering uncertainty in all three components of the value function: the costs, transition probabilities, and the value function at the next time step. We obtained 10 000 posterior samples of each of these three components. Specifically, we obtained 10 000 posterior samples of the costs by sampling from the posterior of the mixed model in which we estimated the yield decline associated with being in the high or medium state, compared to the low state, at each time interval. We obtained 10 000 posterior samples of the transition probabilities by sampling from the posterior of the ordered logistic regression with which we estimated the transition probabilities corresponding to that time-state-action combination. Our parameterization included the cost of yield loss in the destination state s′ during the next time interval in the ca,t (s') term. This means that the cost of yield loss due to *L. hesperus* at $t = 8$ is included in the cost term that is considered when selecting the optimal action at $t = 7$; therefore, we set $V_8(s) = 0\ s \in S$ and worked backwards using dynamic programming to determine the optimal policy for each state at each time. We stored 10 000 posterior samples of the value of each state at each time, and used these stored samples when accounting for uncertainty in, $V_{t+1}(s')$ the value at the next time step, which is a component of $\sum_{s' \in S} Q_{a,t}(s, s') \left[c_{a,t}(s') + V_{t+1}(s')\right]$ Our reasons for accounting for uncertainty in the model parameters, instead of using point estimates, are twofold. First, it is misleading to consider the costs and transition probabilities as known, as we used statistical models to estimate these parameters from the data. The values of these parameters can substantially affect the recommended policy. Unless every posterior distribution is normally distributed, which is unlikely, using maximum likelihood estimates to obtain point estimates of the value function may lead to a different optimal policy than the one obtained using our approach of accounting for uncertainty in all of the model parameters to obtain posterior samples of the value function. To provide growers with useful recommendations, we need to quantify the statistical uncertainty about the values to quantify our confidence in the optimal policy.

Second, propagating uncertainty in all the model parameters provided us a more in-depth quantification of the value differences between the two actions than we would have otherwise obtained. If we had instead used point estimates for the costs and transition probabilities, we would have only obtained point estimates for the value of each action at each time and state, and, therefore, only a point estimate of the difference in value between the two actions. With our Bayesian method, we obtained posterior distributions of the difference in value between the two actions. These distributions facilitate the exploration of a comprehensive set of questions about the differences between the two actions, such as which action leads to the lowest variance in value, which minimizes extreme losses, and which leads to the highest median value.

Optimality Criteria

Different optimality criteria can be considered when choosing the optimal action. To determine the profit-maximizing strategy, we selected the action that maximized the posterior mean of the value function (Eq. 1). In addition, we determined the lowest risk strategy by exploring the likelihood, under each possible action, of net revenue (income from crop yield minus pesticide costs) falling below the minimum revenue required for a profitable harvest. We determined the minimum revenue required for a farmer to break even using the most recent estimates of total cotton production costs published by the UC-Davis Department of Agricultural and Resource Economics: $1800 per acre for both Acala and Pima cotton (Hutmacher **2012a,b**). When selecting the optimal action for each state at each time, we selected the action that minimized the likelihood of revenue falling below this threshold. Again, separate policies were identified for both Acala and Pima cotton due to the substantial difference in value of the species. This criterion can be mathematically formulated as follows, where r_{crit} is the critical revenue threshold:

$$\pi_t(s) = \underset{a}{\text{argmin}} \left\{ P\left(\sum_{s' \in S} Q_{a,t}(s,s') \left[C_{a,t}(s') + V_{t+1}(s') \right] < r_{\text{crit}} \right) \right\} \; t \in [1,7]$$

Sensitivity Analyses

We performed three different sensitivity analyses to quantify the robustness of our results.

Missing data

First, we randomly divided the data set into three subsets with approximately the same number of observations. One at a time, we removed one-third of the data set, repeated the MDP-solving procedure (using the posterior mean to determine optimal policy), and calculated the proportion of time-state combinations in which the optimal policy was the same as when we used the full data set.

Noise

Second, we added stochastic noise to all of the variables in the model (the pest densities at each time interval, and yield), repeated the MDP-solving procedure (again using the posterior mean to select the optimal policy), and calculated the proportion of time-state combinations where the optimal policy was the same as that when we used the real data to fit the model. To each variable, we added normally distributed random noise with mean zero, and standard deviation of 0.1, 0.5, 1, and 2 times the sample standard deviation of the original values for that variable.

Mislabeling

Finally, we assessed our model's robustness to errors in the records of actions (applying or not applying pesticides) at each time point. We randomly selected a certain percentage of observed actions in each time interval, and switched the recorded action (i.e., we labeled applications as no applications, and no applications as applications). We repeated this for various percentages, refit the model, and calculated the proportion of time-state combinations in which the optimal policy was the same as the one selected using the original data.

RESULTS

Effects of *L. hesperus* on yield

We used a Bayesian linear mixed model to estimate the change in yield associated with being in the high and medium pest density states vs. being in the low state, for all eight time intervals (Fig. 2). There was a trend for the difference in yield between fields in the high vs. low state to increase throughout the growing season; a similar trend was noticed for the difference in yield between fields in the medium and low state. We quantified this trend with a Bayesian linear regression of the yield differences between the high and low state, and between the medium and low state, vs. time. The slope when regressing the yield difference between the high and low state vs. time had a posterior mean of 30.3 kg/(hectare × week), with a 95% HPDI of (11.2 kg/(hectare × week), 55.7 kg/(hectare × week)); the slope when regressing the yield difference between the medium and low state vs. time had a posterior mean of 19.0 kg/(hectare × week) with a 95% HPDI of (5.6 kg/(hectare × week), 32.9 kg/(hectare × week)). These entirely positive 95% HPDIs suggest that cotton is more susceptible to yield loss from *L. hesperus* herbivory early in the growing season. Interestingly, the posterior means for the yield differences between high and low and between medium and low states were positive for the 24 July–30 July time interval (the final time interval in our analysis), and the 95% HPDI for the difference in yield between the high and low state was entirely positive.

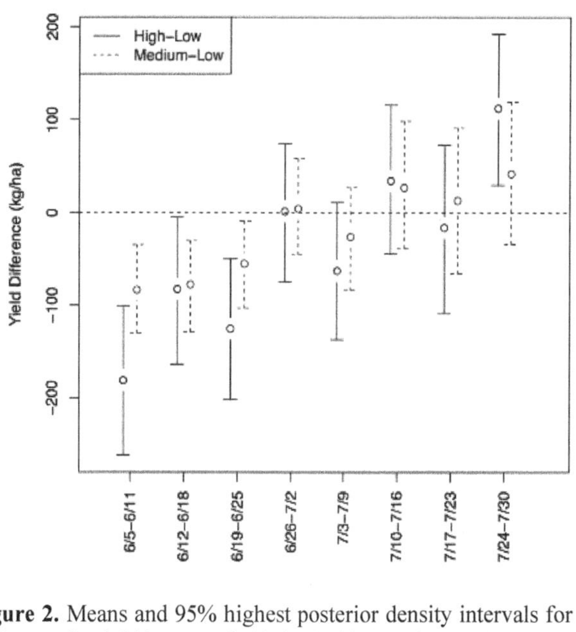

Figure 2. Means and 95% highest posterior density intervals for the difference in yield between fields in the high vs. low, and medium vs. low, pest density states during 8-week-long intervals.

Optimal Policies

Highest Mean

When the optimal policy is determined by selecting the action resulting in the highest posterior mean of the value function, the optimal policies for Pima and Acala cotton are presented in Tables 1and 2, respectively. For Pima cotton, the optimal policy only called for a pesticide application to target *L. hesperus* in the medium and high states during week 1, and the high state during week 2. The optimal policy for Acala cotton only called for pesticide application in the medium and high states during week 1. Tables 3 and 4 display the difference in posterior mean of the value function between the optimal and suboptimal action, for Pima and Acala cotton, respectively. These values can be interpreted as the expected increase in profit when the optimal policy is followed compared to when the suboptimal policy (i.e., applying a pesticide when not recommended or not applying a pesticide when recommended) is followed (assuming the optimal policy is followed at all subsequent time steps).

Table 1. The optimal *L. hesperus* management policy in Pima cotton, for each state at each time. The optimal policy is defined by the management decision, either apply (A) or not apply (NA) pesticides, with the lowest posterior mean of expected long-term costs

	5–11 Jun	**12–18 Jun**	**19–25 Jun**	**26 Jun–2 Jul**	**3–9 Jul**	**10–16 Jul**	**17–23 Jul**
Low	NA	NA	NA	NA	NA	NA	NA
Medium	A	NA	NA	NA	NA	NA	NA
High	A	A	NA	NA	NA	NA	NA

Table 2. The optimal *L. hesperus* management policy in Acala cotton, for each state at each time. The optimal policy is defined by the management decision, either apply (A) or not apply (NA) pesticides, with the lowest posterior mean of expected long-term costs

	5–11 Jun	**12–18 Jun**	**19–25 Jun**	**26 Jun–2 Jul**	**3–9 Jul**	**10–16 Jul**	**17–23 Jul**
Low	NA	NA	NA	NA	NA	NA	NA
Medium	A	NA	NA	NA	NA	NA	NA
High	A	NA	NA	NA	NA	NA	NA

Table 3. The mean difference (dollars/acre) of the posterior means of the value function evaluated at the optimal policy and the suboptimal policy for each state and time, for Pima cotton

	5–11 Jun	**12–18 Jun**	**19–25 Jun**	**26 Jun–2 Jul**	**3–9 Jul**	**10–16 Jul**	**17–23 Jul**
Low	$14.50	$17.51	$20.81	$19.79	$4.04	$54.70	$35.50
Medium	$45.59	$2.79	$18.75	$18.55	$22.46	$17.95	$45.61
High	$33.81	$2.64	$20.06	$12.96	$23.89	$19.99	$42.99

Table 4. The mean difference (dollars/acre) of the posterior means of the value function evaluated at the optimal policy and the suboptimal policy for each state and time, for Acala cotton

	5–11 Jun	**12–18 Jun**	**19–25 Jun**	**26 Jun–2 Jul**	**3–9 Jul**	**10–16 Jul**	**17–23 Jul**
Low	$16.18	$18.15	$20.36	$20.18	$3.26	$44.23	$30.52
Medium	$25.43	$8.45	$19.53	$19.58	$21.73	$18.49	$37.86
High	$18.04	$4.43	$20.11	$15.36	$22.69	$19.99	$35.75

Lowest Risk

When the optimal policy is determined by selecting the action resulting in the lowest likelihood of yield falling below the threshold required for a profitable harvest, the optimal policy for Pima and Acala cotton is presented in Table 5. Coincidentally, when the optimal policy is defined in this way, the optimal policy is the same for both species of cotton. In contrast to the mean-maximizing policy, the risk-minimizing policy never calls for pesticide applications.

Table 5. The optimal *L. hesperus* management policy, for each state at each time, when the optimal policy is defined by the management decision, either apply (A) or not apply (NA) pesticides, that minimizes the likelihood of an unprofitable harvest. This policy was, coincidentally, the same for both Acala and Pima cotton

	5–11 Jun	12–18 Jun	19–25 Jun	26 Jun–2 Jul	3–9 Jul	10–16 Jul	17–23 Jul
Low	NA	NA	NA	NA	NA	NA	NA
Medium	NA	NA	NA	NA	NA	NA	NA
High	NA	NA	NA	NA	NA	NA	NA

Sensitivity Analyses

The optimal policy when using two-thirds of the data was the same as the policy when using the entire data set for 87% of state-time combinations, indicating that our results are robust to data removal. Adding noise to the variables with the same standard deviation as the original data resulted in the same optimal policy in more than 80% of time-state combinations (Fig. 3A), suggesting that our model is robust to moderate measurement error in the underlying data. Even when 25% of the actions were mislabeled, the optimal policy remained unchanged in more than 80% of state-action combinations, suggesting that our model is robust to errors in the records of pest management practices (Fig. 3B).

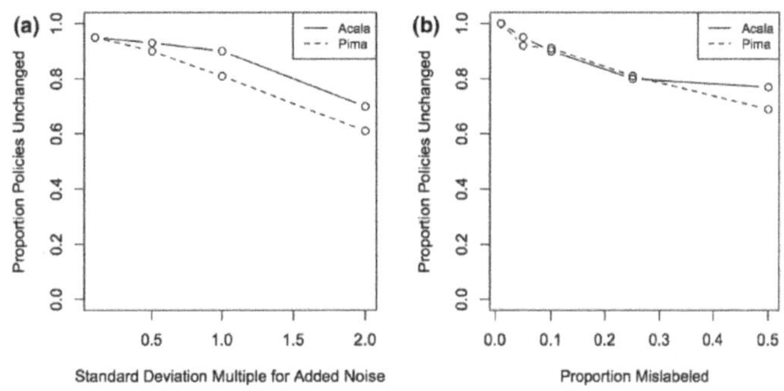

Figure 3. The proportion of time-state instances in which the optimal policy remained unchanged, compared to using the original data, when all the variables used in the model fitting were perturbed with the addition of stochastic variation that was normally distributed with mean set to 0 and standard deviation set to various multiples of the sample standard deviation for each variable (a); the proportion of time-state instances in which the optimal policy remained unchanged when various proportions of the actions were mislabeled (b).

Current Practices

For every state at each time interval, we examined the instances in our data set that were in that state that time, and calculated the proportion of those instances in which the optimal policy was followed (Tables 6 and 7). While growers generally followed the optimal policy during the final weeks of the season, only 2% and 6% of Pima instances followed the optimal policy during week 1 in the medium and high states, respectively. Only 8% and 18% of Acala instances followed the optimal policy during week 1 in the medium and high states, respectively. In other words, most farmers we observed did not apply pesticides targeting *L. hesperus* in this period, despite evidence that*L. hesperus* during this period is associated with decreased yield. On average, farmers followed the optimal policy 92% of the time when a pesticide application was not recommended, but only 10% of the time when an application was recommended.

Table 6. The proportion of Pima cotton instances in the database that followed the optimal policy (defined by maximizing the posterior mean), for each state at each time

	5–11 Jun	12–18 Jun	19–25 Jun	26 Jun–2 Jul	3–9 Jul	10–16 Jul	17–23 Jul
Low	0.98	0.97	0.98	1.00	0.94	1.00	0.85
Medium	0.02	0.86	0.95	0.91	0.98	0.94	0.94
High	0.06	0.21	0.67	0.83	0.81	0.79	0.80

Table 7. The proportion of Acala cotton instances in the database that followed the optimal policy (defined by maximizing the posterior mean), for each state at each time

	5–11 Jun	12–18 Jun	19–25 Jun	26 Jun–2 Jul	3–9 Jul	10–16 Jul	17–23 Jul
Low	0.97	0.98	0.95	0.97	0.99	0.99	1.00
Medium	0.08	0.89	0.89	0.96	0.98	0.97	0.97
High	0.18	0.67	0.74	0.83	0.88	0.90	0.87

Aggregated across both species and all states and times, there were an average of 0.52 pesticide applications per field when an application was not recommended, compared to 0.36 failures to apply pesticides when an application was recommended. So, on average, after considering the fact that applications were only infrequently recommended, the cotton farmers we observed applied pesticides when not recommended more frequently than they failed to apply pesticides when applications were recommended. About 46% of the observations in our data set involved at least one pesticide application that was not recommended.

DISCUSSION

Using a large observational data set from commercial cotton fields, we used a Markov decision process to model the pest management decisions made by farmers throughout eight key weeks of the growing season. Using this model, we were able to determine, for each pest density at each time interval, whether or not a farmer should apply pesticides to treat *L. hesperus* in order to maximize his/her expected profits over the course of the growing season.

Our observation that *L. hesperus* was only associated with decreased yield during the early part of the growing season, and was in fact associated with increased yield during the last week of July, highlights the limitations of adopting a pest management strategy that involves pest suppression when pests exceed a certain threshold. Our results suggest that using a single threshold for*L. hesperus* management throughout the season is undesirable, as the sensitivity of cotton yield to*L. hesperus* pressure decreased steadily throughout the growing season. Based on our results, cotton growers should focus their efforts toward detection and suppression of *L. hesperus* on the earlier part of the growing season (the first 2 weeks of June, in particular). Furthermore, they should avoid pesticide applications later in the growing season, where the yield loss due to *L. hesperus*herbivory (which in some cases was actually a yield increase) does not justify the costs of these applications.

One limitation of the Markov decision process model we implemented is that it involved discretizing pest density, which is a continuous variable, into a finite state space with three states. Doing so implicitly assumes that the value function is a constant value across all pest densities comprising a single state. This is unlikely to be exactly the case, especially for the high state, which had an infinite upper boundary. Despite this limitation, it was necessary to discretize the state space into discrete states in order to implement this model. With more data, it would be possible to accurately estimate yield effects associated with more, finer resolution states.

A second limitation of our approach stems from the fact that our work with observational data restricts our analysis to the existing variation in the data. As our data set did not contain any observations of extremely high *L. hesperus* densities, our recommendation of not applying pesticides during most weeks, even when in the high pest density state, may not actually be optimal in a field with a severe *L. hersperus* outbreak at densities beyond the range of densities that we analyzed. Extremely high *L. hesperus* densities can inflict crop damage so severe that a harvest is abandoned altogether, so it is likely that pesticide applications actually are economically optimal during these severe infestations. However, without data under such conditions, our analysis was unable to consider this possibility.

Interestingly, we found that the policy that minimizes the likelihood of yield falling below the level required for farmers to be financially profitable never involved the application of pesticides. Commercial farmers, particularly when operating under narrow profit margins, may tend to be financially risk averse; some growers may find it more desirable to minimize the risk of losing money rather than to maximize expected profits. We are not sure why pesticide applications were associated with increased risk for unprofitable harvests. However, one possible explanation is that pesticide applications may increase the risk of secondary pest outbreaks, possibly due to their detrimental effects on nontarget, beneficial species (Gross and Rosenheim **2011**).

As noted in the results section, during the 24 July–30 July time interval fields in the high and medium pest density states were associated with higher yield than were fields in the low pest density state. While we do not have data to specifically test this hypothesis, we hypothesize that the increased yield observed in fields with higher *L. hesperus* densities in late July results from *L. hesperus*preferentially attacking young flower buds, which, this late in the season, may be too late to mature before harvest. Herbivory on these young flower buds may result in the cotton plant devoting more resources to the more mature, existing flower buds, which do contribute to yield. So, by removing young flower buds that will not contribute to yield and focusing the plant's resources on harvestable fruits, higher *L. hesperus* densities in late July may contribute to increased yield.

Our finding that high and medium *L. hesperus* densities are associated with the largest yield decline at the beginning of June, as well as our related finding that the optimal policy only involves suppression of *L. hesperus* at the beginning of

June, are both consistent with previous research suggesting that cotton is more susceptible to *L. hesperus* damage early in the growing season. Simulation models (Gutierrez et al. **1975**, Mangel et al. **1985**), experimental studies (Falcon et al. **1971**), and a previous analysis of this data set (that was focused only on estimating yield loss) (Rosenheim and Meisner **2013**) have all provided evidence for increased cotton sensitivity to*L. hesperus* herbivory early in the growing season.

Our current study extends on this existing body of work in several ways. First, we have quantified how much yield loss can be expected from different pest densities at different times of the growing season. Second, we have quantified the trend for cotton sensitivity to *L. hesperus* herbivory to decrease over the growing season. Third, we have quantified how pesticide applications affect the likelihood of pest suppression at different times of the growing season. Fourth, we have integrated information on crop yield loss due to herbivory, the economic cost of that yield loss, pesticide efficacy, and the cost of pesticide applications in order to construct a holistic model of farmer pest management decision-making. This MDP model allowed us to determine what pest management decision a profit-maximizing farmer should make, based on the time in the growing season and the density of *L. hesperus* in a field, and it allowed us to quantify exactly how much increase in profit a farmer should expect when following the optimal, compared to the suboptimal, policy.

This study serves as an example of how data-driven decision-making in agriculture can confer significant advantages. Only 5% of farmers were following the optimal policy when in the medium pest state during week one, so, in some situations, current grower behavior appears to be suboptimal. Following the optimal policy has the potential to enhance commercial cotton production in several ways. First, by more actively suppressing *L. hesperus* in early June, farmers could help reduce yield loss associated with *L. hesperus* at this time. This could increase yield and farmer profits, both of which are desirable in order to sustain the economic vitality of agriculture and generate sufficient resources for a growing population. Second, by eliminating unnecessary pesticide applications in July, farmers could save money and help avoid the adverse effects of pesticides on the environment and human health.

There are several reasons why our ecoinformatics approach, which involved collecting large amounts of historical data from commercial farms, is a valuable approach for helping derive insights into optimal pest management policy. First, collecting data sets of this nature can be an affordable and efficient way to obtain data sets much larger than those that could be obtained experimentally. Replicating an experiment at the scale of a commercial plot hundreds of times requires resources rarely, if ever, available to agricultural researchers. Yield losses of just 1 or 2% can be economically significant to farmers, and it is very difficult to resolve yield effects of this magnitude without a large amount of data. Second, as our data set consists of data collected from commercial farms, it captured the true spatial and temporal scale of commercial agriculture. Experimental studies often rely on small plots, which may not be of sufficient size to provide a realistic picture of commercial cotton fields, which can be hundreds of acres in size. In particular, small plots may fail to capture the spatial dynamics of highly mobile pests. Third, collecting data from actual farms allowed us to quantify how closely, and at which times of the growing season, current farmer decision-making matches the optimal decisions; this information could be useful in guiding outreach and extension efforts that will have the most beneficial impact of cotton production.

While our ecoinformatics approach has several advantages, there are limitations to this approach. First, as our data set consists of observational data, inferences about causal relationships are not possible without making strong and untestable assumptions. When a trend is observed in a controlled experiment, one can be confident that the treatment manipulated by the researcher caused the observed change in response variable; however, in an observational data set, it is impossible to prove that some external factor did not affect both the treatment assignment and the response variable, thus spuriously suggesting a treatment effect. Second, our approach limits us to analyzing the range of variation already present in the data set. For example, we are not able to estimate the yield loss due to extremely high pest densities, as none of the farmers from whom we gathered data allowed the pest densities to reach these levels. However, in an experimental study, the researcher can decide which levels of a treatment he or she wishes to investigate.

Thus, ecoinformatics will not replace or diminish the value of experimental research; rather, it will serve as a complementary approach for generating data-driven crop management recommendations and promising hypotheses for future experimental investigation.

ACKNOWLEDGMENTS

We sincerely thank the growers and crop consultants who generously donated their data and time to make this research possible. We also thank the California State Support Committee of Cotton Incorporated, University of California Statewide IPM Program, USDA-NRICGP (Grant 2006-01761), California Department of Pesticide Regulation, UC-Davis Opportunity Fund (Grant 07427) and National Science Foundation GRFP (Grant DGE-1148897) for financial support.

REFERENCES

1. Agresti, A. 2010. Analysis of ordinal categorical data. John Wiley & Sons, New York, New York, USA.
2. Bauerle, R., and U. Rieder. 2011. Markov decision processes with applications to finance. Springer, New York, New York, USA.
3. Falcon, L. A., R. van den Bosch, J. Gallagher, and A. Davidson. 1971. Investigation of the pest status of Lygus hesperus in cotton in central California. Journal of Economic Entomology 64:56–61.
4. Gelman, A., and J. Hill. 2009. Data analysis using regression and multilevel/hierarchical models. Cambridge University Press, Cambridge, UK.
5. Godfray, H. C. J., J. R. Beddington, I. R. Crute, L. Haddad, D. Lawrence, J. F. Muir, J. Pretty, S. Robinson, S. M. Thomas, and C. Toulmin. 2010. Food security: the challenge of feeding 9 billion people. Science 327:812–818.
6. Godfrey, L. D., P. B. Goodell, E. T. Natwick, D. R. Haviland and V. M. Barlow. 2013. UC IPM pest

management guidelines: cotton. University of California Division of Agriculture and Natural Resources. http://www.ipm.ucdavis.edu/PMG/selectnewpest.cotton.html

7. Gross, K., and J. A. Rosenheim. 2011. Quantifying secondary pest outbreaks in cotton and their monetary cost with causal-inference statistics. Ecological Applications 21:2770–2780.

8. Gutierrez, A. P., L. A. Falcon, W. Loew, P. A. Leipzig, and R. van den Bosch. 1975. An analysis of cotton production in california: a model for acala cotton and the effects of defoliators on its yields. Environmental Entomology 4:125–146.

9. Hutmacher, R. B. 2012a. Sample costs to produce cotton: acala variety. University of California Cooperative Extension. http://cottoninfo.ucdavis.edu/files/150401.pdf

10. Hutmacher, R. B. 2012b. Sample costs to produce cotton: pima variety. University of California Cooperative Extension. http://cottoninfo.ucdavis.edu/files/150403.pdf

11. Mallet, J. 1989. The evolution of insecticide resistance: Have the insects won? Trends in Ecology and Evolution 4:336–340.

12. Mangel, M., S. E. Stefanou, and J. E. Wilen. 1985. Modeling Lygus hesperus injury to cotton yields. Journal of Economic Entomology 78:1009–1014.

13. Puterman, M. L. 2005. Markov decision processes: discrete stochastic dynamic programming. John Wiley & Sons, New York, New York, USA.

14. Rosenheim, J. A., and M. H. Meisner. 2013. Ecoinformatics can reveal yield gaps associated with crop-pest interactions: a proof-of-concept. PLoS ONE 8:e80518.

15. Rosenheim, J. A., K. Steinmann, G. Langellotto, and A. Zink. 2006. Estimating the impact of Lygus hesperus on cotton: the insect, plant, and human observer as sources of variability. Environmental Entomology 35:1141–1153.

16. Rosenheim, J. A., S. Parsa, A. A. Forbes, W. A. Krimmel, Y. H. Law, M. Segoli, M. Segoli, F. S. Sivakoff, T. Zaviezo and K. Gross. 2011. Ecoinformatics for integrated pest management: expanding the applied insect ecologist's tool-kit. Journal of Economic Entomology 104:331–342.

17. Rosner, D., and G. Markowitz. 2013. Persistent pollutants: a brief history of the discovery of the widespread toxicity of chlorinated hydrocarbons. Environmental Research 120:126–133.

18. Roush, R. T. 1987. Ecological genetics of insecticide and acaricide resistance. Annual Review of Entomology 32:361–380.

INDEX

N

O

P

Q

R

S

T

V

W